BIOMECHANICS
VIII-A

International Series on Biomechanics,
Volume 4A

BIOMECHANICS VIII-A

Proceedings of the Eighth International Congress of Biomechanics Nagoya, Japan

EDITORS

Hideji Matsui, D.M.S.
Research Center of Health,
  Physical Fitness and Sports
Nagoya University
Nagoya, Japan

Kando Kobayashi, Ph.D.
Research Center of Health,
  Physical Fitness and Sports
Nagoya University
Nagoya, Japan

HUMAN KINETICS PUBLISHERS
Champaign, Illinois

Publications Director
Richard D. Howell

Production Director
Margery Brandfon

Editorial Staff
John Sauget, Copyeditor
Dana Finney, Proofreader

Typesetters
Sandra Meier
Carol McCarty

Text Layout
Denise Peters
Lezli Harris

Library of Congress Catalog Card Number: 82-84703

ISSN: 0360-344X
ISBN: 0-931250-42-0 (Two-vol. set)
ISBN: 0-931250-43-9 (Vol. A)

Human Kinetics Publishers, Inc.
Box 5076, Champaign, Illinois 61820

CONTENTS

CONTRIBUTORS

Adrian, Marlene J. (869, 883, 903), Department of Physical Education for Women, Biomechanics Laboratory, Washington State University, Pullman, Washington 99164, U.S.A.

Ae, Michiyoshi (648, 737, 762), Institute of Health and Sport Science, The University of Tsukuba, Sakuramura, Niihari-gun, Ibaraki 305, Japan

Amano, Yoshihiro (498, 663), Aichi University of Education, 1, Hirosawa, Igaya-cho, Kariya-shi, Aichi 448, Japan

Amidror, Itzhak (1207), Information and Computer Sciences, Toyohashi University of Technology, 1-1, Hibarigaoka, Tempaku-cho, Toyohashi, Aichi 440, Japan

Andersson, G.B.J. (386, 543), Departments of Clinical Neurophysiology and Orthopedic Surgery, Sahlgren Hospital, S-413 45 Göteborg, Sweden

Andrews, James G. (923, 939), Institute for Developmental Research, Aichi Prefectural Colony, Kamiyacho, Kasugai, Aichi 480-03, Japan

Antonsson, E. (1104), Department of Mechanical Engineering, Massachusetts Institute of Technology, Cambridge, Massachusetts 02139, U.S.A.

Aoki, Hisashi (223, 413), Department of Orthopaedic Surgery, Akita University, School of Medicine, 1-1-1, Hondo, Akita 010, Japan

Arai, Michio (380), Department of Orthopaedic Surgery, Akita University School of Medicine, Akita, Japan

Asami, Toshio (695), University of Tokyo, 3-8-1, Komaba, Meguro-ku, Tokyo 153, Japan

Azuma, Akira (35), Institute of Interdisciplinary Research, Faculty of Engineering, University of Tokyo, 4-6-1, Komaba, Meguro-ku, Tokyo 153, Japan

Balsevich, Vadim Konstantinovich (1032), Department of Biomechanics, Institute of Physical Culture, Maslennikova 144, 644063 Omsk, U.S.S.R.

Bates, Barry T. (251, 574, 635), Biomechanics/Sports Medicine Laboratory, Department of Physical Education, University of Oregon, Eugene, Oregon 97403, U.S.A.

Bauer, Wilhelm L. (801), Sensomotorik-Labor, Universität Bremen, Badgasteiner Strasse, 2800 Bremen 33, B.R.D.

Baumann, Wolfgang (722), Institut für Biomechanik, Deutsche Sporthochschule Köln, Carl-Diem-Wag, D-5000 Köln, B.R.D.

Bechtold, Joan E. (403), Department of Orthopaedics, Mayo Clinic, Rochester, Minnesota 55901, U.S.A.

Bengtzelius, Ulf (861), Chalmer's Institute of Technology, Göteborg, Sweden

Björk, Roland (386), Departments of Clinical Neurophysiology and Orthopedic Surgery, Sahlgren Hospital, S-413 45 Göteborg, Sweden

Bober, Tadeusz (244, 1144), Academy of Physical Education, Biomechanics Laboratory, Al. Olimpijska 35, 51-612 Wrocław, Poland

Bobet, J. (1239), Department of Kinesiology, University of Waterloo, Waterloo, Ontario, Canada N2L 3G1

Boon, K.L. (363), Department of Anatomy and Biomechanics, Vrije Universiteit, 1007 MC Amsterdam, Pustbue 7161, The Netherlands

Bouisset, Simon (615), Université de Paris-Sud, Laboratoire de Physiologie de Mouvement, 91405 Orsay, France

Boukes, R.J. (363), Department of Anatomy and Biomechanics, Vrije Universiteit, 1007 MC Amsterdam, Pustbue 7161, The Netherlands

Bourassa, Paul (582), Department of Mechanical Engineering, Faculty of Applied Science, University of Sherbrooke, Sherbrooke, Province of Quebec, Canada J1K 2R1

Bourgeois, Marc (978), Unité de Recherche de Biomécanique du Mouvement, Université Libre de Bruxelles ISEPK-Laboratoire de l'Effort, C.P, 168, 28 av. Paul Heger 1050 Bruxelles, Belgium

Broeck, M. Van Den (951), Instituut voor Morfologie, Experimental Anatomy, Faculty of Medicine, Vrije Universiteit, Brussels, Belgium

Brouwer, Hendrik Rogier (171), Faculty of Medicine-Faculty of Applied Science, Vrije Universiteit Brussel, Pleinlaan 2, B-1050 Brussels, Belgium

Brüggemann, Peter (793), Institut für Sport und Sportwissenschaften, Ginnheimer Landstr. 39, 6 Frankfurt-Main 90, B.R.D.

Bunch, Richard P. (1089), Biomechanics Laboratory, The Pennsylvania State University, University Park, Pennsylvania 16802, U.S.A.

Burstein, A.H. (1075), Biomechanics Department, Hospital for Special Surgery, Cornell University, Medical College, 535 East 70 Street, New York 10021, U.S.A.

Cahill, Byron P. (403), Department of Orthopedics, Mayo Clinic, Rochester, Minnesota 55901, U.S.A.

Cardon, A. (171), Faculty of Medicine-Faculty of Applied Science, Vrije Universiteit Brussel, Pleinlaan 2, B-1050 Brussels, Belgium

Cappozzo, Aurelio (669, 1067), Laboratorio di Biomeccanica, Instituto di Fisiologia, Umana Università degli Studi, Roma, Italy

Cavanagh, Peter R. (641, 928, 1081, 1089), Biomechanics Laboratory, Pennsylvania State University, University Park, Pennsylvania 16802, U.S.A.

Cerretelli, Paolo (703), Department de Physiologie, Université de Genève, 20, rue de l'Ecole de Médecine, 1211 Genève 4, Switzerland

Chao, Edmund Y.S. (403, 490), Department of Orthopedics, Mayo Clinic, Rochester, Minnesota 55901, U.S.A.

Chiba, G. (467, 485), Department of Orthopaedic Surgery, Nagasaki University, School of Medicine, 7-1, Sakamoto-machi, Nagasaki-shi, Nagasaki 852, Japan

Chikama, H. (97, 110), Department of Orthopaedic Surgery, Faculty of Medicine, Kyushu University, 3-1-1, Maidashi, Higashi-ku, Fukuoka, 812, Japan

Cho, K. (648), Institute of Health and Sport Science, The University of Tsukuba, Sakura-mura, Niihari-gun, Ibaraki 305, Japan

Clarys, Jan P. (951), Instituut voor Morfologie—Experimental Anatomy —Faculty of Medicine, Vrije Universiteit, Laarbeeklaan 103, B-1090 Brussels, Belgium

Conati, E. (1104), Department of Mechanical Engineering, Massachusetts Institute of Technology, Cambridge, Massachusetts 02139, U.S.A.

Cornelis, Jan (986), Department of Electronics, Vrije Universiteit Brussel, Pleinlaan 2, 1050 Brussels, Belgium

Cotton, C.E. (553), Department of Kinanthropology, University of Ottawa, Ottawa, Ontario, Canada K1N 6N5

Crowninshield, R.D. (1023), Biomechanics Laboratory, Department of Orthopaedic Surgery, University of Iowa Hospitals, Iowa City, Iowa, U.S.A.

Dainty, David A. (553), Department of Kinanthropology, University of Ottawa, Ottawa, Ontario, Canada K1N 6N5

Dalrymple, G. (1104), Department of Mechanical Engineering, Massachusetts Institute of Technology, Cambridge, Massachusetts 02139, U.S.A.

Davis, Ken (915), School of Education, Deakin University, Victoria 3217, Australia

Docter G.L. (363), Department of Anatomy and Biomechanics, Vrije Universiteit, 1007 MC Amsterdam, Pustbue 7161, The Netherlands

Doi, T. (105), Department of Orthopaedic Surgery, Faculty of Medicine, University of Tokyo, 7-3-1, Hongo, Bunkyo-ku, Tokyo 113, Japan

Ebashi, Hiroshi (895), Physical Fitness Research Institute, Meiji Foundation of Health and Welfare, 1-1-18, Shiroganedai, Minato-ku, Tokyo 108, Japan

Ekblom, Berit (567), Department of Human Movement Studies, University of Queensland, St. Lucia Queensland 4067, Australia

Ekström, Hans (861), Department of Mechanical Engineering, Linköping Institute of Technology, Linköping University, S-581 83 Linköping, Sweden

Engin, Ali Erkan (125), Department of Engineering Mechanics, The Ohio State University, Boyd Laboratory, 155 West Woodruff Avenue, Columbus, Ohio 43210, U.S.A.

Enoka, R.M. (301), Department of Kinesiology, University of Washington, Seattle, Washington 98195, U.S.A.

Ensink, J. (363), Department of Anatomy and Biomechanics, Vrije Universiteit, 1007 MC Amsterdam, Pustbue 7161, The Netherlands

Fabian, D.F. (1075), Biomechanics Department, Hospital for Special Surgery, Cornell University, Medical College, 535 East 70 Street, New York 10021, USA

Felici, F. (669), Laboratorio di Biomeccanica, Instituto di Fisiologia Umana, Università degli Studi, Città Universitaria - 00100 Roma, Italy

Fidelus, Kazimirez (1175), Institute of Sport, Ceglowska Str., 68-70, 01-809 Warsaw, Poland

Figura, F. (669), Laboratorio di Biomeccanica, Institute di Fisiologia Umana, Università degli Studi, Città Universitaria - 00100 Roma, Italy

Fucci, S. (851), Institute of Sports

Medicine of C.O.N.I., Via dei Campi Sportivi 46, 00197 Roma, Italy

Fuchimoto, T. (754), Laboratory of Exercise Physiology, Osaka College of Physical Education, 1-1, Gakuen-cho, Ibaraki-shi, Osaka 567, Japan

Fujimaki, Etsuo (141, 162), Department of Orthopaedic Surgery, School of Medicine, Showa University, 1-5-8, Hatanodai, Shinagawa-ku, Tokyo 142, Japan

Fujimatsuo, H. (604), Laboratory for Exercise Physiology and Biomechanics, School of Physical Education, Chukyo University, 101, Tokodate, Kaizu-cho, Toyota, Aichi 470-03, Japan

Fujita, K. (604), Laboratory for Exercise Physiology and Biomechanics, School of Physical Education, Chukyo University, 101, Tokodate, Kaizu-cho, Toyota, Aichi 470-03, Japan

Fujita, Masaaki (467, 485), Department of Orthopaedic Surgery, Nagasaki University, School of Medicine, 7-1, Sakamoto-machi, Nagasaki-shi, Nagasaki 852, Japan

Fujiwara, Katsuo (209), Institute of Health and Sport Science, The University of Tsukuba, Sakura-mura, Niihari-gun, Ibaraki 305, Japan

Fukashiro, Senshi (258), Laboratory for Exercise Physiology, Faculty of Education, University of Tokyo, 7-3-1, Hongo, Bunkyo-ku, Tokyo 113, Japan

Fukubayashi, Toru (105), Department of Orthopaedic Surgery, Faculty of Medicine, University of Tokyo, 7-3-1, Hongo, Bunkyo-ku, Tokyo 113, Japan

Fukunaga, Tetsuo (676, 959), Department of Exercise Physiology and Sports Science, College of General Education, University of Tokyo, 3-8-1, Komaba, Meguro-ku, Tokyo 153, Japan

Funk, Sandy (869), Department of Physical Education for Women, Biomechanics Laboratory, Washington State University, Pullman, Washington 99164, U.S.A.

Furukawa, Ryozoh (503), Department of Orthopedic Surgery, Aichi Medical University, Nagakute-cho, Aichi-gun, Aichi 480-11, Japan

Furuya, K. (1198), Department of Orthopaedics Surgery, Tokyo Medical and Dental University, 1-5-45, Yushima, Bunkyo-ku, Tokyo 113, Japan

Gheluwe, B. Van (876, 986), Vrije Universiteit Brussel, H.I.L.O.K. Pleinlaan 2, 1050 Brussels, Belgium

Gollnick, Philip D. (9), Department of Physical Education for Men, Washington State University, Pullman, Washington 99164, U.S.A.

Goto, H. (419), Osaka City University, Osaka 558, Japan

Goto, Yukihiro (1097), Department of Physical Education, Osaka City University, Sumiyoshi, Osaka 558, Japan

Goya, Toshiaki (683), Aichi University of Education, 1, Hirosawa, Igaya-cho, Kariya-shi, Aichi 448, Japan

Grainger, J. (1239), Department of Kinesiology, University of Waterloo, Waterloo, Ontario, Canada N2L 3G1

Grieve, D.W. (527), Royal Free Hospital, School of Medicine, University of London, Biomechanics Laboratory, Department of Anatomy, Clinical Sciences Building, The Royal Free Hospital, Pond Street, London NW3 2QG, England

Hamill, J. (251, 635), Biomechanics/

Sports Medicine Laboratory, Department of Physical Education, University of Oregon, Eugene, Oregon 97403, U.S.A.

Hang, Yi-Shiong (70), National Taiwan University Hospital, Chang-te Street, Taipei, Taiwan, R.O.C.

Hasegawa, Tatsuhiko (89), Century Research Center Corporation, 4-68, Kitakyutaro, Higashi-ku, Osaka 541, Japan

Hashihara, T. (737), Institute of Health and Sport Science, The University of Tsukuba, Sakura-mura, Niihari-gun, Ibaraki 305, Japan

Hashimoto, Fujio (157, 440), College of General Education, Osaka Electro-Communication University, 18-8, Hatsu-machi, Neyagawa-shi, Osaka 572, Japan

Hatano, Izumi (157), Department of Orthopaedic Surgery, Kansai Medical University, 1, Fumizono-cho, Moriguchi-shi, Osaka 570, Japan

Hattori, Tomokazu (503), Department of Orthopedic Surgery, Aichi Medical University, Nagakute-cho, Aichi-gun, Aichi 480-11, Japan

Hay, James G. (923, 939), Department of Physical Education, University of Iowa, Iowa City, Iowa 52242, U.S.A.

Hayashi, Ryoichi (597, 971), Institute of Equilibrium Research, Gifu University School of Medicine, Tsukasa-machi 40, Gifu 500, Japan

Hayashi, S. (1198), Department of Orthopedic Surgery, Tokyo Medical and Dental University, 1-5-45, Yushima, Bunkyo-ku, Tokyo 113, Japan

Hayashi, T. (467, 485), Department of Orthopaedic Surgery, Nagasaki University, School of Medicine, 7-1,

Sakamoto-machi, Nagasaki-shi, Nagasaki 852, Japan

Hennig, Ewald M. (1081, 1089), Biomechanics Laboratory, The Pennsylvania State University, University Park, Pennsylvania 16802, U.S.A.

Hermann, H. (1053), Deutsche Hochschule für Körperkultur, 7010 Leipzig, F.-L.-Jahn-Allee 59, D.D.R.

Himeno, Shinkichi (132), Department of Orthopedic Surgery, Fukuoka Children's Hospital, 2-5-1, Tojin-machi, Chuo-ku, Fukuoka 810, Japan

Hinrichs, Richard N. (641), Department of Physical Education, University of South Carolina, Columbia, South Carolina 29208, U.S.A.

Homma, Saburo (189, 444), Department of Physiology, School of Medicine, Chiba University, 1-8-1, Inohana, Chiba 280, Japan

Honjoh, Hiroshi (503), Department of Orthopedic Surgery, Aichi Medical University, Nagakute-cho, Aichi-gun, Aichi 480-11, Japan

Hori, Masami (503), Department of Orthopedic Surgery, Aichi Medical University, Nagakute-cho, Aichi-gun, Aichi 480-11, Japan

Hoshikawa, Tamotsu (498, 663, 683), Aichi Prefectural University, 3-28, Takada-cho, Mizuho-ku, Nagoya 467, Japan

Howard, A. (1223), Department of Human Movement Studies, University of Queensland, St. Lucia Queensland 4067, Australia

Hutton, R.S. (301), Department of Kinesiology, University of Washington, Seattle, Washington 98195, U.S.A.

Hyodo, K. (959), Department of Exercise Physiology and Sports Science, College of General Education, University of Tokyo, 3-8-1, Komaba, Meguro-ku, Tokyo 153, Japan

Igarashi, Hisato (787), Department of Health, Physical Education and Recreation, University of Oklahoma, 151 West Brooks, Norman, Oklahoma 73019, U.S.A.

Ikegami, Yasuo (963), Research Center of Health, Physical Fitness and Sports, Nagoya University, Furo-cho, Chikusa-ku, Nagoya 464, Japan

Inoue, Yoshimitsu (1097), Kobe University, School of Medicine, 7-12, Kusunoki-cho, Ikuta-ku, Kobe 650, Japan

Insall, J.N. (1075), Biomechanics Department, Hospital for Special Surgery, Cornell University, Medical College, 535 East 70 Street, New York 10021, U.S.A.

Ishii, Kihachi (239, 773), Laboratory of Physiological and Kinesiological Performance, Nippon College of Health and Physical Education, 7-1-1, Fukazawa, Setagaya-ku, Tokyo 158, Japan

Ishii, Nobuko (773), Laboratory of Physiological and Kinesiological Performance, Nippon College of Health and Physical Education, 7-1-1, Fukazawa, Setagaya-ku, Tokyo 158, Japan

Ishiko, Toshihiro (816), Juntendo University, 5-4-54, Fujisaki, Narashino-shi, Chiba 275, Japan

Ito, Akira (754), Laboratory of Exercise Physiology, Osaka College of Physical Education, 1-1, Gakuen-cho, Ibaraki-shi, Osaka 567, Japan

Ito, Masami (1129), Automatic Control Laboratory, School of Engineering, Nagoya University, Furo-cho, Chikusa-ku, Nagoya 464, Japan

Iwai, Akira (380), Miyagi Education

College, Aoba, Aramaki, Sendai-shi, Miyagi 980, Japan

Iwai, Takeshi (773), Laboratory of Physiological and Kinesiological Performance, Nippon College of Health and Physical Education, 7-1-1, Fukazawa, Setagaya-ku, Tokyo 158, Japan

Iwata, Kazuaki (1160), Department of Production Engineering, Faculty of Engineering, Kobe University, Rokko, Nada-ku, Kobe 657, Japan

Jack, Martha L. (889), P.O. Box 776, Richland, Washington 99352, U.S.A.

Jonsson, Bengt (561), Work Physiology Division, National Board of Occupational Safety and Health, Box 6104, S-900 06 Umeå, Sweden

Kakeno, Hidetatsu (591), Toyota Technical College, 2-1, Eisei-cho, Toyota-shi, Aichi 471, Japan

Kameyama, Osamu (157), Department of Orthopaedic Surgery, Kansai Medical University, 1, Fumizono-cho, Moriguchi-shi, Osaka 570, Japan

Kanehisa, Hiroaki (258), Laboratory for Exercise and Biomechanics, Faculty of Education, University of Tokyo, 7-3-1, Hongo, Bunkyo-ku, Tokyo 158, Japan

Kaneko, Masahiro (754), Laboratory of Exercise Physiology, Osaka College of Physical Education, 1-1, Gakuen-cho, Ibaraki-shi, Osaka 567, Japan

Karpeev, A.G. (1032), Department of Biomechanics, Institute of Physical Culture, Maslennikova 144, 644063 Omsk, U.S.S.R.

Kashiwase, Toshio (1011), Department of Electrical Engineering, Chiba University, 1-33, Yayoi-cho, Chiba 260, Japan

Katamoto, Shizuo (816), Juntendo

University, 5-4-54, Fujisaki, Narashino-shi, Chiba 275, Japan

Kato, Hisao (1167), Nagoya Municipal Industrial Research Institute, 3-24, Rokuban-cho, Atsuta-ku, Nagoya 456, Japan

Katoh, Yoshihisa (490), Department of Orthopedics, Mayo Clinic, Rochester, Minnesota 55901, U.S.A.

Kawabe, Shoko (231), Laboratory of Human Movements, Faculty of Letters, Nara Women's University, Nara 630, Japan

Kawachi, S. (1198), Department of Orthopedics Surgery, Tokyo Medical and Dental University, 1-5-45, Yushima, Bunkyo-ku, Tokyo 113, Japan

Kawahats, Kiyonori (289), Human Performance Laboratory, College of Liberal Arts, Kyoto University, Yoshida-Nihonmatsu-cho, Sakyo-ku, Kyoto 606, Japan

Kawai, Tadahiko (132), Institute of Industrial Science, Tokyo University, 7-22-1, Roppongi, Minato-ku, Tokyo 106, Japan

Kawano, Tsuneo (1160), Department of Production Engineering, Faculty of Engineering, Kobe University, Rokko, Nada-ku, Kobe 657, Japan

Kazei, Nobuyuki (419), Bukkyo University, 96, Kitahananobo-cho, Murasakino, Kita-ku, Kyoto 603, Japan

Kijima, Akira (239), Department of Literature, Laboratory of Physical Education, University of Kanto Gakuin, 4834, Mutsuura-cho, Kanazawa-ku, Yokohama-shi 236, Japan

Kinoshita, Hiroshi (574), Biomechanics/Sports Medicine Laboratory, Department of Physical Education, University of Oregon, Eugene, Oregon 97403, U.S.A.

Kira, H. (485), Department of Orthopaedic Surgery, Nagasaki University School of Medicine, 7-1, Sakamoto-nachi, Nagasaki 852, Japan

Kito, Nobukazu (498), Aichi University of Education, 1, Hirosawa, Igaya-cho, Kariya-shi, Aichi 448, Japan

Klinger, Anne K. (882), Clatsop Community College, 16th and Jerome Streets, Astoria, Oregon 97103, U.S.A.

Kojima, Takeji (321), Department of Physical Education, University of Tokyo, Komaba, Meguro-ku, Tokyo 153, Japan

Kondo, M. (959), Department of Exercise Physiology and Sports Science, College of General Education, University of Tokyo, 3-8-1, Komaba, Meguro-ku, Tokyo 153, Japan

Kornecki, Stefan (244), Academy of Physical Education, Biomechanics Laboratory, Al. Olimpijska 35, 51-612 Wrocław, Poland

Kulig, Kornelia (1144), Academy of Physical Education, Biomechanics Laboratory, Al. Olimpijska 35, 51-612 Wroclaw, Poland

Kumamoto, Minoyori (157, 419, 440, 809, 828), College of Liberal Arts, Kyoto University, Yoshida-Nihonmatsu-cho, Sakyo-ku, Kyoto 606, Japan

Kuriyama, Setsuro (141), Department of Orthopaedic Surgery, School of Medicine, Showa University, 1-5-8, Hatanodai, Shinagawa-ku, Tokyo, Japan

Kurokawa, Takao (294), Department of Biophysical Engineering, Faculty of Engineering Science, Osaka University, Machikaneyama-cho, 1-1, Toyonaka, Osaka 560, Japan

Kurosawa, H. (105), Department of Orthopaedic Surgery, Faculty of Medicine, University of Tokyo, 7-3-1 Hongo, Bunkyo-ku, Tokyo 113, Japan

LaFortune, Mario A. (928), Biomechanics Laboratory, The Pennsylvania State University, University Park, Pennsylvania 16802, U.S.A.

Landjerit, B. (455), Laboratoire de Biomecanique, École Nationale Supérieure des Arts et Metiers, Paris, France

Lanshammar, Håkan (397, 1123), Institute of Technology, Uppsala University, Box 256, S-751 21, Uppsala, Sweden

Laughlin, Cynthia K. (903), Department of Physical Education for Women, Biomechanics Laboratory, Washington State University, Pullman, Washington 99164, U.S.A.

Laughman, R. Keith (403, 490), Department of Orthopedics, Mayo Clinic, Rochester, Minnesota 55901, U.S.A.

Leemputte, M.F. Van (264, 997, 1138), Laboratory for Biomechanics, Institut voor Lichamelijke Opleiding Katholieke Universiteit te Leuven Tervuursevest 101 B 3030 Heverlee, Belgium

Leo, Tommaso (1067), Instituto di Automatica, Universita degli Studi, Ancona, Italy

Leonardo, Maria (851), Institute of Sports Medicine of C.O.N.I., Via dei Campi Sportivi 46, 00197 Roma, Italy

Lewillie, Léon (978), Unité de Recherche de Bioméchanique du Movement, Université Libre de Bruxelles ISEPK—Laboratoire de l'Effort, C.P. 168, 28 av. Paul Héger B-1050 Bruxelles, Belgium

Macellari, Velio (1067), Laboratorio di Technologie Biomediche, Instituto Superiore di Sanità, Roma, Italy

Macmillan, Norman H. (1081, 1089), Materials Research Laboratory, Pennsylvania State University, University Park, Pennsylvania 16802, U.S.A.

Maeshima, T. (816), Senshu University, 3-8, Kanda Jinbo-cho, Chiyoda-ku, Tokyo 101, Japan

Mann, Robert W. (1104, 1181), Department of Mechanical Engineering, Massachusetts Institute of Technology, 77 Massachusetts Avenue, Rm. 3-144, Cambridge, Massachusetts 02139, U.S.A.

Mano, T. (281), Department of Physiology, Hamamatsu University, School of Medicine, Hamamatsu 431-31, Japan

Marchetti, Marco (669), Laboratorio di Biomeccanica, Instituto di Fisiologia Umana, Università degli Studi, Città Universitaria—00100 Roma, Italy

Marhold, Gert (1053), Deutsche Hochschule für Körperkultur, 7010 Leipzig, F.-L.-Jahn-Allee 59, D.D.R.

Martin, E.E. (1032), Department of Biomechanics, Institute of Physical Culture, Maslennikova 144, 644063 Omsk, U.S.S.R.

Maruyama, Hirotake (419), Seibo Junior College, Kyoto 612, Japan

Mason, Michael (553), Department of Kinanthropology, University of Ottawa, Ottawa, Ontario, K1N 6NS Canada

Massez, C. (951), Instituut voor Morfologie—Experimental Anatomy—Faculty of Medicine, Vrije Universiteit, Laarbeeklaan 103, B-1090 Brussels, Belgium

Masuda, Makoto (423), Department of Physiology, The Jikei University,

School of Medicine, 3-25-8, Nishishinbashi, Minato-ku, Tokyo 105, Japan

Matake, Tomokazu (1115), Department of Mechanical Engineering, Nagasaki University, 1-14, Bunkyo-cho, Nagasaki 852, Japan

Matoba, Hideki (217), Laboratory of Biomechanics and Physiology, College of General Education, Yamaguchi University, 1-1677, Yoshida, Yamaguchi 753, Japan

Maton, Bernard (455), Laboratoire de Physiologie du Travail, CHU Pitié-Salpetrière, 91 Bd. de l'Hôpital, Paris 75634 cedex 13, France

Matsui, Hideji (3, 498, 683), Research Center of Health, Physical Fitness and Sports, Nagoya University, Furo-cho, Chikusa-ku, Nagoya 464, Japan

Matsuo, Akifumi (676, 959), Department of Exercise Physiology and Sports Science, College of General Education, University of Tokyo, 3-8-1, Komaba Meguro-ku, Tokyo 153, Japan

Matsuo, T. (373), Shinko-En Children Hospital, Kaminofu, Shingu, Kasuya 811-01, Japan

Matsusaka, N. (467, 485), Department of Orthopaedic Surgery, Nagasaki University, School of Medicine, 7-1, Sakamoto-machi, Nagasaki 852, Japan

McGill, S. (553), Department of Kinanthropology, University of Ottawa, Ottawa, Ontario, Canada K1N 6N5

Miki, Shunichiroh (503), Department of Orthopedic Surgery, Aichi Medical University, Nagakute-cho, Aichi-gun, Aichi 480-11, Japan

Miller, Doris I. (822), Department of Kinesiology, University of Washing-

ton, Seattle, Washington 98195, U.S.A.

Mimatsu, K. (223, 413), Institute for Developmental Research, Aichi Prefectural Colony, 713-8, Kamiya-cho, Kasugai, Aichi 480-03, Japan

Misaki, Norimasa (1160), Department of Production Engineering, Faculty of Engineering, Kobe University, Rokko, Nada, Kobe 657, Japan

Mishima, Ken (294), Department of Biophysical Engineering, Faculty of Engineering Science, Osaka University, 1-1, Machikaneyama-cho, Toyonaka, Osaka 560, Japan

Mita, Katsumi (223, 413), Institute for Developmental Research, Aichi Prefectural Colony, 713-8, Kamiya-cho, Kasugai, Aichi 480-03, Japan

Mita, Tsutomu (1011), Department of Electrical Engineering, Chiba University, 1-33, Yayoi-cho, Chiba 260, Japan

Mitarai, Genyo (281), Department of Aerospace Physiology, Research Institute of Environmental Medicine, Nagoya University, Furo-cho, Chikusa-ku, Nagoya 464, Japan

Miyake, Akihide (597, 971), Institute of Equilibrium Research, Gifu University, School of Medicine, Tsukasa-machi 40, Gifu 500, Japan

Miyamura, Miharu (963), Research Center of Health, Physical Fitness and Sports, Nagoya University, Furo-cho, Chikusa-ku, Nagoya 464, Japan

Miyanaga, Yutaka (105), Department of Orthopaedic Surgery, Faculty of Medicine, University of Tokyo, 7-3-1, Hongo, Bunkyo-ku, Tokyo 113, Japan

Miyashita, Mitsumasa (180, 258, 480, 629, 842), Laboratory for Biomechanics and Exercise Physiology, Faculty of Education, University of

Tokyo, 7-3-1, Hongo, Bunkyo-ku, Tokyo 113, Japan

Miyozaki, M. (444, 467), Department of Orthopaedic Surgery, Nagasaki University, School of Medicine, 7-1, Sakamoto-machi, Nagasaki 852, Japan

Mizuno, Yoshio (597), Institute of Equilibrium Research, Gifu University, School of Medicine, 40, Tsukasa-machi, Gifu 500, Japan

Mizutani, Shiro (663), Mie University, 1515, Uehama-cho, Tsu-shi 514, Japan

Mizutani, Y. (380), Department of Orthopaedic Surgery, Akita University, School of Medicine, 1-1-1, Hondo, Akita 010, Japan

Monte, A. Dal (851), Institute of Sports Medicine of C.O.N.I. Via dei Campi Sportivi 46, 00197 Roma, Italy

Morecki, Adam (341), International Federation for the Theory Machines and Mechanisms, Central Office, Al. Neipodleglosci, 222 R. 206, 00-663 Warszawa, Poland

Mori, Takemi (180), Department of Orthopaedic Surgery, Tokyo Welfare-Pension Hospital, 23, Tsukudo-cho, Shinjuku-ku, Tokyo 162, Japan

Morimoto, Shigeru (423), Department of Physiology, The Jikei University School of Medicine, 3-25-8, Nishishinbashi, Minato-ku, Tokyo 105, Japan

Moritani, Toshio (312, 432), Bio-dynamics Laboratory, Department of Physical Education, University of Texas at Arlington, Arlington, Texas 76019, U.S.A.

Moriwaki, Toshimichi (1160), Department of Production Engineering, Faculty of Engineering, Kobe University, Rokko, Nada, Kobe 657, Japan

Morrey, B.F. (490), Department of

Orthopedics, Mayo Clinic, Rochester, Minnesota 55901, U.S.A.

Morrison, W. (553), Department of Kinanthropology, University of Ottawa, Ottawa, Ontario, Canada K1N 6N5

Munro, A.R. (306), Department of Human Movement and Recreation Studies, University of Western Australia, Nedlands, W.A. 6009, Australia

Murakami, Naotoshi (217), Department of Physiology, Yamaguchi University, School of Medicine, 1144, Oguchi, Ube-shi, Japan

Murakami, Teruo (97, 110), Department of Mechanical Engineering, Kyushu University, 6-10-1, Hakozaki, Higashi-ku, Fukuoka 812, Japan

Muraki, Yukito (762), Institute of Health and Sport Science, University of Tsukuba, Sakura-mura, Niihari-gun, Ibaraki 305, Japan

Murase, Ken-ichi (1023), Department of 2nd Physiology, Fukushima Medical College, 5-75, Sugizuma-cho, Fukushima 960, Japan

Muro, Masuo (312, 432), Tokyo College of Pharmacy, 1432-1, Horinouchi, Hachioji-shi, Tokyo 192-03, Japan

Mutoh, Yoshiteru (165, 180), Department of Orthopaedic Surgery, Tokyo Welfare-Pension Hospital, 23, Tsukudo-cho, Shinjuku-ku, Tokyo 162, Japan

Nachemson, A.L. (543), Departments of Clinical Neurophysiology and Orthopedic Surgery, Sahlgren Hospital, Göteborg, Sweden

Nagata, Akira (412, 432), Laboratory of Bio-dynamics, Faculty of Science, Tokyo Metropolitan University, 1-1-1, Yakumo, Meguro-ku, Tokyo 146, Japan

Nakagawa, Hiroshi (419), Osaka

University of Economics, Higashiyodogawa-ku, Osaka 533, Japan

Nakagawa, N. (809), Osaka University of Economics, Osaka 533, Japan

Nakamura, Yoshio (157, 180), Laboratory for Exercise, Physiology and Biomechanics, Faculty of Education, University of Tokyo, 7-3-1, Hongo, Bunkyo-ku, Tokyo 113, Japan

Narikiyo, Tatsuo (1129), Automatic Control Laboratory, School of Engineering, Nagoya University, Furocho, Chikusa-ku, Nagoya 464, Japan

Nicol, Klaus (1231), Institut für Leibesübungen, Westfälische Wilhelms—Universität Münster, Horstmarer Landweg 62B, 4400 Münster, B.R.D.

Nigg, Benno M. (1041), Biomechanics Laboratory, Faculty of Physical Education, The University of Calgary, 2500 University Drive N.W., Calgary, Alberta, Canada T2N 1N4

Niinomi, Shigeru (1215), Labor Accident Prosthetic and Orthotics Center, 1-10-5, Komei, Minato-ku, Nagoya 455, Japan

Nishio, A. (97, 110), Department of Orthopaedic Surgery, Faculty of Medicine, Kyushu University, 3-1-1, Maidashi, Higashi-ku, Fukuoka 812, Japan

Nishizaki, Hiromi (97, 110), Department of Orthopaedic Surgery, Faculty of Medicine, Kyushu University, 3-1-1, Maidashi, Higashi-ku, Fukuoka 812, Japan

Nissinen, Mauno A. (781), Institut für Sport und Sportwissenschaften, Ginnheimer Landstr. 39, 6 Frankfurt/ Main 90, B.R.D.

Niwa, Shigeo (503), Department of Orthopaedic Surgery, Aichi Medical University, 21, Nagakute-cho, Aichi-gun, Aichi 480-11, Japan

Nomura, Haruo (1160), Department of Health and Physical Education, Faculty of Liberal Arts, Kobe University, Rokko, Nada, Kobe 657, Japan

Nomura Takeo (842), Institute of Sports Science, The University of Tsukuba, Sakura-mura, Niihari-gun, Ibaraki 305, Japan

Norimatsu, Toshiharu (467, 485), Department of Orthopaedic Surgery, Nagasaki University, School of Medicine, 7-1, Sakamoto-machi, Nagasaki 852, Japan

Norman, Robert William (1239), Department of Kinesiology, University of Waterloo, Waterloo, Ontario, Canada N2L 3G1

Notte, Volker (695), Institut für Biomechanik, Deutsche Sporthochschule, Carl-Diem-Wag, D-5000, Köln, B.R.D.

Nowacki, Zbigniew (1144), Academy of Physical Education, Biomechanics Laboratory, Al. Olimpijska 35, 51-612, Wrocław, Poland

Nyssen, M. (986), Medical Informatics, Vrije Universiteit Brussel, Pleinlaan 2, 1050 Brussels, Belgium

Ohenheimer, D. (1104), Department of Mechanical Engineering, Massachusetts Institute of Technology, Cambridge, Massachusetts 02139, U.S.A.

Ohkuwa, Tetsuo (963), Nagoya Institute of Technology, Gokiso-cho, Showa-ku, Nagoya 460, Japan

Ohmichi, Hitoshi (258, 480), Laboratory for Biomechanics and Exercise Physiology, Faculty of Education, University of Tokyo, 7-3-1, Hongo, Bunkyo-ku, Tokyo 113, Japan

Ohtsuki, N. (97, 110), Department of

Mechanical Engineering, Kyushu University, 6-10-1, Hakozaki, Higashi-ku, Fukuoka 812, Japan

Ohtsuki, Tatsuyuki (231), Laboratory of Human Movements, Faculty of Letters, Nara Women's University, Nara 630, Japan

Oka, Hideo (157, 419, 809), Osaka Kyoiku University, High School, 1-5-1, Midorigaoka, Ikeda-shi, Osaka 563, Japan

Okada, Morihiko (209), Institute of Health and Sport Science, The University of Tsukuba, Sakura-mura, Niihari-gun, Ibaraki 305, Japan

Okamoto, Tsutomu (157, 419, 809, 829), Kansai Medical School, 18-89, Uyamahigashi-cho, Hirakata-shi, Osaka 573, Japan

Okawa, Yoshikuni (147), Faculty of Engineering, Gifu University, 3-1, Naka-monzen-cho, Kakamigahara, Gifu 504, Japan

Okumura, H. (485), Department of Radiation Biophysics, Atomic Disease Institute, Nagasaki University, School of Medicine, 7-1, Sakamoto-machi, Nagasaki 852, Japan

Okumura, Shinji (1198), Department of Orthopedic Surgery, Tokyo Medical and Dental University, 1-5-45, Yushima, Bunkyo-ku, Tokyo 113, Japan

Oonishi, Hironobu (89), Department of Orthopedic Surgery, Osaka-Minami National Hospital, 677-6, Kido-cho, Kawachinagano, Osaka, 586, Japan

Örtengren, Roland (386, 543), Departments of Clinical Neurophysiology and Orthopedic Surgery, Sahlgren Hospital, S-413 45 Göteborg, Sweden

Osternig, L.R. (251, 635), Biomechanics/Sports Medicine Laboratory, Department of Physical Education,

University of Oregon, Eugene, Oregon 97403, U.S.A.

Otis, James C. (1075), Biomechanics Department, Hospital for Special Surgery, Cornell University, Medical College, 535 East 70 Street, New York 10021, U.S.A.

Ottonsson, Stig (861), Department of Mechanical Engineering, Linköping Institute of Technology, Linköping University, S-581 83 Linköping, Sweden

Payne, Andrew H. (746), Physical Education Department, University of Birmingham, Birmingham, P.O. Box 363, Birmingham B15 2TT, England

Pedersen, D.R. (1023), Biomechanics Laboratory, Department of Orthopaedic Surgery, University of Iowa Hospitals, Iowa City, Iowa 52242, U.S.A.

Peeraer, L. (1138), Laboratory for Biomechanics, Instituut voor Lichamelijke Opleiding, Katholieke Universiteit te Leuven, Tervuursevest, 101, B-3030 Heverlee, Belgium

Peres, G. (455), Laboratoire de Physiologie du Travail du CNRS, CHU Pitie-Salpetriere 91, Boulevard de l'Hôpital, Paris 756 34 cedex 13, France

Persyn, U. (833), Institute of Physical Education, Katholieke Universiteit, Leuven, Tervuursevest 101, B-3030 Heverlee, Belgium

Philippe, C. (455), Laboratoire de Biomécanique, Ecole Nationale Superieure des Arts et Métiers, Paris, France

Piette, G. (951), Instituut voor Morfologie — Experimental Anatomy — Faculty of Medicine, Vrije Universiteit, Laarbeeklaan 103, B-1090 Brussels, Belgium

Prampero, Pietro E. di (703), Centro Studi di Fisiologia del Lavoro Muscolare del C.N.R., Milano, Italy

Purcell, Michael (869), Department of Physical Education for Women, Biomechanics Laboratory, Washington State University, Pullman, Washington 99164, U.S.A.

Putnam, Carol A. (688), School of Physical Education, Dalhousie University, Halifax, Nova Scotia, Canada B3H 3J5

Pyke, F.S. (306), Department of Sports Studies, Camberra College of Advanced Education, University of Western Australia, Nedlands, Western Australia 6009, Australia

Rau, Günter (513), Helmholtz-institute for Biomedical Engineering, Aachen, B.R.D.

Robeaux, R. (951), Instituut voor Morfologie — Experimental Anatomy — Faculty of Medicine, Vrije Universiteit, Laarbeeklaan 103, B-1090 Brussels, Belgium

Rossignol, P.F. (306), Department of Human Movement and Recreation Studies, University of Western Australia, Nedlands, Western Australia 6009, Australia

Rowell, Derek (1104, 1181), Department of Mechanical Engineering, Massachusetts Institute of Technology, 77 Massachusetts Avenue, Rm. 3-144, Cambridge, Massachusetts 02139, U.S.A.

Ryushi, T. (959), Department of Exercise Physiology and Sports Science, College of General Education, University of Tokyo, 3-8-1, Komaba, Meguro-ku, Tokyo 153, Japan

Saeki, C. (373), Shinko-En Children's Hospital, Kaminofu, Shingu, Kasuya 811-01, Japan

Sagawa, Kazunori (239), Laboratory of Physiological and Kinesiological Performance, Nippon College of Health and Physical Education, 7-1-1, Fukazawa, Setagaya-ku, Tokyo 158, Japan

Saibene, Franco (703), Centro Studi di Fisiologia de Lavoro Muscolare del C.N.R., Milano, Italy

Saito, Mitsuru (963), Toyota Technological Institute, 12-2, Hisakata Tempaku-ku, Nagoya 468, Japan

Saito, Shinichi (648, 762), Institute of Health and Sport Science, University of Tsukuba, Sakura-mura, Niiharigun, Ibaraki 305, Japan

Saito, Susumu (1023), Department of 2nd Physiology, Fukushima Medical College, 5-75, Sugizuma-cho, Fukushima 960, Japan

Sakamoto, T. (762), Institute of Health and Sport Science, University of Tsukuba, Sakura-mura, Niiharigun, Ibaraki 305, Japan

Sakurai, Shinji (629), Laboratory for Exercise Physiology and Biomechanics, Faculty of Education, University of Tokyo, 7-3-1, Hongo, Bunkyo-ku, Tokyo 113, Japan

Sawai, Kazuhiko (503), Department of Orthopedic Surgery, Aichi Medical University, Nagakute-cho, Aichi-gun, Aichi 480-11, Japan

Sawhill, J.A. (251, 635), Biomechanics/Sports Medicine Laboratory, Department of Physical Education, University of Oregon, Eugene, Oregon 97403, U.S.A.

Schandevijl, H. Van (876), Vrije Universiteit Brussel H.I.L.O.K. Pleinlaan, 2, 1050 Brussels, Belgium

Schneider, Erich (403, 490), M.E.M. Institut für Biomechanik, der Univeristät Bern, Murtenstrasse 35, CHO-3010 Bern, Switzerland

Schultz, A.B. (543), Department of Materials Engineering, University of Illinois at Chicago Circle, Chicago, Illinois, U.S.A.

Sellars, I.E. (116), Biomedical Engineering, Medical School, University of Cape Town, Observatory, 7925, Cape Town, South Africa

Seluyanov, V.N. (1152), State Central Institute of Physical Education, Department of Biomechanics, Syrenevyi Blvd. 4, 105008 Moscow, U.S.S.R.

Shibayama, Hidetaro (895), Physical Fitness Research Institute, Meiji Life Foundation of Health and Welfare, 1-1-18, Shiroganedai, Mibato-ku, Tokyo 108, Japan

Shibukawa, Kanji (648, 737, 762), Institute of Health and Sport Science, University of Tsukuba, Sakura-mura, Niihari-gun, Ibaraki 305, Japan

Shimada, Masatoshi (1097), Department of Physics, Osaka Kyoiku University, Tennoji, Osaka 543, Japan

Shirasaki, Y. (105), Department of Orthopaedic Surgery, Faculty of Medicine, University of Tokyo, 7-3-1, Hongo, Bunkyo-ku, Tokyo 113, Japan

Shitama, S. (373), Department of Control Engineering, Kyushu Institute of Technology, 1-1, Sensui, Tobata-ku, Kitakyushu 804, Japan

Spaepen, A.J. (264, 997, 1138), Laboratory for Biomechanics, Instituut voor Lichamelijke Opleiding, Katholieke Universiteit te Leuven, Tervuursevest 101, B-3030 Heverlee, Belgium

Stijnen, V.V. (264, 997, 1138), Laboratory for Biomechanics, Instituut voor Lichamelijke Opleiding, Katholieke Universiteit te Leuven, Tervuursevest 101, B-3030 Heverlee, Belgium

Strandberg, Lennart (397, 1123), National Board of Occupational Safety and Health, Accident Research Section, Arbetarskyddsstyrelsen, S-171 84, Solna, Sweden

Sugiura, Yasuo (165), Division of Orthopaedic Surgery, Nishio Municipal Hospital, 2-1, Hananoki-cho, Nishio-shi, Aichi 445, Japan

Suzuki, Kenji (380), Department of Orthopaedic Surgery, Akita University, School of Medicine, 1-1-1, Hondo, Akita 010, Japan

Suzuki, Masayasu (239), Laboratory of Physiological and Kinesiology Performance, Nippon College of Health and Physical Education, 7-1-1, Fukazawa, Setagaya-ku, Tokyo 158, Japan

Suzuki, Ryohei (57, 467, 485), Department of Orthopaedic Surgery, Nagasaki University, School of Medicine, 7-1, Sakamoto-machi, Nagasaki 852, Japan

Suzuki, Shuji (301, 444), Department of Physiology, School of Medicine, Kyorin University, 6-20-2, Shinkawa, Mitaka-shi, Tokyo 181, Japan

Suzuki, Yoshitaka (1215), Labor Accidents Prosthetics and Orthotics Center, 1-10-5, Komei, Minato-ku, Nagoya 455, Japan

Tachi, Susumu (1181), Mechanical Engineering Laboratory, Ministry of International Trade and Industry, Tsukuba Scientific Research City, Ibaraki 305, Japan

Tada, Shigeru (648, 737), Institute of Health and Sport Science, The University of Tsukuba, Sakura-mura, Niihari-gun, Ibaraki 305, Japan

Tagawa, Yoshihiko (1005), Educational Center for Information Processing, Kurume Institute of Technology, 2228-66, Mukaino, Kurume 830, Japan

Takahama, M. (380), Department of Orthopaedic Surgery, Akita University, School of Medicine, 1-1-1, Hondo, Akita 010, Japan

Takahashi, Goro (842), Institute of Sports Science, The University of Tsukuba, Sakura-mura, Niihari-gun, Ibaraki 305, Japan

Takata, Kazuyuki (591), Toyota Technical College, 2-1, Eisei-cho, Toyota-shi, Aichi 471, Japan

Takeuchi, Norio (132), Department of Orthopedic Surgery, Fukuoka Children's Hospital, 2-5-1, Tojin-machi, Chuo-ku, Fukuoka 810, Japan

Takeuchi, Shinya (591), Aichi University of Education, 1, Hirosawa, Igaya-cho, Kariya-shi, Aichi 448, Japan

Takeuchi, T. (1198), Department of Orthopedic Surgery, Tokyo Medical and Dental University, 1-5-45, Yushima, Bunkyo-ku, Tokyo 113, Japan

Tamura, Hiroshi (294), Department of Biophysical Engineering, Faculty of Engineering Science, Osaka University, 1-1, Machikaneyama-cho, Toyonaka, Osaka 560, Japan

Taniguchi, Takao (373), Kyushu Institute of Technology, 1-1, Sensui, Tobata-ku, Kitakyushu 804, Japan

Tashiro, Yoshihisa (141), Department of Orthopaedic Surgery, School of Medicine, Showa University, 1-5-8, Hatanodai, Shinagawa-ku, Tokyo 142, Japan

Tatara, Yoichi (471), Department of Engineering, Shizuoka University, 3-5-1, Johoku, Hamamatsu-shi, Shizuoka 432, Japan

Tateishi, Tetsuya (105), Mechanical Engineering Laboratory, 1-2, Namiki, Sakura-mura, Niihari-gun, Ibaraki 305, Japan

Tetewsky, A.T. (1104), Department of Mechanical Engineering, Massachusetts Institute of Technology, Cambridge, Massachusetts 02139, U.S.A.

Tezuka, Masataka (869), Meiji University, 1-9-1, Eifuku, Suginami-ku, Tokyo 168, Japan

Therrien, R. (582), Department of Mechanical Engineering, Faculty of Applied Sciences, University of Sherbrooke, Sherbrooke, Province of Quebec, Canada J1K 2R1

Tillberg, Bengt (567), Department of Human Work Science, University of Luleå, S-951 87, Luleå, Sweden

Tokuhara, Yasuhiko (440, 809), College of General Education, Teikoku Women's University, 6-173, Fujita-cho, Moriguchi-shi, Osaka 570, Japan

Toyonaga, T. (97, 110), Department of Orthopaedic Surgery, Faculty of Medicine, Kyushu University, 3-1-1, Maidashi, Higashi-ku, Fukuoka 812, Japan

Toyooka, Jiro (754), Laboratory of Exercise Physiology, Osaka College of Physical Education, 1-1, Gakuen-cho, Ibaraki-shi, Osaka 567, Japan

Toyoshima, Shintaro (683), Aichi Prefectural University, 3-28, Takada-cho, Mizuho-ku, Nagoya 467, Japan

Trozzi, V. (851), Institute of Sports Medicine of C.O.N.I., Via dei Campi Sportivi 46, 00197 Roma, Italy

Tsuchiya, Kazuo (1215), Labor Accident Prosthetic and Orthotics Center, 1-10-5, Komei, Minato-ku, Nagoya 455, Japan

Tsujino, Akira (1097), Department of Physical Education, Osaka Kyoiku University, 3-1-1, Jonan, Ikeda, Osaka 563, Japan

Tsukahara, Susumu (1023), Department of 2nd Physiology, Fukushima Medical College, 5-75, Sugizuma-cho, Fukushima 960, Japan

Uemura, Shokichi (141), Department of Orthopaedic Surgery, School of Medicine, Showa University, 1-5-8, Hatanodai, Shinagawa-ku, Tokyo 142, Japan

Ueya, Kiyomi (654), Department of Physical Education, Yamanashi University 4-4-37, Takeda, Kofu-shi 400, Japan

Umazume, Yoshiki (423), Department of Physiology, The Jikei University, School of Medicine, 3-25-8, Nishishinbashi, Minato-ku, Tokyo 105, Japan

Uemoto, Kazumi (971), Institute of Equilibrium Research, Gifu University, School of Medicine, Tsukasa-machi 40, Gifu 500, Japan

Usui, Shiro (1207), Information and Computer Sciences, Toyohashi University of Technology, 1-1, Hibarigaoka, Tempaku-cho, Toyohashi, Aichi 440, Japan

Vaughan, C.L. (116, 923, 939), Biomedical Engineering, University of Cape Town, South Africa

Verhetsel, D. (833), Institute of Physical Education, Katholieke Universiteit, Leuven te Tervuursevest 101, B-3030 Heverlee, Belgium

Veraecke, H. (833), Institute of Physical Education, Katholieke Universiteit Leuven te Tervuursevest 101, 3030 Heverlee, Belgium

Viitasalo, Jukka T. (271), Department of Biology of Physical Activity, University of Jyväskylä, FIN-40100, Jyväskylä 10, Finland

Watanabe, Kazuhiko (856), Laboratory of Physiology and Sports Biomechanics, School of Education,

Hiroshima University, 2-17, Midori, Fukuyama-shi, Hiroshima 720, Japan

Watanabe, Satoru (597, 971), Institute of Equilibrium Research, Gifu University, School of Medicine, Tsukasa-machi 40, Gifu 500, Japan

Watanabe, Shiroh (444), Department of Physiology, School of Medicine, Kyorin University, 6-20-2, Shinkawa, Mitaka-shi, Tokyo 181, Japan

Watanabe, Yosaku (591), Toyota Technical College, 2-1, Eisei-cho, Toyota-shi, Aichi 471, Japan

Welch, W. (171), Faculty of Medicine, Faculty of Applied Science, Free University of Brussels (V.U.B.), Pleinlaan 2, B-1050 Brussels, Belgium

Wielki, Czeslaw (1190), Université de Louvain, Fac. de Médicine, Inst. d'Educ. Phys. Lab. JECO, Place Pierre de Coubertain, B-1348 Louvain-La-Neuve, Belgium

Willems, E.J. (264, 997, 1138), Laboratory for Biomechanics, Instituut voor Lichamelijke Opleiding, Katholieke Universiteit te Leuven, Tervuursevest 101, B-3030 Heverlee, Belgium

Williams, Keith R. (641), Department of Physical Education, University of California at Davis, Davis, California 95616, U.S.A.

Wilson, Barry D. (1223), Department of Human Movement Studies, University of Queensland, St. Lucia, Queensland 4067, Australia

Winkel, Jörgen (567), Department of Human Work Science, University of Luleå, S-951 87, Luleå, Sweden

Winter, David A. (329, 1239), Faculty of Human Kinetics and Leisure Studies, Department of Kinesiology, University of Waterloo, Waterloo, Ontario, Canada N2L 3G1

Wit, Andrzej (1175), Institute of Sport, 01-809 Warsaw, Ceglowska Str. 68/70, Poland

Wood, G.A. (306), Department of Human Movement and Recreation Studies, University of Western Australia, Nedlands, Western Australia 6009, Australia

Yabe, Kyonosuke (223, 413), Institute for Developmental Research, Aichi Prefectural Colony, Kamiya-cho, Kasugai, Aichi 480-03, Japan

Yamaguchi, K. (467, 485), Department of Orthopaedic Surgery, Nagasaki University, School of Medicine, 7-1, Sakamoto-machi, Nagasaki-shi, Nagasaki 852, Japan

Yamaguchi, Toru (1011), Department of Electrical Engineering, Chiba University, 1-33, Kamiya-cho, Kasugai, Aichi 480-03, Japan

Yamamoto, Takashi (604), Laboratory for Exercise Physiology and Biomechanics, School of Physical Education, Chukyo University, 101, Tokodate, Kaizu-cho, Toyota, Aichi 470-03, Japan

Yamashita, Noriyoshi (440, 809), College Liberal Arts, Kyoto University, Yoshida-Nihonmatsu-cho, Sakyo-ku, Kyoto 606, Japan

Yamashita, Tadashi (373, 1005), Department of Control Engineering, Kyushu Institute of Technology, 1-1, Sensui, Tobata-ku, Kitakyushu 804, Japan

Yamazaki, Setsumasa (503), Department of Orthopedic Surgery, Aichi Medical University, Nagakute-cho, Aichi-gun, Aichi-ken, 480-11, Japan

Yamazaki, Yoshihiko (281), Department of Health and Physical Education, Nagoya Institute of Technology, Gokiso, Showa-ku, Nagoya 466, Japan

Yanagida, Yasuyoshi (1160), Department of Health and Physical Education, Faculty of Liberal Arts, Kobe University, Rokko, Nada, Kobe 657, Japan

Yata, H. (959), Department of Exercise Physiology and Sports Science College of General Education, University of Tokyo, 3-8-1, Komaba, Meguro-ku, Tokyo 153, Japan

Yoshida, Akira (842), Institute of Sports Science, The University of Tsukuba, 1-1-1, Tennodai, Sakuramura, Niihari-gun, Ibaraki 305, Japan

Yoshizawa, Masatada (809, 828), College Education, Fukui University, 3-9-1, Bunkyo, Fukui 910, Japan

Zatsiorsky, Vladimir M. (1152), State Central Institute of Physical Education, Department of Biomechanics, Syrenevyi Blvd. 4, 105008 Moscow, U.S.S.R.

Zawadzki, Jerzy (244), Academy of Physical Education, Biomechanics Laboratory, Al. Olimpijska 35, 51-612 Wrocław, Poland

Zetterberg, Carl (386), Departments of Clinical Neurophysiology and Orthopedic Surgery, Sahlgren Hospital, S-413 45 Göteborg, Sweden

OPENING
SESSION

Welcome by the Congress Chairman

It is my pleasure to meet with the special guests and friends from all over the world here today to open the 8th International Congress of Biomechanics.

I remember that the only representative from Japan to attend the first meeting in Zurich which was held as a working group of the ICSPE was Dr. Ikai who passed away in 1973. After he returned home, Dr. Ikai strongly stressed that we should also initiate the study of biomechanics in Japan.

The study or research in biomechanics is gaining in popularity year by year. And we received over 200 manuscripts for the current meeting. The fields of study are wide-spread and are divided into seven sections in this meeting.

I think that the principal aim of science during the last half of the 20th century and in the 21st century is to discover and produce new energy, and to obtain or maintain healthy life as human beings. Biomechanics is a science which has a role under the latter condition. Our knowledge of human beings has external implications, and it should always be fed back to improve the lives of human beings.

Biomechanics is such a science, and should be fed back to us. Today is the first day of the Congress, and it will continue for four days until July 24. I hope that this Congress will become a great stepping stone for the further development of biomechanics research. Please make valuable presentations out of your research work and produce active discussion. The memorial lectures for Dr. Wartenweiler and Dr. Ikai have been scheduled during this meeting.

Dr. Wartenweiler devoted his life to establishing the International Society of Biomechanics. Today, Mrs. Wartenweiler from Switzerland is attending this Opening Ceremony.

Dr. Ikai was the pioneer in this area of research in Japan. We also have Mrs. Ikai from Tokyo here today.

I would like to introduce Mrs. Wartenweiler and Mrs. Ikai.

Thank you very much.

Hideji Matsui D.M.S.
Chairman of VIIIth I.C.B.

Welcome by the Honorary Advisor

It is our great pleasure that the VIIIth International Congress of Biomechanics will be held in Nagoya, under the contribution of the researchers of our country.

Biomechanics is a newly developed science in the basic and interdisciplinary research area for human adaptation to modern industrialized society, and there is no doubt that biomechanics has an important role to play in that society.

Exchanges of research results at the international level are indispensable for the development of science. As an example, Nagoya University, with the aid of the Japan Society for the Promotion of Science, had previously invited Prof. R.C. Nelson, the President of the International Society of Biomechanics, as an invited professor for research cooperation.

I believe that the VIIIth International Congress of Biomechanics will lead to a further great step in biomechanics research through the meaningful exchange among the participants from almost 30 different countries.

I heartily welcome all of the participants to the Congress at Nagoya.

Naotaka Ishizuka, M.D.
Honorary Advisor of VIIIth I.C.B.
President of Nagoya University

Welcome by the Governor of the Aichi Prefecture

Mr. President, Vice Chairman Rau, ladies and gentlemen, and our distinguished guests:

You are most welcome to Nagoya City. From all over the world prominent scientists in biomechanics were invited, and the 8th International Congress of Biomechanics is ready to open in this City of Nagoya. We heartily welcome it.

The progress of science in the 20th century is really eye-opening. It was 80 years ago, in the early years of this century that the Wright Brothers made their first flight. Today we fire rockets to Mars and other planets far out in space as well as to the moon. Clues to solving the secret of the cosmos are now in our hands. Our challenge to science is fast and endless.

On the other hand, science is still powerless in certain respects. The human species is a creation of the Almighty, and its physical mechanism is itself a supreme work of art in this world. Even today's most up-to-date science is still helpless in explaining the riddle of our physical constitution and system. Complexity and exquisiteness are functions of the human body and logical elucidation of the same is attempted in this branch of science called biomechanics, whose future achievements are eagerly anticipated. Day and night, you are making assiduous researches, for which I offer my deep respect.

As you are already aware, the three prefectures of Tokai region centering on Nagoya are now drawing the keen attention of the world as a prospective site for the 1988 Olympics. You are actively engaged in researches of sport science, hygiene, physical education, and like studies, and you must, in my thinking have a profound interest in the Olympics. In politics, economics, culture, sports, and many other fields, this city is striving to be internationalized. I ask our guests to observe these efforts for themselves.

In human history an unprecedented and phenomenal development has been achieved in this 20th century. Toward the 21st century, energy, resources, and other new and important problems await our increased efforts for solution, and our future is not necessarily optimistic.

May this conference of biomechanics produce successful achievements

5

through your endeavors and may you continue your noble studies for the prosperity of us all hereafter as well. I again extend you my hearty welcome.

Thank you very much.

Yoshiaki Nakaya
Honorary Advisor of VIIIth I.C.B.
Governor of Aichi Prefecture

WARTEN-WEILER MEMORIAL LECTURE

Muscle Characteristics as
a Foundation of Biomechanics

Philip D. Gollnick
Washington State University, Pullman, Washington, U.S.A.

All human movement and athletic activity is a result of an intricate control of the skeletal muscles by the central nervous system. This is possible since skeletal muscles are aggregates of motor units with different contractile and metabolic properties which endow them with capabilities for a wide variety of contractile patterns. The size, that is the number of fibers per motor unit, their composition, that is the type of fibers, and the type of innervation of the individual motor units determines how each functions during activity. The control that the central nervous system has over motor activity extends from the capacity to activate a single motor unit to an activation of all motor units in a given muscle. The multiplicity of movement patterns that humans produce in daily life, extending from the most delicate and precise movements to massive displays of strength, is testimony of the ability to control the muscles and of their ability to adapt to a wide variety of constant patterns of use. The purpose of this article is to give a synopsis of the characteristics of the motor units found in muscle and to briefly cover some adaptations that occur in them with training.

Properties of Motor Units

Contractile Properties

Skeletal muscles contain two broad types of fibers that can be identified from their contractile properties (see Figure 1). One type of fiber requires a short and the other a long time to attain peak tension after activation (Bárány et al., 1965; Bárány, 1967; Barnard et al., 1971; Buchthal and Schmalbruch, 1980). On this basis these units can be identified simply as

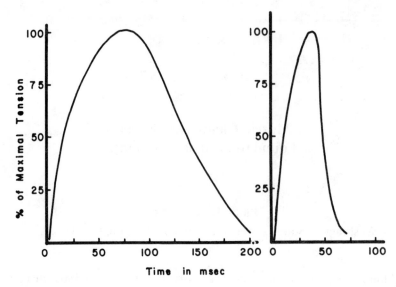

Figure 1 — A schematic representation of the time to peak tension and the return to resting tension in a slow-twitch (ST) muscle (left panel) and a fast-twitch (FT) muscle (right panel).

fast-twitch (FT) and slow-twitch (ST). Though there are differences in the absolute times to reach peak tension between animal species (Buchthal and Schmalbruch, 1980), the general pattern is that FT motor units usually attain peak tension in about ½ the time required for ST units. Similarly, the time for relaxation — the time most commonly measured is the ½ relaxation — is also about twice as long for ST as compared with FT motor units. These observations are true of humans as well as the other animals (Eberstein and Goodgold, 1968; Buchthal and Schmalbruch, 1969; Buchthal and Schmalbruch, 1970; Buchthal et al., 1973; Garnett et al., 1979).

The biochemical basis for the differences in contractile characteristics of the two major fiber types lies in the biochemical composition of the contractile elements. Bárány (1965, 1967) was one of the first to demonstrate that major differences exist in the rate that myosin splits ATP, that is, its ATPase activity. The maximal ATPase activity of myosin is positively correlated with contractile speed in a wide variety of animal species, including humans (Bárány, 1967; Barnard et al., 1971). For the human, the myosin ATPase activity of ST and FT muscle has been estimated to be 0.16 and 0.48 umoles of ATP split $\times$ mg^{-1} $\times$ min^{-1}, respectively (Essén et al., 1975). In the rat, a commonly studied animal, the maximal activity of myosin when activated by actin is about 0.50 and 1.40 umoles of ATP split $\times$ mg^{-1} $\times$ min^{-1} for the ST soleus and FT white vastus muscles, respectively (see Figure 2) (Watrus, 1980). This higher activity is consistent with the fact that the time to peak tension is shorter in

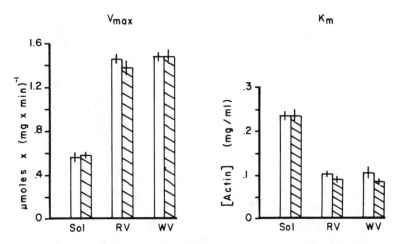

Figure 2—The maximal actin activated ATPase activity of myosin purified from the slow-twitch soleus (Sol), the fast-twitch red vastus (RV), and the fast-twitch white vastus (WV) of the rat. The open bars represent myosin prepared from control animals and the cross-hatched bars are values for myosin prepared from trained animals. The Km for myosin (the concentration of actin that produces ½ of the Vmax) is also given. Data are from Watrus (1980).

rat as compared to human skeletal muscle (Buchthal and Schmalbruch, 1970).

The myosin from ST and FT muscle also possesses a different capacity for resistance to loss of ATPase activity when treated with alkali or acid (Brooke and Kaiser, 1970). This is illustrated in Figure 3. This property serves as the basis for the histochemical identification of the fiber types in skeletal muscle (Brooke and Kaiser, 1970, 1974; Barnard et al., 1971; Peter et al., 1972; Essén et al., 1975; Saltin et al., 1977). For the histochemical identification of fiber types, cross-sections of muscle are fresh frozen and pre-incubated at pHs of 10.4, 4.6, or 4.2 (Brooke and Kaiser, 1970, 1974; Saltin et al., 1977). Following a standard histochemical method for the staining of myofibrillar ATPase, the ST fibers have a low or negative staining intensity at the high pH, whereas the FT fibers stain intensely. After pre-incubation at pH 4.6 some FT fibers lose staining for myofibrillar ATPase, whereas it is retained by others. There is also an intense staining of the ST fibers. Following a pre-incubation at pH 4.2, almost all FT fibers lose staining for myofibrillar ATPase with the ST fibers still staining moderately. A commonly used nomenclature for identifying the fibers after their pre-incubations is simply ST for the fibers that are alkaline labile, FTa for the FT fibers that are ATPase negative after the treatment at pH 4.6, FTb for the FT fibers that retain stainable ATPase after pre-incubation at pH 4.6, and FTc for the small percent of the FT fibers that stain after all pre-incubations (Saltin et al.,

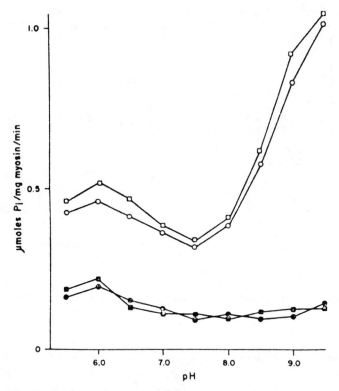

Figure 3 — The influence of pH on the maximal ATPase activity of myosin purified from fast-twitch muscles (open symbols) or from slow-twitch muscles (filled symbols). Adapted from Bárány et al. (1965).

1977). An alternate nomenclature is type I for the ST fibers, type IIa for FTa, type IIb for FTb, and type IIc for FTc fibers (Brooke and Kaiser, 1970, 1974). The ST and FT convention is preferred by this writer since it conveys some physiological meaning to the identification scheme.

The differences in the ATPase of the myofibrillar complex is the result of the existence of polymorphoric forms of the proteins in the contractile element (Dhoot and Perry, 1979, 1980; Gauthier, 1979, 1980; Gauthier and Lowey, 1979; Weeds, 1980). For the protein myosin this includes variations in the heavy chain, including the "rod" and "head" portions of the molecule (see Figure 4). There are also differences in the low molecular weight proteins that are associated with the myosin head, the head is the part of the protein where the ATPase activity is located. There are isozymes of all of the proteins that combine to form the contractile complex with the possible exception of actin (see Table 1). Such data demonstrate that the entire contractile complex exists in polymorphoric form. This appears to be the result of a specific expression of

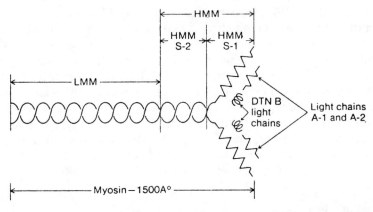

Figure 4—A schematic representation of the myosin complex. This shows the rod portion (LMM) and the head portion (HMM). The light chains are identified as DTNB and A-1 and A-2.

Table 1

Subunits of the Contractile Complex[a]

Component	Fast-Twitch		Slow-Twitch
Myosin Complex			
Heavy chain	200,000 d		200,000 d
Light chain			
LC1 (Alkali 1)	25,000 d	Ia	27,500 d
		Ib	26,500 d
LC2 (DTNB)	18,000 d		–
LC3 (Alkali 2)	16,000 d		19,000 d
Actin Complex			
Actin	alpha		alpha
Tropomyosin	alpha		beta
Troponin			
C	fast		slow
I	fast		slow
T	fast		fast

[a]These data are a compilation of results from Desmedt and Godaux, 1977b; Dhoot and Perry, 1979; Weeds, 1980.

the genes that assemble these proteins (Dhoot and Perry, 1980). The usual expression of these genes is such that for a given fiber type only one set of contractile proteins is included in the contractile unit.

The principal factor that determines the specific gene expression appears to be the nerve. On this basis, it should be noted that all of the fibers in a given motor unit are the same type (Edström and Kugelberg,

1968; Kugelberg, 1973; Pette and Spamer, 1979). Moreover, the conversion of one type of fiber to another must involve a change of all of the isozymes in the contractile unit. Such conversion can be induced by a number of experimental perturbations including cross-innervation (Bárány and Close, 1971; Sréter et al., 1975), chronic electrical stimulation (Lømo et al., 1980), and the chronic stress of elimination of a synergistic muscle or stretch (Ianuzzo et al., 1976; Holly et al., 1980). In all cases, it must be stressed that the identification of a type of fiber or motor unit is based on the properties of the contractile proteins contained in it. Moreover, the subclasses of FT fibers is also a result of different isozymes of the contractile proteins (Gauthier, 1980).

Metabolic Properties

The metabolic properties of the motor units in skeletal muscle can be identified either from the histochemical staining for oxidative and glycolytic enzymes (Edström and Kugelberg, 1968; Barnard et al., 1970, 1971; Peter et al., 1972; Kugelberg, 1973; Brooke and Kaiser, 1974; Buchthal and Schmalbruch, 1980) or by the direct biochemical analysis of muscles with a homogeneous fiber population or analysis of individual fibers (Barnard et al., 1971; Baldwin et al., 1972, 1973; Essén et al., 1975; Lowry et al., 1978; Pette and Spamer, 1979).

Generally ST motor units are well endowed with mitochondrial enzymes for end-terminal oxidation of fats and carbohydrates. Conversely, they have low enzyme activities (concentrations) for degrading glycogen to lactate. The FTa fibers are similar to ST fibers in that they have a relatively high oxidative potential. They also have a high concentration of enzymes for the Embden-Meyerhof pathway, however. FTb fibers are the antithesis of ST fibers in that they have a low concentration of oxidative enzymes and a high concentration of enzymes for glycolysis.

Though the above statements are generally true, some additional information should be added. First, there is wide variation in the absolute activity of the enzymes for both the aerobic and anaerobic pathways in the fibers. This includes fibers with similar contractile properties (Lowry et al., 1978; Pette and Spamer, 1979). However, within a single motor unit the activities for the different energy producing pathways are homogeneous (Edström and Kugelberg, 1968; Kugelberg, 1973). Another important property of the individual fiber is that the activities of the enzymes are consistent over the entire length of the fiber (Pette et al., 1980). Thus, for either histochemical staining or biochemical analysis it does not matter where a cross-sectional segment is taken from the fiber.

The metabolic properties of the fibers have also been incorporated into schemes for identifying motor units. Thus, Peter et al. (1972) have used a combination of contractile and metabolic characteristics to identify fibers as slow-twitch-oxidative (SO), fast-twitch-oxidative glycolytic

(FOG), or fast-twitch-glycolytic (FG). Baldwin and co-workers (1972, 1973) have identified fibers simply as slow-twitch red, fast-twitch red, or fast-twitch white. Burke and associates (1971, 1974) have used a system based on the fatigue properties of the different motor units. In this system there are slow-twitch (S) (these are fatigue resistant), fast-twitch fatigue resistant (FFR), and fast-twitch fast fatigue (FF) units.

Patterns of Motor Unit Use

A system of control exists in the central nervous system by which the different motor units can be recruited to produce the force and type of contraction needed for a particular activity. The systematic use of the different motor units in response to varying physiological demands is the result of an orderly procedure within the central nervous system for motor unit activation. This system is based upon the existence of motoneurons and cell bodies of the nerves such that their ease of activation is different (Hursh, 1939; Rushton, 1951; Hennemon et al., 1965, 1965, 1974; Buchthal and Schmalbruch, 1980). The functional nature of this system has been elaborated by Henneman and co-workers (1965, 1965, 1974). In this scheme the ST motor units are innervated by small, low threshold motoneurons. FT motor units have larger, more difficult to activate motor nerves. However, within each type of motor unit there are differences such that a continuum exists for motor unit activation. This allows for the progressive addition of motor units in response to added tension demands. Evidence for the orderly recruitment of motor units in man during a variety of physical activities has come from both histochemical and electrophysiological methods (Gollnick et al., 1972b, 1973b, 1974a, 1974b; Armstrong et al., 1974; Hannerz, 1974; Desmedt and Godaux, 1977a, 1977b; Garnett et al., 1979; Kugelberg and Lindegren, 1979).

The general pattern of motor unit recruitment is that for low intensity exercise there is a primary reliance upon the ST motor units. However, when such activity is carried out for prolonged periods of time there is a progressive recruitment of additional motor units until all motor units, including the FTb units, have been used if the exercise is continued to exhaustion (Gollnick et al., 1973b, 1974a; Armstrong et al., 1974). As the force-generating requirement of a muscular contraction increases there is a progressive involvement of additional motor units with those of the FT type being involved at high tensions.

The existence of fibers with different contractile speeds could lead one to conclude that at high rates of contraction there would be a preferential use of FT motor units. Electrophysiological data do not support this possibility for most normal muscular activity (Desmedt and Godaux, 1977a, 1977b). However, the relative contribution to the force developed by the different types of motor units may be influenced by the speed of the movement. This topic is covered in depth in an article by Komi in this volume.

The general pattern of fiber use as described is that for man. Some caution must be exercised when comparisons are made between man and most other animal species. This extends to the size, metabolic properties, and patterns of use of the different motor unit types. In most animals the FTa (FOG) fibers possess the highest oxidative potential. They also appear to be the first recruited during exercise (Armstrong et al., 1974, 1977). Regarding fiber size, in man the fibers are fairly uniform in cross-sectional area (Edström and Ekblom, 1972; Gollnick et al., 1972a, 1973a; Costill et al., 1976, 1979). In most other mammals the FTb fibers are more than twice the size of the ST (SO) and FTa (FOG) fibers (Armstrong et al., 1974; Ianuzzo et al., 1976; Gonyea et al., 1977; Keens et al., 1978; Gonyea, 1980). Usually the ST fibers are the smallest. These differences must be considered when studies conducted with animals are extrapolated to man.

Fatigue Properties

The fatigability of the different motor units in muscle is related to their oxidative capacity (Edström and Kugelberg, 1968; Kugelberg, 1973; Garnett et al., 1979; Kugelberg and Lindegren, 1979). Conversely, it can be suggested that since the most fatigue-resistant fibers are also the most easily recruited, that the maintenance of a high level of oxidative enzymes is the function of a chronic high level of activity of the fibers. The biochemical basis for the resistance to fatigue lies in the fact that with a highly developed enzyme system for end-terminal oxidation for both fats and carbohydrates, the fibers are capable of an effective use of the energy stores in the muscle and of the total body. This conclusion is based on the observation that exhaustion of man during prolonged exercise is closely associated with the depletion of the glycogen stores of the exercising muscles (Christensen and Hansen, 1939; Bergström and Hultman, 1967; Bergström et al., 1967; Gollnick et al., 1973b; Hermansen et al., 1967; Karlsson et al., 1974). By use of fatty acids and a complete oxidation of the glycogen rather than the production of lactate, the glycogen stores of the muscles can be conserved. Thus, when glycogen is oxidized it provides at least 12 times more ATP than when it is degraded to lactate and escapes from the muscle into the blood. Viewed from another standpoint, the rate of glycogen depletion would occur much more slowly when it is oxidized to CO_2 and H_2O as compared to lactate production. This basic difference in the metabolism extends to the heart and diaphragm where the major fuel for the chronic activity of these muscles is fat.

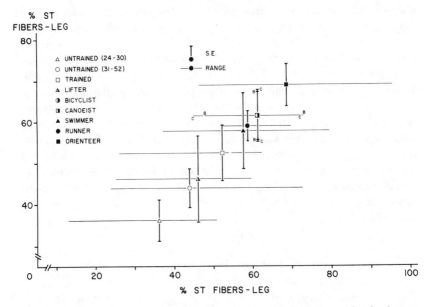

Figure 5—A plot of the percentage of ST fibers in the vastus lateralis muscle of sedentary subjects and of individuals who excel in a variety of athletic events. Data are from Gollnick et al. (1972a).

There are also differences in the fatigue properties of motor units contained in the skeletal muscles of man as compared with those in other animal species. Thus, according to the studies of Burke et al. (1971, 1974), the FF (FG) motor units will become exhausted after about 1 min of stimulation, the FFR (FOG) units will continue contracting for about 5 min before a decline in tension is noted. After 60 min of contraction these units produce only about 10% of their initial tension. The S (SO) units continue at a constant level of tension development for more than 1 hr. In contrast, the FTb motor units retain up to 40% of their initial tension developing capacity after 50 min of stimulation where 3000 stimuli (60/min) have been delivered to the motor unit. Moreover, in man the ST and FTa motor units have very similar fatigue characteristics (Garnett et al., 1979). This is in striking contrast to the S and FFR motor units of the cat.

Fiber Composition and Performance

Early studies of the properties of the skeletal muscles revealed that some athletes possessed fiber compositions (see Figure 5) in their skeletal muscles that could be interpreted as being desirable for success in the event in which they excelled (Gollnick et al., 1972a). For example, long distance runners and other athletes who participated in endurance events were found to have high percentages of ST, aerobic fibers in their muscles, whereas sprinters had high percentages of the FT, anaerobic fibers. These initial findings were subsequently confirmed by several laboratories (Costill et al., 1976, 1979; Saltin et al., 1979). Findings such as these led some to conclude that the type of training an athlete engaged in was the determining factor for the development of muscle that contained either the ST fibers with high aerobic capacity needed for endurance or the FT fibers needed for the speed events. However, such conclusions were unwarranted from the cross-sectional studies from which these data were obtained. Subsequent longitudinal studies have not demonstrated that any major changes in the basic characteristic related to the speed of the muscle fiber occur with training (Gollnick et al., 1973a).

Hereditary Factors in Muscle Composition

The basic characteristic of skeletal muscle, that is, its contractile speed, appears to be genetically determined. The observation that the composition in the skeletal muscles of monozygotic twins is nearly identical supports this thesis (Keens et al., 1978). Moreover, within a population of young (aged 16 years) males and females there is a near normal distribution (bell-shaped curve) of fiber compositions in the skeletal muscles (Saltin et al., 1977). When these data are extended to the general population, this normal distribution of fiber compositions could provide for an adequate number of individuals with the unusually high percentages of the different fiber compositions to explain the distributions seen in elite endurance or sprint athletes. This would negate the need for converting one fiber type to another with training. If fiber composition alone is the determining factor for success in an athletic event, these data would suggest that many individuals undoubtedly exist with great athletic potential. Whether they would be successful if they had either an interest or an opportunity to participate is unknown.

Adaptations in Muscle with Training

Aerobic Metabolism

Although the contractile properties of skeletal muscle appear to be unaltered by training, major changes do occur in the metabolic potential of the muscle and in the individual fibers with training. The most dramatic manifestation of this adaptation is an increase in the concentration of mitochondria and concurrently of the enzymes for oxidative

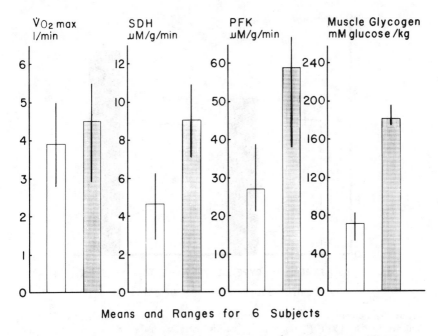

Figure 6 — A comparison of the maximal oxygen uptake ($\dot{V}O_2$ max), succinate dehydrogenase (SDH), phosphofructokinase (PFK), and concentration of muscle glycogen in six subjects before (open bars) and after (shaded bars) a six month program of endurance training. All post-training values are significantly higher than the pre-training values.

metabolism (see Figure 6) that occurs with endurance training (Morgan et al., 1971; Baldwin et al., 1972; Gollnick et al., 1973a; Hoppeler et al., 1973; Costill et al., 1976; Gollnick and Sembrowich, 1977; Terjung, 1976). These increases in oxidative potential occur in both the high oxidative ST and the low oxidative FT fibers. The magnitude of the adaptation in muscle is related to the type and duration of the exercise used in the training program. The biggest gains in oxidative potential are induced by prolonged endurance training, whereas only modest changes occur after high intensity, sprint-type training (Gollnick et al., 1973a; Saltin et al., 1976; Terjung, 1976).

The higher oxidative potential of endurance-trained muscle would appear to increase its capacity for using the energy reserves, both those stored intracellularly and those transported to it by the blood. The elevated aerobic capacity also increases the ability to use fatty acids transported to the muscle by the blood. The combination of these processes exerts a glycogen-sparing effect on the muscle and results in a greater endurance capacity since depletion of intracellular glycogen reserves of muscle and termination of moderately severe exercise occur simultaneously (Bergström and Hultman, 1967; Bergström et al., 1967; Gollnick et al., 1973b; Hermansen et al., 1967).

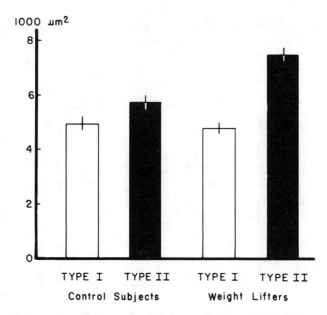

Figure 7 — A comparison of the cross-sectional area of slow-twitch (identified as TYPE I) and fast-twitch (TYPE II) fibers in the vastus lateralis muscles of control subjects and weight lifters. Data were compiled from Edström and Ekblom (1972) and Gollnick and associates (1972a).

Anaerobic Metabolism

The influence of training on the anaerobic capacity of skeletal muscle is not as clear as that of the aerobic potential. Several published studies have failed to demonstrate any major changes in the activities of key enzymes for lactate production (Gollnick and Hermansen, 1973; Saubert et al., 1973; Staudte et al., 1973). In other studies there were significant increases in these enzymes (Gollnick et al., 1973a; Costill et al., 1979). This may be dependent upon the type and intensity of the training program. This also may be attributable either to the fact that the studies were improperly conducted or that the normally high concentration of these enzymes in skeletal muscle obviates any need to augment their concentration to enable the muscle to meet the energy demands of very heavy exercise. Further investigations may lead to a clarification of this issue.

Muscular Enlargement

Muscular strength is closely related to the total cross-sectional area of a muscle. Heavy resistance training increases both the strength and the total cross-sectional area of skeletal muscle. Evidence exists which demonstrates that weight lifting increases the cross-sectional area of FT fibers (see Figure 7) of human muscle (Edström and Ekblom, 1972;

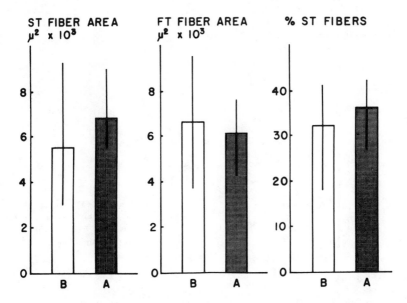

Figure 8—The effect of endurance training on the fiber composition and fiber area. The difference between the pre-training value (B) and post-training (A) value for ST fiber area is statistically significant (P < 0.05).

Gollnick et al., 1972a; Costill et al., 1979). In contrast, endurance training results in an increase in the cross-sectional area of ST fibers (see Figure 8) (Gollnick et al., 1973a). Successful participants in endurance events, however, do not always possess unusually large ST fibers. By changing fiber size it is possible to somewhat alter the basic composition of a muscle. Comparisons of the area of muscle fibers from infants and adults (Moss, 1968; Edström and Ekblom, 1972; Gollnick et al., 1972a, 1973a; Colling-Saltin, 1978; Costill et al., 1979) clearly demonstrate that considerable hypertrophy of the fibers occurs with normal growth and development. Whether the individual muscle fibers possess the potential for enlargement to an extent that would produce the very large muscles seen in competitive weight lifters and body builders has been questioned, however. An additional mechanism for increasing muscle bulk has been suggested to be an increase in fiber number (Van Linge, 1962; Reitsma, 1969; Gonyea et al., 1977; Schiaffino et al., 1979; Gonyea, 1980).

Gollnick and co-workers (1981) have developed methods whereby all of the fibers in skeletal muscle could be individually teased free, examined over their entire length for branching points, and counted. In these studies, there was no change in fiber number in enlarged as compared to control skeletal muscles in the rat. Muscular enlargement ranging from 10 to 110% was induced in the penniform plantaris and extensor digitorum longus muscles and the parallel fibered soleus muscle by surgical

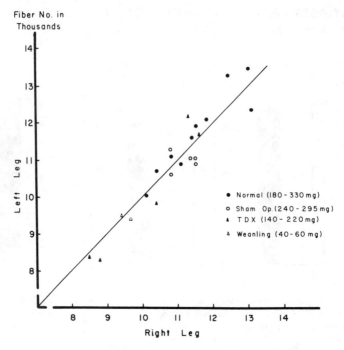

Figure 9A — The number of fibers in the plantaris muscles of the right and left legs from a number of experimental groups (see symbols on Figure). The line is the line of identity. These data demonstrate that though the number of fibers in the plantaris muscle is similar for the two legs of the same animal, there is considerable variation between animals.

ablation of a synergistic muscle and the combination of surgical ablation of a synergistic muscle and treadmill running. Muscles were examined from 4 to 40 weeks after the surgical ablation of the synergistic muscle. This time interval was allowed so that any possible effect of inflammation would have passed. Although considerable variation existed in the number of fibers per muscle between animals, the total fiber number in the right and left limb muscle for the same animal was remarkably similar (see Figure 9). The difference in the dry weight of individual fibers in the enlarged muscle as compared to those from the normal muscle demonstrated that the increase in weight was the result of hypertrophy and not hyperplasia. The incidence of branched fibers was similar for normal and enlarged muscles. The observation (see Figure 10) that branch points can occur anywhere along the length of the fiber, however, illustrates the technical difficulty of using histological cross-sections to evaluate its frequency (Van Linge, 1962; Reitsma, 1969; Gonyea et al., 1977; Schiaffino et al., 1979; Gonyea, 1980). The apperance of branched fibers in all muscles was interpreted as an indication that these abnormal fibers are normally present in small percentages and do not represent an

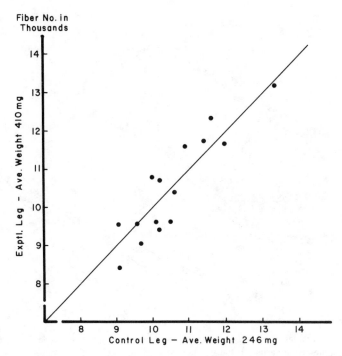

Figure 9B—A plot of the number of fibers in the plantaris muscle of a control leg and a leg where muscular enlargement had been induced by the surgical ablation of the gastrocnemius muscle. The line is the line of identity. These data illustrate that there is no difference in fiber number between a control and enlarged muscles. The variation between animals is similar to that of the control groups of Figure 9A.

active process of a longitudinal division of fibers.

The findings of Gollnick and associates (1981) support the older concept that the number of fibers in skeletal muscle is established early in life and that the exceptional enlargement that occurs in response to a variety of overloads is a true hypertrophy (Morpurgo, 1897; Seibert, 1928). Such data support the conclusions of Hall-Craggs and co-workers (1970a, 1970b, 1972) and of James (1973, 1976) that if a splitting of fibers does occur in skeletal muscle during enlargement it is of minimal importance in the overall increase in muscular size. James (1976) has in fact concluded that the changes in fiber cross-sectional area during hypertrophy of the mouse extensor digitorum longus muscle are such that a hypoplasia must have occurred.

Conversion of Fiber Types

As indicated previously, there is a question of whether one type of fiber

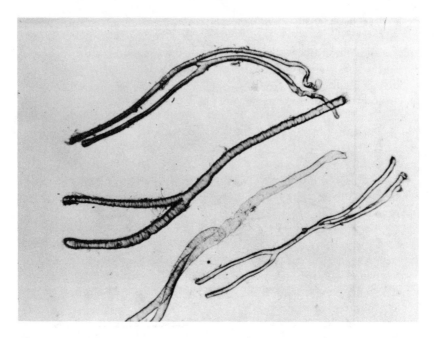

Figure 10—Examples of bifurcated fibers teased from a rat plantaris muscle. These fibers illustrate some of the variation that can be found in such fibers.

can be converted to another by normal physiological factors such as those associated with training. In this case the logical conversions would be for an increase in the ST fibers in response to endurance type activity and the appearance of a greater percentage of FT fibers for those activities where short-explosive contractions predominate the activity. In several early studies with animals, fibers were identified simply as "red" or "white" on the basis of histochemical staining intensities for oxidative enzymes. With such methods it was reported that an increase of red fibers occurred as a result of endurance training (Barnard et al., 1970; Faulkner et al., 1972a, 1972b). Termination of the training resulted in a reversion to the control situation. This was interpreted as evidence for the interconvertibility of fiber types. A similar report also exists from a longitudinal study of the effect of training on the skeletal muscle of man (Morgan et al., 1971). Such alterations in staining intensity for mitochondrial enzymes is consistent with the biochemical studies demonstrating that endurance training induces increases in mitochondrial protein concentrations and concomitantly in the activities of enzymes associated with the mitochondria (see previous statement). However, the increase in oxidative potential occurs in all types of fibers when the intensity and duration are sufficient to produce a recruitment of all motor unit types in the muscle. Conventional histochemical methods are not sensitive enough to distinguish this increase in the normally high oxidative fibers and since it appears as an increase in the normally low staining fibers they now become difficult to identify. With the short-heavy resistance exercise of weight lifting there is a decline in the oxidative potential of skeletal muscle. This should, on the basis of the conclusion from standard staining for mitochondrial enzymes, be viewed as an increase in the "white" fibers. As indicated initially, the criterion that has been established to identify a fiber type is the myofibrillar protein. On this basis, changes in oxidative potential do not represent changes in fiber types.

Significant amounts of data are available on the fiber composition of muscles of athletes. The majority of these data have come from studies where the fiber composition was determined from needle biopsy samples. As a result, there is appreciable descriptive information on the fiber distribution of several muscles from athletes specializing in a variety of athletic events. In most of these studies the fiber composition was assessed from the histochemical identification of fibers on the basis of the histochemical demonstration of myofibrillar ATPase. In some instances this was combined with procedures to establish the metabolic profile of the fibers (Edström and Ekblom, 1972; Gollnick et al., 1972a; Saltin et al., 1977; Costill et al., 1979, 1976). From a cross-sectional standpoint the fiber composition of athletes falls within the range for normal subjects. This is not surprising since the range encompassed by normal subjects is rather large.

In spite of the fact that most athletic subgroups possess fiber distributions in their muscles that are within the range of normal subjects, there are some indications of selective fiber distributions in certain athletic groups. Most notable is the fact that those individuals who excel in events requiring high endurance usually possess a predominance of ST fibers in their muscles. Conversely, there is a tendency for sprinters to have slightly higher than average percentages of FT fibers. There are, however, very few studies to substantiate this. Gollnick and co-workers (1973a) did examine the fiber composition of biopsy samples from the vastus lateralis muscle of six subjects before and after a six-month training program of bicycle exercise. Since this type of exercise depends to a large extent on the quadriceps muscle group, it should have undergone extensive use and any change that would occur should have been evident in this muscle. This was verified by a large increase in the oxidative potential of the muscle as a result of the training. The fiber composition, as determined by staining procedures that identified only FT and ST fibers, was not altered by the training (see Figure 8). This occurred in spite of the fact that the muscles of these subjects were initially high in the percentage of FT fibers. Since the FT fibers were not differentiated into the FTa, FTb, and FTc sub-populations, there was no indication of a shift in the distribution of this fiber type. If a major shift in the percentage of the two major fiber types did occur, however, it should have been detectable with the methods used. Saltin et al. (1976) also failed to detect any interconversion of fibers when comparisons were made between legs of subjects where combinations of no training, endurance training, and sprint training were practiced with only one leg. These data suggest that the presence of high percentages of one or the other fiber type in the muscles of specialized athletes is the result of a natural endowment and these individuals found their way into the activity as a result of this endowment.

References

ARMSTRONG, R.A., Saubert, C.W., IV, Sembrowich, W.L., Shepherd, R.E., and Gollnick, P.D. 1974. Glycogen depletion in rat skeletal muscle fibers at different intensities and duration of exercise. Pfluegers Arch. 352:243-256.

ARMSTRONG, R.A., Marum, P., Saubert, C.W., IV, Seeherman, H.J., and Taylor, C.R. 1977. Muscle fiber activity as a function of speed and gait. J. Appl. Physiol.: Respirat. Environ. Exercise Physiol. 43:672-677.

BALDWIN, K.M., Klinkerfuss, G.H., Terjung, R.L., Molé, P.A., and Holloszy, J.O. 1972. Respiratory capacity of white, red, and intermediate muscle: adaptive response to exercise. Am. J. Physiol. 222:373-378.

BALDWIN, K.M., Winder, W.W., Terjung, R.L., and Holloszy, J.O. 1973. Glycolytic enzymes in different types of skeletal muscle: adaptation to exercise. Am. J. Physiol. 225:962-966.

BÁRÁNY, M., Bárány, K., Reckard, T., and Volpe, A. 1965. Myosin of fast and slow muscles of the rabbit. Arch. Biochem. Biophy. 109:185-191.

BÁRÁNY, M. 1967. ATPase activity of myosin correlated with speed of muscle shortening. J. Gen. Physiol. 50:197-218.

BÁRÁNY, M., and Close, R.I. 1971. The transformation of myosin in cross-innervated rat muscles. J. Physiol. 213:455-474.

BARNARD, R.J., Edgerton, V.R., Furukawa, T., and Peter, J.B. 1971. Histological, biochemical, and contractile properties of red, white, and fibers. Am. J. Physiol. 220:410-414.

BARNARD, R.J., Edgerton, V.R., and Peter, J.B. 1970. Effect of exercise on skeletal muscle. I. Biochemical and histological properties. J. Appl. Physiol. 28:762-766.

BERGSTRÖM, J., and Hultman, E. 1967. A study of the glycogen metabolism during exercise in man. Scand. J. Clin. Lab. Invest. 19:218-228.

BERGSTRÖM, J., Hermansen, L., Hultman, E., and Saltin, B. 1967. Diet, muscle glycogen and physical performance. Acta Physiol. Scand. 71:140-150.

BROOKE, M.H., and Kaiser, K.K. 1970. Three "myosin adenosine triphosphatase" systems: the nature of their pH lability and sulfhydryl dependence. J. Histochem. Cytochem. 18:670-672.

BROOKE, M.H., and Kaiser, K.K. 1974. The use and abuse of muscle histochemistry. Ann. N.Y. Acad. Sci. 228:121-144.

BUCHTHAL, F., and Schmalbruch, H. 1969. Spectrum of contraction times of different fibre bundles in the brachial biceps and triceps muscle of man. Nature 222:89-91.

BUCHTHAL, F., Dahl, K., and Rosenfalch, D. 1973. Rise time of the spike potential in fast and slowly contracting muscle of man. Acta Physiol. Scand. 87:261-269.

BUCHTHAL, F., and Schmalbruch, H. 1970. Contraction time and fibre types in intact human muscle. Acta Physiol. Scand. 79:435-452.

BUCHTHAL, F., and Schmalbruch, H. 1980. Motor unit of mammalian muscle. Physiol. Rev. 60:90-142.

BURKE, R.E., Levine, D.N., Zajac, F.E., III, Tsairis, P., and Engel, W.K. Mammalian motor units: physiological-histochemical correlation in three types of cat gastrocnemius. Science 174:708-712.

BURKE, R.E., and Tsairis, P. 1974. The correlation of physiological properties with histochemical characteristics in single motor units. Ann. N.Y. Acad. Sci. 228:145-158.

CHRISTENSEN, E.-H., and Hansen, O. 1939. Arbeitsfähigkeit und Ehrnährung. Skand. Arch. Physiol. 81:160-171.

COLLING-SALTIN, A.-S. 1978. Enzyme histochemistry of skeletal muscle of the human foetus. J. Neurol. Sci. 39:169-185.

COSTILL, D.L., Coyle, E.F., Fink, W.F., Lesmes, G.R., and Witzmann, F.A. 1979. Adaptations in skeletal muscle following strength training. J. Appl. Physiol.: Respirat. Environ. Exercise Physiol. 46:96-99.

COSTILL, D.L., Daniels, J., Evans, W., Fink, W., Krahenbuhl, G., and Saltin, B. 1976. Skeletal muscle enzymes and fiber composition in male and female track athletes. J. Appl. Physiol. 40:149-154.

DESMEDT, J.E., and Godaux, E. 1977a. Fast motor units are not preferentially activated in rapid voluntary contractions in man. Nature 267:717-719.

DESMEDT, J.E., and Godaux, E. 1977b. Ballistic contractions in man: characteristic recruitment pattern of single motor units of the tibialis anterior muscle. J. Physiol. 64:673-693.

DHOOT, G.K., and Perry, S.V. 1979. Distribution of polymorphoric forms of troponin components and tropomyosin in skeletal muscle. Nature 278:714-718.

DHOOT, G.K., and Perry, S.V. 1980. Factors determining the expression of the genes controlling the synthesis of the regulatory proteins in striated muscle. In: D. Pette (ed.), Plasticity of Muscle, pp. 255-267. Walter de Gruyter and Co., Berlin.

EBERSTEIN, E., and Goodgold, J. 1968. Slow and fast twitch fibers in human skeletal muscle. Am. J. Physiol. 215:535-541.

EDSTRÖM, L., and Kugelberg, E. 1968. Histochemical composition, distribution of fibres and fatigability of single motor units. J. Neurol. Neurosurg. Psychiat. 31:424-433.

EDSTRÖM, L., and Ekblom, B. 1972. Differences in sizes of red and white muscle fibres in vastus lateralis of musclulus quadriceps of normal individuals and athletes. Relation to physical performance. Scand. J. Clin. Lab. Invest. 30:175-181.

ESSÉN, B., Jansson, E., Henriksson, J., Taylor, A.W., and Saltin, B. 1975. Metabolic characteristics of fibre types in human skeletal muscles. Acta Physiol. Scand. 95:153-165.

FAULKNER, J.A., Maxwell, L.C., Brook, D.A., and Lieberman, D.A. 1972a. Adaptation of guinea pig plantaris muscle fibers to endurance training. Am. J. Physiol. 221:291-297.

FAULKNER, J.A., Maxwell, L.C., and Lieberman, D.A. 1972b. Histochemical characteristics of muscle fibers from trained and detrained guinea pigs. Am. J. Physiol. 222:836-840.

GARNETT, R.A.F., O'Donovan, M.J., Stephens, J.A., and Taylor, A. 1979. Motor unit organization of human medial gastrocnemius. J. Physiol. 287:33-43.

GAUTHIER, G.F. 1979. Ultrastructural identification of muscle fiber types by immunocytochemistry. J. Cell Biol. 82:391-400.

GAUTHIER, G.F., and Lowey, S. 1979. Distribution of myosin isoenzymes among skeletal muscle fiber types. J. Cell Biol. 81:10-25.

GAUTHIER, G.F. 1980. Distribution of myosin isoenzymes in adult and developing muscle fibers. In: D. Pette (ed.), Plasticity of Muscle. pp. 83-96. Walter de Gruyter and Co., Berlin.

GOLLNICK, P.D., Armstrong, R.B., Saubert, C.W., IV, Piehl, K., and Saltin, B. 1972a. Enzyme activity and fiber composition in skeletal muscle of untrained and trained men. J. Appl. Physiol. 33:312-319.

GOLLNICK, P.D., Armstrong, R.B., Sembrowich, W.L., Shepherd, R.E., and Saltin, B. 1972b. Glycogen depletion pattern in human skeletal muscle fibers after heavy exercise. J. Appl. Physiol. 34:615-618.

GOLLNICK, P.D., Armstrong, R.B., Saltin, B., Saubert, C.W., IV, Sembrowich, W.L., and Shepherd, R.E. 1973a. Effect of training on enzyme activity and fiber composition of human skeletal muscle. J. Appl. Physiol. 34:107-111.

GOLLNICK, P.D., Armstrong, R.B., Saubert, C.W., IV, Sembrowich, W.L., Shepherd, R.E., and Saltin, B. 1973b. Glycogen depletion patterns in human skeletal muscle fibers during prolonged work. Pfluegers Arch. 344:1-12.

GOLLNICK, P.D., and Hermansen, L. 1973. Biochemical adaptations to exercise: anaerobic metabolism. In: J.H. Wilmore (ed.), Exercise and Sports Science Reviews. Vol. 1: pp. 1-43. Academic, New York.

GOLLNICK, P.D., Piehl, K., and Saltin, B. 1974a. Selective glycogen depletion pattern in human muscle fibres after exercise of varying intensity and at varying pedalling rates. J. Physiol. 241:45-57.

GOLLNICK, P.D., Karlsson, J., Piehl, K., and Saltin, B. 1974b. Selective glycogen depletion in skeletal muscle fibers of man following sustained contractions. J. Physiol. 241:59-67.

GOLLNICK, P.D., and Sembrowich, W.L. 1977. Adaptations in human skeletal muscle. In: E.A. Amsterdam, J.H. Wilmore, and A.N. DeMaria (eds.), Exercise in Cardiovascular Health and Disease. pp. 70-94. Yorke, New York.

GOLLNICK, P.D., Timson, B.F., Moore, R.L., and Riedy, M. 1981. Muscular enlargement and the number of fibers in skeletal muscle of rats. J. Appl. Physiol.: Respirat. Environ. Exercise Physiol. 50:936-943.

GONYEA, W., Ericson, G.C., and Bonde-Petersen, F. 1977. Skeletal muscle fiber splitting induced by weight-lifting exercise in cats. Acta Physiol. Scand. 99:105-109.

GONYEA, W.J. 1980. Role of exercise in inducing increases in skeletal muscle fiber number. J. Appl. Physiol.: Respirat. Environ. Exercise Physiol. 48:421-426.

HALL-CRAGGS, E.C.B., and Lawrence, C.A. 1970a. Longitudinal fiber division in skeletal muscle: a light and electronmicroscopic study. Z. Zellforsch. 109:481-494.

HALL-CRAGGS, E.C.B. 1970b. The longitudinal division of fibres in overloaded rat skeletal muscle. J. Anat. 107:459-470.

HALL-CRAGGS, E.C.B. 1972. The significance of longitudinal fibre division in skeletal muscle. J. Neurol. Sci. 15:27-33.

HANNERZ, J. 1974. Discharge properties of motor units in relation to recruitment order in voluntary contraction. Acta Physiol. Scand. 91:374-384.

HENNEMAN, E., and Olson, C.B. 1965. Relations between structure and function in the design of skeletal muscle. J. Neurophysiol. 28:581-598.

HENNEMAN, E., Somjen, G., and Carpenter, D. 1965. Functional significance of cell size in spinal motoneurons. J. Neurophysiol. 28:650-680.

HENNEMAN, E., Clamann, H.P., Gilles, J.D., and Skinner, R.D. 1974. Rank-order of motoneurons within a pool: law of combination. J. Neurophysiol. 37:1338-1349.

HERMANSEN, L., Hultman, E., and Saltin, B. 1967. Muscle glycogen and prolonged severe exercise. Acta Physiol. Scand. 71:129-139.

HOLLY, R.G., Barnett, J.G., Ashmore, C.R., Taylor, R.G., and Molé, P.A. 1980. Stretch-induced growth in chicken wing muscles: a new model of stretch hypertrophy. Am. J. Physiol. 238:C62-C71.

HOPPELER, H., Lüthi, P., Classen, H., Weibel, E.R., and Howald, H. 1973. The ultrastructure of the normal human skeletal muscle. A morphometric analysis on untrained men, women, and well-trained orienteers. Pfluegers Arch. 344:217-232.

HURSH, J.B. Conduction velocity and diameter of nerve fibers. 1939. Am. J. Physiol. 127:131-139.

IANUZZO, C.D., Gollnick, P.D., and Armstrong, R.B. 1976. Compensatory adaptations of skeletal muscle to long-term functional overload. Life Sci. 19:1517-1524.

JAMES, N.T. 1973. Compensatory hypertrophy in the extensor digitorum longus muscle of the rat. J. Anat. 116:57-65.

JAMES, N.T. 1976. Compensatory muscular hypertrophy in the extensor digitorum longus muscle of the mouse. J. Anat. 122:121-131.

KARLSSON, J., Noredesjö, L.-O., and Saltin, B. 1974. Muscle glycogen utilization during exercise after physical training. Acta Physiol. Scand. 90:210-217.

KEENS, T.G., Chen, V., Patel, P., O'Brien, P., Levison, H., and Ianuzzo, C.D. 1978. Cellular adaptations of the ventilatory muscles to a chronic respiratory overload. J. Appl. Physiol.: Respirat. Environ. Exercise Physiol. 44:905-908.

KOMI, P.V., Vittasalo, T., Havu, M., Thorstesson, A., and Karlsson, J. 1976. Physiological and structural performance capacity: effect of heredity. In: P.V. Komi (ed.), Biomechanics V-A, pp. 118-123, University Park Press, Baltimore, MD.

KUGELBERG, E. 1973. Histochemical composition, contraction speed and fatigability of rat soleus motor units. J. Neurol. Sci. 20:177-198.

KUGELBERG, E., and Lindegren, B. 1979. Transmission and contraction fatigue of rat motor units in relation to succinate dehydrogenase activity of motor unit fibres. J. Physiol. 288:285-300.

LØMO, T., Westergaard, R.H., and Engebretsen, L. 1980. Different stimulation patterns affect contractile properties of denervated rat soleus muscles. In: D. Pette, (ed.), Plasticity of Muscle. pp. 297-309. Walter de Gruyter & Co., Berlin.

LOWRY, C.V., Kimmey, J.S., Felder, S., Chi, M.M.-Y., Kaiser, K.K., Passonneau, P.N., Kirk, K.A., and Lowry, O.H. 1978. Enzyme patterns in single human muscle fibers. J. Biol. Chem. 253:8269-8277.

MORGAN, T.E., Cobb, L.A., Short, F.A., Ross, R., and Gunn, D.R. 1971. Effect of long-term exercise on human muscle mitochondria. In: B. Pernow and B. Saltin, (eds.), Muscle Metabolism During Exercise. pp. 87-95. Plenum, New York.

MORPURGO, B. 1897. Uber Aktivitat-Hypertrophie der willkuerlichen Muskeln (Concerning activity-hypertrophy of voluntary muscles). Virchow's Arch. Path. Anat. 150:522-554.

MOSS, F.P. 1968. The relationship between dimensions of the fibers and the number of nuclei during normal growth of skeletal muscle in the domestic fowl. Am. J. Anat. 122:555-564.

PETER, J.B., Barnard, R.J., Edgerton, V.R., Gillespie, C.A., and Stempel, K.E. 1972. Metabolic profiles of the three fiber types of skeletal muscle in guinea pigs and rabbits. Biochemistry. 11:2627-2633.

PETTE, D., and Spamer, C. 1979. Metabolic subpopulations of muscle fibers: a quantitative study. Diabetes 28: Suppl. 1:25-29.

PETTE, D., Wimmer, M., and Nemeth, P. 1980. Do enzyme activities vary along muscle fibres. Histochemistry 67:225-231.

REITSMA, W. 1969. Skeletal muscle hypertrophy after heavy exercise in rats with surgically reduced muscle function. Am. J. Phys. Med. 48:237-258.

RUSHTON, W.A.H. 1951. A theory of the effects of fiber size in medullated nerve. J. Physiol. 115:101-122.

SALTIN, B., Nazar, K., Costill, D.L., Stein, E., Jansson, E., Essén, B., and Gollnick, P.D. 1976. The nature of the training response; peripheral and central adaptations to one-legged exercise. Acta Physiol. Scand. 96:289-305.

SALTIN, B., Hendriksson, J., Nygaard, E., and Andersen, P. 1977. Fiber types and metabolic potentials of skeletal muscles in sedentary men and endurance runners. Ann. N. Y. Acad. Sci. 301:3-29.

SAUBERT, C.W., IV, Armstrong, R.B., Shepherd, R.E., and Gollnick, P.D. 1973. Anaerobic enzyme adaptations to sprint training in rats. Pfluegers Arch. 341:305-312.

SCHIAFFINO, S., Bormioli, S., and Aloisi, M. 1979. Fiber branching and formation of new fibers during compensatory muscle hypertrophy. In: A. Mauro (ed.), Muscle Regeneration. pp. 177-188. Raven Press, New York.

SIEBERT, W.E. 1928. Untersuchungen uber Hypertrophy des Skeletmuskels (Investigation of hypertrophy of skeletal muscles). Z. klin. Med. 109:350-359.

SOLA, O.M., Christensen, D.L., and Martin, A.W. 1973. Hypertrophy and hyperplasia of adult chicken anterior latissimus dorsi muscles following stretch with and without denervation. Expt. Neurol. 41:76-100.

SPAMER, C., and Pette, D. 1979. Activities of malate dehydrogenase, 3-hydroxyacyl-CoA dehydrogenase and fructose-1, 6-diphosphatase with regard to metabolic subpopulations of fast- and slow-twitch fibers in rabbit muscles. Histochemistry 60:9-19.

SRÉTER, F.A., Luft, A.R., and Gergely, J. 1975. Effect of cross-reinnervation on physiological parameters and on properties of myosin and sarcoplasmic reticulum of fast and slow muscles of the rabbit. J. Gen. Physiol. 66:811-821.

STAUDTE, H.W., Exner, G.U., and Pette, D. 1973. Effects of short-term, high intensity (sprint) training on some contractile and metabolic characteristics of fast and slow muscle of the rat. Pfluegers Arch. 344:159-168.

TERJUNG, R.L. 1976. Muscle fiber involvement during training of different intensities and durations. Am. J. Physiol. 230:946-950.

VAN LINGE, B. 1962. The response to strenuous exercise. J. Bone Joint Surg. 44:711-721.

WATRUS, J.M. 1980. Influence of chronic exercise on myosin from cardiac and skeletal muscle of hamsters and rats. Ph.D. Thesis, Washington State University, Pullman, Washington.

WEEDS, A. 1980. Myosin light chains, polymorphism and fibre types in skeletal muscles. In: D. Pette (ed.), Plasticity of Muscle. pp. 55-68. Walter de Gruyter and Co., Berlin.

IKAI
MEMORIAL
LECTURE

Biomechanical Aspects of Animal Flying and Swimming

Akira Azuma
University of Tokyo, Tokyo, Japan

Living creatures have many ways of flying and swimming. The way used depends greatly on body size and shape and the animal's mode of life in its environment. Animals with small bodies utilize the resistive force or drag exclusively, whereas larger and faster animals rely on reactive force or lift.

In flying, thin and laterally wide plates, namely wings, are used by almost all animals. In order to generate a lifting force against the force of gravity the wing must be advanced parallel to the wing surface with a small angle of attack, whereas in order to produce a propulsive force against the drag of the body, the wing must be moved in the vertical direction in what is called beating motion. Both the lifting and propulsive forces must act through the center of gravity to maintain stable flight.

In swimming, the presence of a buoyant force eliminates the need for generating a lifting force and, therefore, the only condition required for stable motion is that the mean propulsive force be directed through the center of gravity. For this reason, a variety of body and wing (fin) motions can be observed in addition to the above method of beating motion. Hydrodynamic characteristics such as performance and efficiency are also discussed for exemplified species performing the above described body and wing motions.

Variety of Motions

Animals, or more generally living creatures, range in body size from small single-celled creatures (only microns long) to large mammalia (with body lengths of some tens of meters). Each moves in its own way, selecting its mode of locomotion from a variety of movements according to its size, shape and living environment. Usually the mode a creature uses is the most economical for its mode of life as it has been developed in

35

response to severe environmental conditions.

Every animal lives in a fluid — some in water, some in air. The density of most animals is approximately equal to that of water, presumably because they lived in the water in their early stages. Animals that move in water can therefore float without making a constant effort to support their weight. Thus, in locomotion they can concentrate on propelling themselves against the resistive and reactive forces caused by their relative motion in water. An animal that moves in air, on the other hand, must first of all support itself utilizing a fluid-dynamic force other than the buoyant force of air which, at 1/820 that of water, is negligibly small. Thus, the modes of locomotion that can be used by air-going animals are limited appreciably.

Before any detailed discussion on the moving behavior of animals, four important non-dimensional parameters which represent the physical characteristics of motion in a fluid should be introduced (Streeter, 1961). These are Reynold's number, Froude number, reduced frequency and Froude efficiency.

The Reynold's number (Re) is the ratio of inertial force to viscous force of a fluid and is thus represented by the product of the speed (U) and the typical length of the body (ℓ) over the kinematic viscosity of the fluid (ν) (which is the ratio of viscosity (μ) to density (φ) of the fluid), i.e., Re = $U\ell/\nu$.

The Froude number (Fr) is the ratio of inertial force to gravity or buoyant force and is represented by the ratio of speed (U) to the square root of gravitational acceleration (g) times the body length (ℓ), i.e., Fr = $U/\sqrt{g\ell}$.

The reduced frequency (k) is a dimensionless angular velocity ($2\pi f$) based on the speed/chord ratio [U/(c/2)], i.e., k = $\pi cf/U$. Unsteady hydrodynamic characteristics are closely related to this nondimensional frequency.

The Froude efficiency (η) is the propulsive efficiency or useful rate of work for a given power. It is defined as the ratio of the mean rate of work of propulsion, which is given by the product of the mean propulsive force (T) and the speed (U), to the mean rate of work needed to maintain the locomotion (P), i.e., η = TU/P.

Snaking

In an animal with an elongated or slender body, swimming is usually performed with an undulatory motion of the body by sending transverse waves either forward or backward. This motion, which resembles that of a snake or eel, may be called "snaking" or "angulliform" motion. Careful observation reveals several noticeable differences among the various species of such elongated animals, however.

As shown in Figure 1 (Hiramoto and Baba, 1978), a spermatozoon of

Figure 1—Swimming of a spermatozoon
(Hiramoto and Baba, 1978).

an echinoderm swims by repeatedly beating a whip-like flagellum and thus sending several transverse waves backward. The waves are characterized by almost the same amplitude along the longitudinal body axis and they stay in the same plane. The locomotion of this system can thus be considered to be the undulatory motion of a circular cylinder having constant diameter.

Since in all micro-organisms the length and diameter of the flagellum are on the order of $10 \sim 1,000 \ \mu m$ and $0.2 \sim 1.0 \ \mu m$ respectively (Holwill, 1977), the hydrodynamic force acting on the flagellum is, because of the very low Reynolds number, mainly a resistive or drag force. Thus the propulsive force is proportional to the total length obtained by multiplying the number of waves (n) by the wavelength (λ), the

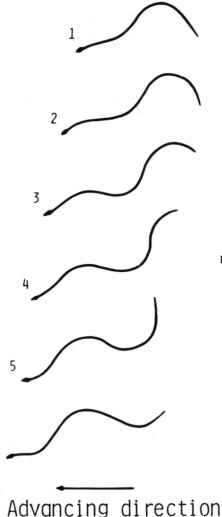

Figure 2—Snaking motion of a snake.

Advancing direction

fluid viscosity (μ), the wave propagation speed (C), and the square of the amplitude-to-wave length ratio [$(\pi a/\lambda)^2$]. The maximum efficiency of the system is very low (a few percent) and is attained at $\pi a/\lambda \simeq 0.6$ and U/C $\simeq 0.15$.

A bacterial flagellum is as small as $0.01 \sim 0.03$ μm in diameter and only a few μm in length. It is also driven to make a spiral motion by a rotary motor embedded in the cell (Berg, 1975). The marine worm Noreis diversicolor also swims by throwing its body into waves which travel from tail to head (Gray, 1939). Then many laterally extended parapodia make a rowing motion.

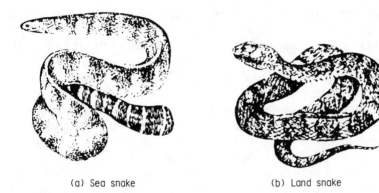

(a) Sea snake   (b) Land snake

Figure 3 — Tail configuration of snakes.

In larger slender animals such as sea snakes, cyclostomata and eels, the waves of transverse displacement are, as shown in a typical example for a snake in Figure 2, characterized by an increase in amplitude as they pass down along the body (Hertel, 1966). Since locomotion in the water is performed at a high Reynolds number such as Re $\simeq$ 10^4, these animals can utilize positively a force which is effective only when the amplitude is increased toward the tail, namely, an inertial force. This force is proportional to the product of the added mass of the surrounding fluid and the square of the wave propagation speed.

The added mass is principally proportional to the square of the height of the body and is almost totally independent of the body width or thickness. Thus, the animal's ability to perform snaking locomotion in a fluid is enhanced either by a tail which is thin but of full height, as in the sea snake, or by dorsal and/or ventral fins, as in slender animals such as eels. For example, in Figure 3 (Azuma, 1980), the tail of a sea snake (a) is seen to be rather flat and wide right up to the tip, while the tail of a land snake (b) is rounded and tapered toward the tip. A similar difference can be found between marine and land iguanas. This suggests a way to distinguish the mode of life of (extinct) animals having unknown ecological behavior.

Jetting

The simplest way to utilize the inertial force of a fluid is 'jet propulsion'. Thus this method of propulsion is widely adopted by a variety of aquatic animals for normal swimming and/or for emergencies.

The best known species of marine animals to use this method are the cephalopods, particularly squids, octopuses and nautiluses, and the medusae such as jellyfish and salps.

Usually the jet propulsion is performed in two phases, the inhalant phase and the exhalant phase. In the inhalant phase the water is sucked into a mantle cavity through a sucking nozzle or the open rim of the mantle so gently that no appreciable positive or negative force is created. In the exhalant phase, the water is expelled rearward by high internal cavity pressure through an exhaust or jet nozzle which may be the same as or separate from the sucking nozzle.

Through these recovery and power strokes, the propulsive force is generated intermittently and the impulse acting on the body through the one operation cycle is proportional to the product of the mass of the fluid sucked into and expelled from the cavity and the jet speed (V_j) minus (or plus) the sucking speed (V_s) when the sucking nozzle is directed rearward (or forward). This impulse is balanced with the momentum of the body, which is the product of the mass of the body (M) and the resultant velocity (U). Since the efficiency of this system is given by U/V_j, the jet speed should be close to the cruising speed obtained except in the case where the system is used in an emergency, at which time the maximum thrust must be produced without consideration of economy.

Another interesting example is the scallop Patinopecten which swims by alternately opening and closing bivalves. The entrapped water in the recovery stroke is expelled in the power stroke through two openings, causing the shell to jump up in the direction opposite to the water jet once each time the stroke is repeated. The valves are connected by a small block of abductin, an elastic protein which forms the inner hinge ligament. The spring action of the abductin can store a part of the power for locomotion which would otherwise be wasted in overcoming the inertial force accompanying the valve motion.

Paddling

If a thin plate, for example a circular disc, is moved in the water with a constant speed (C) in the direction normal to the plate surface, then the plate generates a drag force which hinders its motion. When the plate is a paddle for rowing a boat, the drag force is, in the power stroke, utilized as the propulsive force for the boat. If the boat is moving with a constant speed (V), the propulsive force is proportional to the product of the area of the plate (S) and the square of the relative speed [$U^2 = (C - V)^2$] and the efficiency in this steady power stroke is given by $\eta = V/C$. As in the case of jet propulsion, the propulsive force grows larger as the relative speed increases, whereas the efficiency approaches 1 as the speed of the boat approaches the rowing speed, or $C = V$, at which time the propulsive force is zero.

When a paddle is in linearly accelerated motion in a fluid, the added mass of the surrounding fluid will accompany the paddle and thus an inertial force will be superimposed on the resistive force as a reaction force. This inertial force is approximately equal to the product of the accelera-

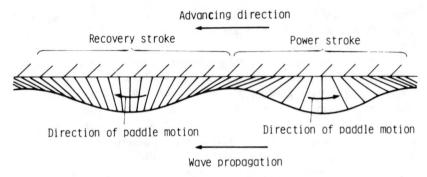

Figure 4—Series of paddling motions.

tion in relative speed and the mass of the fluid contained in a circular cylinder of a diameter equal to the width of the paddle for a long paddle or in a sphere of a diameter equal to the diameter of the paddle for a circular paddle (Streeter, 1961).

In the locomotion of an aquatic bird swimming on the water surface, the webbed feet of the bird can be considered to be a pair of paddles which are driven either alternately or simultaneously in the power stroke and returned to the initial position after being laid level so as to minimize both inertial and drag force in the recovery stroke. The moment generated around the center of gravity in the power stroke is opposed by any moment generated by a shift in the center of buoyancy due to slight changes in the body attitude. In the fully submerged state, the webs are driven simultaneously in a rear position as in the swimming of a frog so as to direct the propulsive force toward the center of gravity without creating any moment. This method is used because the hydro-static and -dynamic pressure distribution makes the restoring moment insufficient.

A pair of rows of paddles called 'pleopods' or 'swimmerets' is used in many species of shrimps, lobsters and prawns. Each paddle is concaved forward and fringed with bristles for catching the water fully during the power stroke and for enabling the water to escape during the recovery stroke. The paddles are also quickly and successively tilted rearward in the power stroke and gradually returned forward in the recovery stroke as if a wave were propagating forward in the manner shown in Figure 4. In an emergency, a tail fan is opened wide and the whole tail is flicked under the body so rapidly that the shrimp shoots backward.

Gliding

A thin plate or wing moving in a fluid with an angle of attack (α) generates an aerodynamic force and moment, both of which are proportional to the product of the air density (ρ), the wing area (S), and the square of the relative speed of the fluid with respect to the wing, (U). The force can be divided into two components, 'lift' ($L = \frac{1}{2}\rho U^2 S C_L$) and 'drag' ($D = \frac{1}{2}\rho U^2 S C_D$), which act normal and parallel to the flow direction, respectively.

If the upper surface of the wing is given a convex curvature, the lift will be larger for a given angle of attack. And if the wing span (b) is much larger than the wing chord (c) or if the ratio ($\mathcal{R} = b^2/S$), called the 'aspect ratio', is made as large as possible, then the ratio of the lift (L) to the drag (D), called the 'lift-to-drag ratio', will become appreciably greater than 1 even when the aerodynamic drag of a body is added to the drag of the wing alone (Perkins and Hage, 1949).

Many species of birds and a few species of fishes can perform powerless flight, i.e., gliding or soaring, by extending their wings or fins in a fixed manner without any active beating motion. The ratio of distance (ℓ) to height (h), or ℓ/h, in steady gliding flight under calm conditions is equal to the lift-to-drag ratio of the whole body, i.e., $\ell/h = L/D$. Thus, many large birds have wings with large spans for making long distance flight as shown in Figure 5 (courtesy of Tanaka and Iwago).

The speed of steady gliding flight is approximately proportional to the square root of the ratio of wing loading (W/S) and the lift coefficient (C_L), or $U \simeq \sqrt{(2/p)(W/S)}\,/CL$. The maximum flight range is attained approximately at $C_L \simeq 0.5$ and in exemplified birds of Figure 5, $(L/D)_{max} \simeq 20$ for the hawk and $\simeq 50$ for the albatross, respectively.

For a leisurely powerless flight the rate of descent (w) must be minimum. Many gliding birds can normally keep the rate of descent around $w \simeq 0.5$ m/sec and thus gain altitude by searching out updrafts that are faster than the sinking speed.

As seen from the typical example of a hawk shown in Figure 5a, land birds that prefer gliding flight have wide wings with a moderate aspect ratio and a tail wing of large wing area in extended state. The primary feathers at the opposite ends of the wings are separated to form 'slotted wings' which not only can sustain a higher load even at slow speed in turbulent air but also have improved load distribution, a property they attain through a clever arrangement that involves slight upward bending of the wings at the tips so as to get a higher lift-to-drag ratio without increasing the bending moment at the wing root (Whitcomb, 1976; Spillman, 1978; Conner, 1978).

The large sea birds typically represented by the albatross shown in Figure 5b have main wings with a very wide span and pointed tips, and an extremely small tail wing. The large aspect-ratio wing guarantees high performance soaring. Actually, these birds can continue powerless flight over the ocean utilizing horizontal wind shear rather than updrafts. This is performed by a method called 'dynamic soaring' in which the high performance bird takes a zigzag course, ascending against and descending with the wind (Azuma and Suzuki, 1981). The pointed tips also ensure a high L/D like the 'elliptic wing' without increasing the bending moment

(a) Chicken hawk [18)

(b) Wandering albatross [19)

Figure 5—Gliding birds (courtesy of Tanaka and Iwago).

at the wing root (Jones, 1950).

Flying squids shown in Figure 6 (courtesy of Iwago) shoot out of the water under acceleration produced by the jetting of water. As seen in this impressive picture, the flight is performed by fins and laterally extended arms like those of a 'canard'. In order to sustain the necessary lift, the main wing has to be a lifting surface which is membranous and is reinforced by five pairs of tentacles, the longest two of which form the leading edge of the membranous wing and maintain the elliptic planform required for high performance flight.

Figure 6—Flying squids in formation flight (courtesy of Iwago).

What is the membrane made of? One possibility is that the squid of this special species has membraned arms. This photograph, however, does not support this possibility. It is obvious that the membranes would have to be spread to the fullest extent during flight in order to provide a smooth wing surface. A close look at the picture shows, however, that the length of the connecting membrane is so short that if it is truly a membrane, the squid would lose its freedom of movement in its normal activities in the water. Therefore, I would like to suggest another possibility—that the squid makes a membranous wing or webs among the arms just before taking off, probably by spurting out a very sticky mucus. An instant wing!

This hypothesis is reinforced by observing the arrangement of the ten-

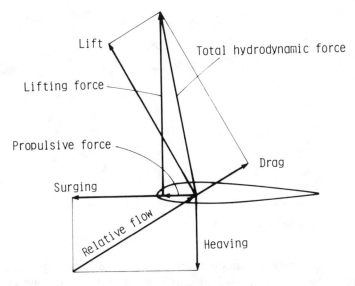

Figure 7—Fluid-dynamic forces acting on an airfoil in motion.

tacles, specifically their tips, which are, as clearly seen in Figure 6, rounded and make narrower gaps than those at the middle parts of the tentacles. This formation prevents the open trailing edge of the mucus membrane from shrinking and maintains the extended wing form.

Beating

As shown in Figure 7, a wing moving in a fluid generates lift and drag in the directions normal and parallel to the relative velocity. Therefore, a wing capable of combining surging (longitudinal) motion and heaving (transverse) motion creates a 'propulsive force' or 'thrust' along the wing surface in addition to the 'lifting force' normal to the surface. Thus, in addition to the two modes of wing locomotion already described, namely, (a) the paddling plate and (b) the fixed wing, there are five other possible modes to be considered. As shown in Figure 8, these are (c) the rotary wing which is rotated in the plane of the wing surface in one direction around a normal axis, (d) the beating wing which rocks around an axis lying in the wing surface at one terminal end of the wing, (e) the fanning wing which rocks around an axis lying in an extension of the wing surface ahead of the wing, (f) the sweeping wing which rocks around an axis normal to the wing surface, and (g) the cyclo-gyro wing which rotates around an axis lying parallel to and under the wing surface.

Each of these wings is also capable of 'feathering motion', a pitch-up or -down motion around a spanwise axis. This axis is usually located at the aerodynamic center or at a quarter chord and the feathering motion

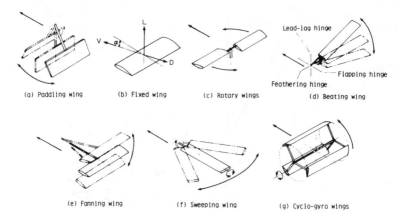

Figure 8 — Modes of wing motion.

is superimposed on other motions with an adequate phase lag. It should be noted, however, that it is impossible to utilize the rotary and cyclo-gyro wings in animal locomotion.

The thrust is proportional to the square of the reduced frequency ($k = \pi c f / U$) and to the square of the ratio of the amplitude of transverse motion to the chord (h/c), i.e., $(kh/c)^2 = (\pi f h / U)^2$. Therefore, most of the thrust is generated at the outer part of the beating wings (Azuma, 1980). Thus, the inner part where the amplitude is small must, together with the body and the tail wing, create the lifting force. In order to attain optimal flight, the beating motion includes, in addition to the flapping motion, the lead-lag motion in the wing surface and the feathering motion along a line connecting the aerodynamic centers of the wing sections. These coupled motions can be performed through the three joints of the wing (Azuma, 1979).

The most powerful motion is the flapping down of the wing in the power stroke, during which the greater part of the thrust is generated. To facilitate this motion, the bird's pectralis major is situated vertically under the shoulder. Thus the wing is in the final analysis a 'shoulder wing' which can be provided close to the center of gravity. The beating wing is the only way to sustain the body in the air against the force of gravity and to propel it against the drag of the body. No other mode of wing locomotion is possible for flight in the air because in all other ways the position of the wings would be separated from the center of gravity, making it almost impossible to cancel out the resulting moment. This will be clearer from the subsequent sections.

The flapping up of the wing in the recovery stroke does not generally generate a propulsive force or a lifting force. This is especially true in low speed flight. The adverse effect of the recovery stroke can be avoided by making a quick return to the upmost position with supination or pitch-up

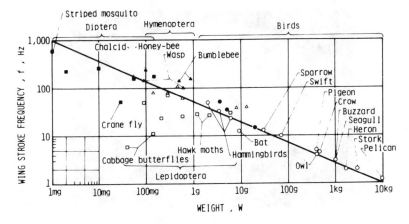

Figure 9 — Beating frequency versus mass of birds and insects.

of feathering and by introducing a phase difference between the flapping motions around the wrist and shoulder hinges so as to reduce the wing span during the recovery stroke (Sato et al., 1979; Azuma, 1980).

As shown in Figure 9, statistical data reveal that the beating frequency increases as the mass of the bird becomes small or more specifically in proportion to the -0.4 power of the mass ($f \propto m^{-0.4}$). However, the power balance between the inertial power and the available power, which are considered to be proportional to $m^{5/3} f^3$ and mf, respectively, gives the relation $f \propto m^{-1/3}$. Thus, as the mass decreases the increase in beating frequency for a given decrease in mass is larger than that theoretically expected for the inertial force. Therefore, it would seem that small birds such as hummingbirds and almost all insects would find it difficult to beat their wings at the required high frequency. This problem is, however, alleviated by introducing a spring device which can store and release the energy by converting the kinetic energy of the wing to elastic energy and vice versa (Azuma, 1979). The exoskeleton of insects is known to have such mechanical characteristics.

Many differences can be found between birds and insects in their biomechanical characteristics. Since insects have very small body size and mass, the wing area can be small even though the wing loading and thus the flight speed are kept small. Then too, the Reynolds numbers are also small, ranging from 10^2 to 10^4 as compared with 10^4 to 10^5 for birds. This fact is closely related to the aerodynamic configuration of insects in the following way:

Almost all insect wings are membranous, of small span (or small aspect ratio) and driven around a single joint, and, except for aquatic insects the body configurations including the wing surfaces are different from that of the birds, in that they are not smooth.

The wing is not required to have a large aspect ratio as is necessary in

(a) Folding fan

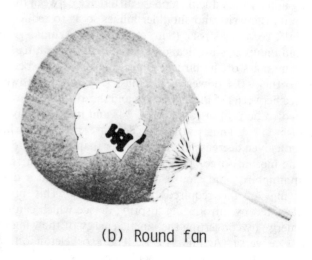

(b) Round fan

Figure 10—Fans.

soaring birds, except in the flight of dispersion or migration during which some species adopt longer feathers than normal for phase variation. The aerodynamic characteristics, for example the maximum lift-to-drag ratio, are improved somewhat by adopting a corrugated wing with a sharper leading edge than the conventional wing having a round leading edge and smooth surfaces (Pope and Harper, 1960).

The thorax of an insect is covered by an exoskeleton of sclerotic cuticle which supports the membranous wing through a single universal joint and beats it either directly through the contraction of inner and outer muscles or indirectly through a 'click mechanism' by distorting the thorax

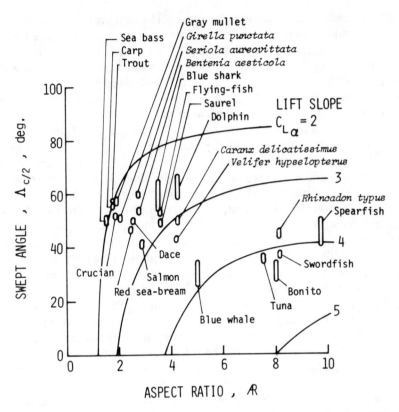

Figure 11—Statistical relation between the swept angle and the aspect ratio of caudal fins.

(Borror et al., 1964). In either case, a spring device is provided to cancel the inertial force generated by the wing motion. The elastic deformation of the membranous wings is thought to compensate for insufficient beating motion due to the single joint.

Fanning

Snaking motion performed by a slender fish generates thrust which is proportional to the square of the maximum height of the body and to the square of the amplitude of the snaking motion (Wu, 1971). If the motion is of a small wave number less than 1 and of large increment in amplitude toward the tail, then the caudal fin can be considered to be a wing in transverse or heaving motion and feathering motion with some phase difference, the latter of which is strongly affected by the elasticity of the fin (Azuma, 1980). Since the motion as well as the planform is similar to that of a fan, either a folding fan called a 'sensu' (see Figure 10a) or a round fan called an 'uchiwa' (see Figure 10b), the word 'fanning' has been used to describe the motion.

Unlike beating, the fanning of a fin is performed uniformly along the fin span and is thus more powerful for the generation of propulsive force only, as in swimming where no lifting force is required on the fin.

A folding fan is used to direct a breeze onto the user by rocking it around a single hinge, the wrist, at a rather high frequency. On the other hand, a round fan with a long handle is used to fan a fire or a person by rocking it around two hinges, the wrist and the elbow, at a lower frequency than that of a folding fan. An important difference between these two kinds of fanning exists in the phase lag between the heaving and feathering motion. In the folding fan, the two motions are performed without any phase lag because the single hinge is not mechanically capable of providing phase lag except through elastic deformation. In the rounded fan, on the other hand, the feathering motion follows the heaving motion with some phase lag which is given mechanically through at least two hinges. It is known from the hydrodynamic considerations (Azuma, 1980) that the phase lag (ψ) is, for optimal efficiency, a function of the reduced frequency (k) and should be $\psi = 90°$ for very low k and $0°$ for high values of k greater than 1. For a cruising fish, the reduced frequency is low (k < 0.3), like that in the cruising flight of birds and insects, so that the fish prefers to swim in the mode of two-hinge locomotion. When starting from rest or in an emergency, the reduced frequency is high (k > 1.0) because of the need for a sudden, large thrust. Therefore, no phase lag is necessary so that the mode of one-hinge locomotion can be used.

Figure 11 shows the statistical relation between the aspect ratio and swept angle of the caudal fin at mid chord ($\Lambda_{c/2}$) for various fish. The figure also shows the lift slope ($C_{L\alpha}$), which is a measure of the performance of a wing. It can be seen that high performance fish which cruise continuously have a fin of large aspect ratio and swept angle, called a 'lunate tail', which allows them to swim fast with high efficiency. On the other hand, fish living near the bottom have a fin of large area but low aspect ratio in the form of a rounded fan or the triangular wing of an airplane which allows them to start and stop quickly.

To maintain steady swimming the mean thrust for each period of oscillation must be directed to the center of gravity. As a primary propulsive device the caudal fin arrangement is well suited not only for the above purpose but also for making possible a locomotive muscle arrangement that preserves the streamlined shape of the body.

Sweeping

The rocking motion of a thin and long blade, which is called 'sweeping' or 'sculling', can be considered as a coupled motion consisting of a lead-lag motion in the surface of the blade and a feathering motion for pitch change.

(a) Ro

(b) Kai

Figure 12 — Oriental sweeping devices.

Since sweeping is effective for the generation of propulsive force near the blade tip where the relative velocity is large, the sweeping blade (or wing) has been used for rowing a boat as with a 'ro' or 'kai' as shown in Figure 12.

In a combined flow generated by the forward motion of the boat and the sweeping of the blade, the main force generated by the blade, aside from drag and inertial forces, is lift, in all but the early and later stages of the motion. Since the lift is directed nearly normal to the blade surface, the propulsive force is given by the sine component of the blade inclination. The cosine component exerts an undesirable force on the rower if it is not counterbalanced by a rope fixed to the bottom of the boat. The ro has such a rope (Takusagawa and Yoshioka, 1976) while the kai does not.

The ro and kai have been widely used in Asia as propulsive devices. The ro is convenient for longer journeys since it can be rowed with both hands and the kai is convenient for rowing a boat while fishing since it can be operated by one hand.

Blade sweeping can be observed in the locomotion of several propelling devices such as the pectoral fins of fish, the caudal fin of the long-tailed shark and the swimming legs of Portunidae. Shown in Figure 13 (Azuma, 1980) is a portunid crab, Neptunus trituberculatus, having a pair of swimming legs, in each of which there are, in addition to one universal joint at the root, three lead-lag hinges and one feathering

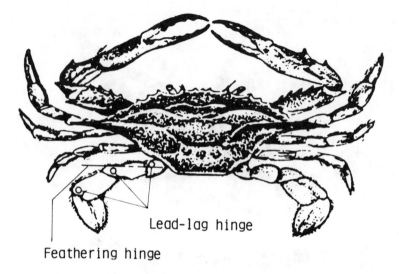

Figure 13 — Swimming legs of a portunid crab.

hinge. Thus the crab can swim laterally by sweeping the legs in exactly the same way that a boat is rowed in the oriental way.

Conclusion

In conclusion, one can express profound wonder and awe at how living creatures have found the ways of locomotion that are most economical for their modes of life and at how they have developed sophisticated mechanisms to perform these locomotions. The flight of the squid is a superb example. A likely mechanism for the flying of this species has been hypothesized, although the evidence is indirect. However, it would be nice to know how the squid has obtained this way of making and utilizing the instant wing for flight without any knowledge of hydrodynamics and flightdynamics, at least in the sense that we humans think of knowledge. The evolutionary process has been long and its results marvelous. Still, one cannot help asking: "Is such an accomplishment really possible simply as the accumulation of natural selection and mutation?"

References

AZUMA, A. 1979. Flying behavior of living creatures. Blue Backs B-378, Kodansha Co. Ltd., Tokyo. (in Japanese)

AZUMA, A. 1980. Swimming behaviour of living creatures. Blue Backs B-412, Kodansha Co. Ltd., Tokyo. (in Japanese)

AZUMA, A. 1980. Unsteady aerodynamics in relation to rotary and beating wings. Proceedings of 11th Annual Meeting of the Japan Society for Aeronautical and Space Sciences, 90-95. (in Japanese)

AZUMA, A., and Suzuki, K. 1981. On the dynamic soaring of the albatross. Proceedings of 12th Annual Meeting of the Japan Society for Aeronautical and Space Sciences, 18-19. (in Japanese)

BERG, H.C. 1975. How bacteria swim. Scientific American, 233:2, 36-44.

BORROR, D.J., Delong, D.M., and Triplehorn, C.A. 1964. An Introduction to the Study of Insects. Holt, Rinehart and Winston, New York.

CONNER, D. 1978. CTOL concepts and technology development. Astronautics & Aeronautics. Vol. 16, No. 7/8, July/August, 29-48.

GRAY, J. 1939. Studies in animal locomotion VIII. The kinetics of locomotion of nereis liversicolor. J. Exp. Biol., 16:9-17.

HERTEL, H. 1966. Structure-Form-Movement. Katz, M.S., Translation ed. Reinhold Publishing Corporation, New York.

HIRAMOTO, Y., and Baba, S.A. 1978. A qualitiative analysis of flagellar movement in echinoderm spermatozoa. J. Exp. Biol., 76:85-104.

HOLWILL, M.E.J. 1977. Low Reynolds number undulatory propulsion in organisms of different sizes. In: T.J. Pedlay (ed.), Scale Effects in Animal Locomotion, Academic Press, London.

JONES, R.T. 1950. The spanwise distribution of life for minimum induced drag of wings having a given lift and a given bending moment. NACA TN 2249.

PERKINS, C.D., and Hage, R.E. 1949. Airplane Performance, Stability and Control. John Wiley & Sons, Inc., New York.

POPE, A., and Harper, J.J. 1960. Low Speed Wind Tunnel Testing. John Wiley & Sons, Inc., New York.

SATO, M., Azuma, A., and Saito, S. 1979. Analysis of wing motion in living creatures. Proceedings of 17th Aircraft Symposium, The Japan Society for Aeronautical and Space Sciences, 14-17. (in Japanese)

SPILLMAN, J.J. 1978. The use of wing tip sails to reduce vortex drag. Paper No. 618, Aeronautical Journal, 82. September.

STREETER, V.L. (ed.). 1961. Handbook of Fluid Dynamics. McGraw-Hill Books, Inc., New York.

TAKUSAGAWA, Z., and Yoshioka, I. 1976. A study of ro (oriental sweep). Bulletin of the Faculty of Eng. Yokohama National Univ., 25:69-75.

WHITCOMB, R.T. 1976. A design approach and selected wind-tunnel results at high subsonic speeds for wing-tip mounted winglets. NASA TN D-8260.

WU, T.Y. 1971. Hydrodynamics of swimming propulsion. Part 3. Swimming and optimum movements of slender fish with side fins. J. Fluid Mech. 46:(3)545-568.

I.
ORTHOPEDICS

Keynote Lectures

Loose Shoulder and Suspension Mechanism of the Gleno-Humeral Joint

Ryohei Suzuki
Department of Orthopaedic Surgery,
Nagasaki University School of Medicine, Japan

Surgical treatment of recurrent dislocation of the shoulder joint is one of the most interesting problems for orthopedic surgeons. Hitherto, more than 100 kinds of surgical techniques have been published, but it is necessary to analyze the etiology and pathomechanics of the disorder in order to select the operative method.

Most cases of recurrent dislocation of the shoulder joint have a history of traumatic dislocation and have been treated successfully by the Bristow operation in our clinic. But there are some cases in which no episode of trauma is found and this Bristow operation does not succeed. In these cases, the cause of dislocation is considered to be instability of the shoulder joint itself. This unstable state was called a "loose shoulder joint" by Endo, a Japanese orthopedic surgeon (Endo et al., 1973).

On the other hand, loose shoulder joint also can be detected among patients without recurrent dislocation who complain of dull pain along the arm and shoulder, especially at the time of carrying heavy objects, or of pain on motion. One can also find people with loose shoulder joint who have no complaints. It is an interesting fact that some of the patients with such loose shoulders can dislocate and reduce their own shoulder joints by voluntary muscular action.

Definition of the Loose Shoulder Joint

According to Endo, the loose shoulder joint is defined as a shoulder with abnormal loosening without distinct abnormalities of the bones constituting the shoulder joint and muscles of the shoulder girdle.

Therefore, in spite of having marked loosening, paralytic shoulder is excluded from this category.

Method of Examination for Loose Shoulder Joint

The simplest method for detection of a loose shoulder joint is to determine instability in the superior-inferior direction.

The patient is asked to relax his shoulders in the erect posture, and the arms are pulled down by the examiner. Inspection or palpation of the abnormal depression in the subacromial region indicates the presence of abnormal laxity of the shoulder joint.

In the same condition a 3 to 5 kg weight is applied to the forearm and an antero-posterior x-ray picture is taken. In the case of loose shoulder joint, abnormal downward shift of the humeral head from the glenoid fossa is observed. Figure 1a shows an x-ray without a weight and Figure 1b shows the joint with a load.

The rate of inferior shift of the humeral head is calculated by B/A (see Figure 2). Using this value, Endo classified the loose shoulder joint into three types. In Type 1, the head of the humerus is present within the scapular joint but a part of the head shifts inferiorly from the joint cavity upon loading with the weight. The portion of B/A is less than 30%. In Type II, B/A is more than 30%. In Type III, a part of the head is already out of the glenoid fossa inferiorly without loading and the degree of shift becomes even greater upon loading.

Incidence of Loose Shoulder in the Population

Based on the examination method described, 14,558 subjects between the ages of 5 and 88 years were studied by Endo and others. The results show that the incidence of loose shoulder joint is highest between the ages of 20 and 30 years and about twice as high in females as in males. The incidence of each type is shown in Table 1.

Role of the Scapula in Loose Shoulder

It is interesting that inferior subluxation by passive traction and voluntary anterior dislocation of the humeral head can be prevented in most cases by the patient's own effort to elevate the scapula, or by passively holding the scapula in an abducted position. This fact can be confirmed by roentgenogram.

On the other hand, on the stress roentgenogram taken with a 5 kg weight suspended in the relaxed posture, there are found inferior sublux-

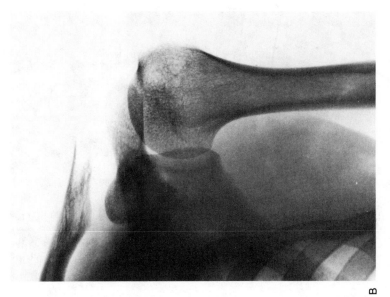

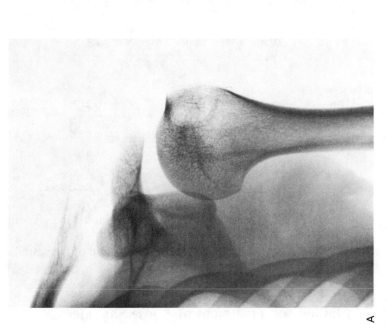

Figure 1 — Roentgenograms of loose shoulder joint in the erect posture. A: Without loading the weight. B: With loading a 5 kg weight. The humeral head shifts downward.

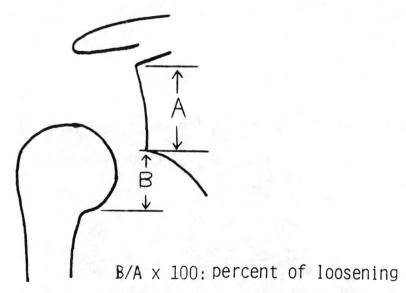

B/A x 100: percent of loosening

Figure 2—Rate of inferior shift of the humeral head is calculated by B/A. A is the longitudinal diameter of the glenoid fossa. B is the length of downward shift of the humerus.

Table 1

**Incidence of Loose Shoulder in 14,558 Subjects
(Males: 7,504; Females: 7,054)**

	Males	Females
Type I	1.6%	2.2%
Type II	0.7%	1.5%
Type III	0.2%	0.5%
Total	2.5%	4.2%

ation of the gleno-humeral joints and downward tilting of the glenoid fossa.

These facts provide evidence that the abducted position of the scapula is very important in order to stabilize the loose joint.

Method for Treatment of Loose Shoulder

In some cases, the Bristow operation was carried out for the prevention of anterior recurrent dislocation due to instability of the shoulder joint.

The aim of this method is to prevent dislocation at the moment of swinging the arm upward by the transfer of the coracobrachialis muscle and the short head of the biceps brachii muscle to the anterior glenoid rim. This operation is very effective for the treatment of recurrent dislocation caused by trauma, but is never useful for loose shoulder.

As previously mentioned, an unstable shoulder joint becomes tight by abducting the scapula. Nobuhara (1979) turned up the glenoid fossa by glenoid osteotomy, and excellent results were obtained. Endo (1979) transferred an inferior half of the pectoralis major muscle to the inferior angle of the scapula combined with anterior capsulorrhaphy.

In our method of treatment, the pectoralis major muscle is completely detached from the humerus, passed through the interval between the thoracic cage and the latissimus dorsi muscle, and sutured to the inferior angle of the scapula. Using this procedure, forward rotation position of the scapula, consisting of abduction and external rotation, is obtained. This operation was performed on 30 joints of 25 patients and the results were satisfactory.

Electromyographical Studies on the Normal Shoulder Joint

Electromyographical studies were performed on normal adults to reveal the role of the shoulder girdle muscles for prevention of inferior dislocation of the humeral head. The examined muscles were the upper portion of the trapezius, the supraspinatus, the middle portion of the deltoideus, the pectoralis major and the rhomboideus. In the relaxed erect posture, EMGs of these muscles were examined with intramuscular fine wire electrodes.

The results were different in detail from case to case but, in general, no action potential was observed without loading the arm. By increasing the load gradually, however, electrical activity began to appear in the shoulder girdle muscles. In most cases, the supraspinatus muscle was activated at first and the contraction became stronger with an increase of load. The trapezius muscle also participated in suspension by positioning the scapula in abduction. In some cases, the pectoralis major and deltoideus took part in suspension of the arm directly (see Figure 3). At any rate, it is clear that the supraspinatus muscle plays the most important role in suspension of the arm.

Electromyographical Analysis of Loose Shoulder

EMG Discharge at Rest

In the relaxed erect posture, loose-shoulder patients with severe com-

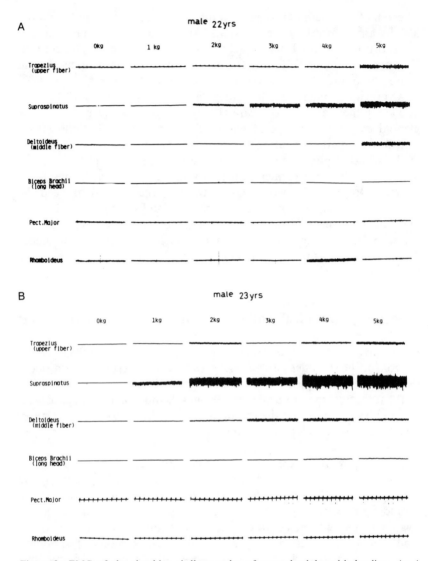

Figure 3—EMG of the shoulder girdle muscles of normal adults with loading. A: A 22-year-old male. B: 23-year-old male. The supraspinatus muscle plays the most important role in the suspension mechanism of the gleno-humeral joint.

plaints tend to have increased discharge from the trapezius and supraspinatus muscles. Figure 4 shows the EMG of an 18-year-old female patient with severe pain and stiffness of the shoulder girdle muscles. After surgery, however, EMG discharge disappeared completely from these muscles and she became free of pain and stiffness.

However, some cases of loose shoulder have been found in which ac-

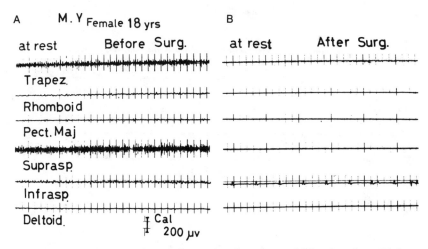

Figure 4—EMG of the shoulder girdle muscles of an 18-year-old female patient with loose shoulder in the relaxed erect posture. A: Before surgery. B: After surgery.

tion potentials were scarcely observed from the shoulder girdle muscles, although they had severe looseness.

Arm Suspension with Loading

On the stress roentgenogram taken with a 5 kg weight suspended, inferior subluxation of the shoulder joint and downward tilting of the glenoid fossa were observed, but the EMG of the trapezius and supraspinatus muscles did not change from that of the unloaded condition. However, the patient was able to prevent subluxation by strong contraction of these muscles.

Inferior subluxation and downward tilting of the glenoid fossa did not occur after surgery even when the weight was added and most cases revealed intensive action potential from the pectoralis major muscle which had scarcely shown any activity before surgery (see Figure 5).

The angle of downward tilt of the glenoid fossa was measured in 15 cases before and after surgery. The angle of tilt decreased 4.5 degrees on the average after surgery.

On the other hand, there was a case with severe loosening in which EMG-discharge scarcely occurred in the unloaded condition as well as in the condition with gradual increase of the load.

EMG and Scapular Motion at Elevation and Depression of the Arm

Compared with the preoperative movement of the scapula in elevation, the postoperative movement showed quicker motion at the beginning of arm elevation. With this postoperative movement the transferred pec-

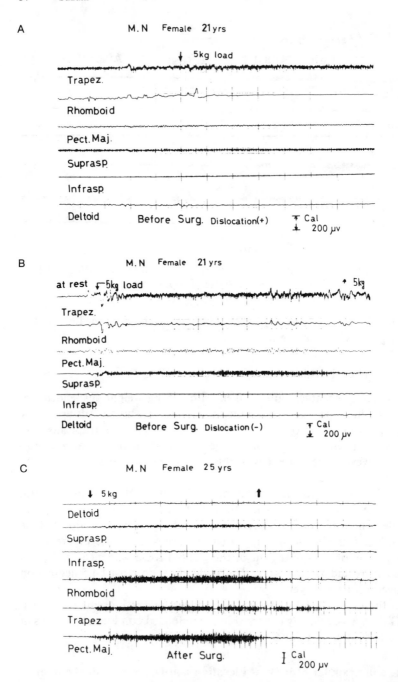

Figure 5—EMG of the shoulder girdle muscles of a 21-year-old patient with loose shoulder. A: Inferior subluxation and downward tilting of the glenoid fossa occur by loading a 5 kg weight. B: Inferior subluxation is prevented by muscular action. C: EMG after surgery with a load. Inferior subluxation is prevented.

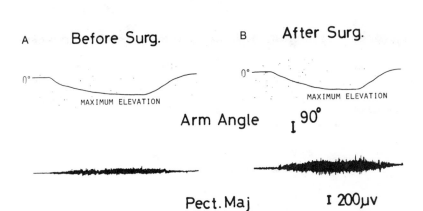

S.H 22 yrs

A   Before Surg.   B   After Surg.

0°   0°

MAXIMUM ELEVATION   MAXIMUM ELEVATION

Arm Angle   I 90°

Pect. Maj   I 200μv

Figure 6—EMG of the pectoralis major muscle of a 22-year-old male patient with loose shoulder recorded simultaneously with arm angle change during elevation and depression of the arm. A: Before surgery. B: After transfer of the pectoralis major.

toralis major muscle, which had scarcely participated in arm elevation before surgery, increased its contraction gradually from the initial stage. At the time of depression, on the contrary, this muscle showed gradually decreasing EMG-discharge (see Figure 6). In this patient, the winged scapula at the time of depression disappeared completely on the operated side.

Voluntary Anterior Dislocation of the Shoulder Joint

In our patients, seven cases had voluntary anterior dislocation before surgery. Muscular activities were different from case to case, but, in general, at the moment of dislocation, intensive action potential appeared in the pectoralis major muscle (see Figure 7). At the same time, the activity of the trapezius muscle was suppressed. On the contrary, at the moment of voluntary reduction, reverse relationship between these muscles was observed. After transfer of the pectoralis major muscle, voluntary dislocation became impossible in almost all of the patients.

Case Presentation

The following are interesting cases:

Case 1.   A 14-year-old girl. Chief complaint was recurrent and volun-

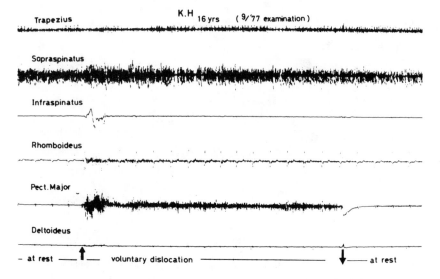

Figure 7—EMG of the shoulder girdle muscles of a 16-year-old male patient with loose shoulder. Intensive action potential appears from the pectoralis major muscle at the time of voluntary anterior dislocation.

tary anterior dislocation of the right shoulder joint after humeral neck fracture, which occurred six months prior to her admission into our clinic. The fracture was treated with a hanging cast immediately after injury. The fractured humerus was completely united leaving moderate humerus varus deformity. After removal of the plaster cast, she noticed that she could dislocate the right shoulder joint by herself.

On examination, range of motion of the shoulder joint was normal and no weakness of the muscle was found. She was able to dislocate the right shoulder joint anteriorly and reduce it by voluntary muscular action. The stress roentgenogram showed loose shoulder only on the right side (see Figure 8a). After pectoralis major transfer, she has never experienced dislocation (see Figure 8b).

Case 2.   A 20-year-old waitress. She never experienced shoulder dislocation, but had dull pain in both shoulders, and edema and numbness of the arms. The stress roentgenogram showed loosening of the shoulders. Pulsation of the radial artery became weaker by arm traction.

Plethismographical findings improved markedly after pectoralis major transfer. The operated scapula was also stabilized and all the pathological symptoms completely disappeared.

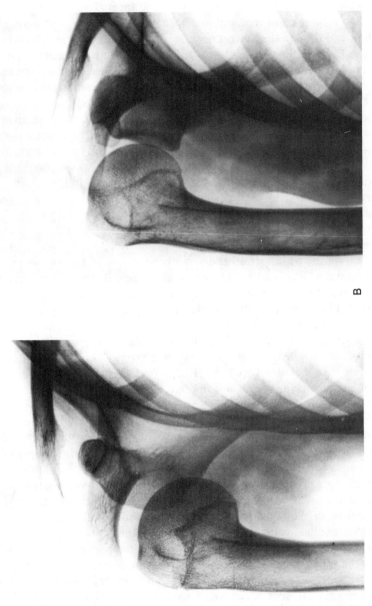

Figure 8—Stress roentgenograms of posttraumatic loose shoulder joint in a 14-year-old female patient. Humerus-varus deformity is observed. A: Inferior subluxation before surgery. B: Inferior subluxation is prevented after surgery.

Discussion

The gleno-humeral joint is comprised of the glenoid fossa and the head of the humerus. As the former is shallow and faces almost laterally, and the latter is large compared with the former, it is impossible to maintain the humeral head in the correct position in the erect posture without soft tissues. The question, then, is how to prevent downward dislocation or subluxation of the humeral head. Basmajian (1967) attached importance to the roles of the slope of the glenoid fossa, the coracohumeral ligament and the supraspinatus muscle in maintaining joint stability.

Electromyographic studies were performed to confirm Basmajian's theory. In normal adults, activity of the shoulder girdle muscles was not found in the relaxed erect posture without loading. But by increasing the load gradually, various muscles were activated. There was no doubt that the supraspinatus muscle played the most important role in joint stability.

On the contrary, in most patients with loose shoulder having severe complaints, the trapezius and the spuraspinatus were active even in unloaded arm hanging. When downward traction was applied to the arm of the patients, downward tilting of the glenoid fossa and inferior subluxation of the humeral head was observed on the roentgenograms. In this state, EMG-discharge from the trapezius and supraspinatus muscles was nearly the same. On the other hand, the patients' own effort to prevent inferior dislocation activated the trapezius and supraspinatus intensively, tilting the glenoid fossa somewhat upwards. Consequently, the coracohumeral ligament became tight. These hyperactivities of the muscles are considered to be a cause of some complaints of the patients.

In rare cases belonging to Type III (see Table 1), the active suspension mechanism of the arm does not operate well.

As for the treatment of loose shoulder, the glenoid fossa is turned somewhat upwards by transfer of the pectoralis major muscle to the inferior angle of the scapula. Consequently, the humerus is brought to the adducted position in relation to the scapula, and the suspension mechanism of the arm can operate well.

In the loose shoulder with posttraumatic humerus-varus deformity, the loosening was considered to be due to approximation of the major tuberculum of the humerus to the superior border of the glenoid fossa, which resulted in insufficient suspension mechanism of the supraspinatus muscle and coracohumeral ligament.

On the other hand, the pectoralis major muscle seems to play an important role in producing voluntary anterior dislocation which is characteristic of loose shoulder. Therefore, transfer of this muscle is reasonable because the dislocation force is weakened, and the abduction force of the scapula against dislocation is increased by this procedure.

In the case with loose shoulder, paresis, and circulation disturbance of the arm, the pectoralis major transfer was also effective. It might be due to widening of the thoraco-clavicular space by forward rotation of the scapula.

References

BASMAJIAN, J.V. 1967. Muscle Alive. 2nd Ed. Williams & Wilkinson Co., Baltimore.

ENDO, H. 1979. Pectoralis major muscle transfer for loose shoulder joint. Shujutsu. 33(8):907-911. (in Japanese)

ENDO, H., Kuge, A., Murata, Y., Takigawa, H., Takada, K., Kuriwaka, Y. and Yamamoto, O. 1973. Fieldwork of loose shoulder. Cent. Jap. J. Orthop. Traumat. 16(1):103-105. (in Japanese)

NOBUHARA, K. 1979. On the glenoid osteotomy. Shujutsu. 33(8):899-906. (in Japanese)

Little League Elbow:
A Clinical and Biomechanical Study

Yi-Shiong Hang
National Taiwan University Hospital, Taipei, Taiwan

Excessive pitching places abnormal stress on the elbow and frequently leads to injury. Trauma to the medial tendons at the elbow joint during the pitching act has been well recognized as a unique throwing injury. The underlying mechanism for this problem seems to be correlated with the extreme valgus position of the elbow during the acceleration phase as documented by cinematographic observations.

Various reports with regard to the incidence of Little League Elbow have been published. It is hypothesized that to throw a breaking pitch requires more forceful flexion and pronation of the wrist (DeHaven and Evarts, 1973) which is thought to impose a greater stress on the elbow, compared to the standard straight pitch (Adams, 1965). The present study was concerned with a clinical and roentgenographic evaluation of the elbows of a competitive group of Little Leaguers in Taiwan as well as a biomechanical study attempting to record forces acting at the elbow joint during the pitching motion and to correlate the elbow loading with pitching style.

A Clinical and Roentgenographic Study

Preadolescent and adolescent boys actively engaged in baseball are apt to injure the elbow (Bennett, 1947; Brogdon, 1960; Adams, 1965; Torg et al., 1972; Tullos and King, 1972; Gugenheim et al., 1976; Larson et al., 1976).

Ever since the Little Leaguers from Taiwan won the World Series in 1969, Little League baseball has become one of the most popular and most competitive sports in this society. Prompted by various reports with regard to incidence of the Little Leaguer's elbow, an organized effort was made to identify the nature and the incidence of the elbow abnormality in an intensely competitive group of the preadolescent baseball players as the first phase of a longitudinal follow-up study in the future.

Material and Method

Little League baseball players (n = 112) from eight regional champion teams who participated in the national championship in 1980 were the subjects of this investigation. This survey was confined to the last eight teams to evaluate the effects of pitching on the preadolescent elbows primarily because the subjects represented the most active group of the players with considerable pitching experience.

Information was obtained regarding age, number of years pitched, symptoms referable to the elbow and shoulder, number of throws per day during practice and the type of breaking pitches thrown by the pitchers examined. All players were examined personally with respect to flexion contractures as well as valgus deformities of the elbow, if present, with a goniometer. Particular attention was directed to comparing the pitching and non-pitching arms since the physical make up of each individual is variable. History of pain, local tenderness, and measures adopted to relieve symptoms were documented. Anteroposterior roentgenological examinations were made of both elbows in one view.

Results

The average age of the players at the time of survey was 11 years 8 months with a range from 10 years 8 months to 11 years 10 months. The period of pitching experience ranged from 1 to 4 years with a mean of 2 years 6 months.

Seventy-two (64%) of the 112 boys reported an episode of soreness of the pitching elbow. When the subjects were divided into Group I-39 pitchers, Group II-16 catchers, and Group III-57 others, it was found that the soreness of the elbow was experienced in 27 (69%) of Group I, 11 (69%) in Group II and 34 (60%) in Group III. Tenderness over the medial epicondylar region was found in 41 boys (37%). When considering the player's position, this was present in 16 (41%) in Group I, 6 (38%) in Group II and 19 (33%) in Group III.

In no instance had any boy stopped throwing or seen a physician. Flexion contracture of the elbow of more than 5° was found in 7 (18%) in Group I, 3 (19%) in Group II and 2 (4%) in Group III.

Valgus deformity of the pitching elbow compared with non-pitching elbow was detected in 3 (8%) boys in Group I and only one in Group III. No catcher was found to have this deformity.

Table 1

Radiological Findings of the Elbow

	Pitcher	Catcher	Others	Total
Number	39	16	57	112
Hypertrophy	39 (100%)	16 (100%)	48 (84%)	103 (91%)
Separation	24 (62%)	14 (88%)	30 (53%)	68 (61%)
Fragmentation	6 (15%)	9 (56%)	12 (21%)	27 (24%)

Roentgenograms of the elbows were obtained and comparisons of the pitching and non-pitching arms were made (see Table 1). The hypertrophy of the medial epicondyle and cortex thickening in the pitching arm were noted in all pitchers and catchers, whereas this change was noted in 84% of the rest of the players.

Separation of the medial epicondyle was defined as an increase in the distance between the medial epicondylar ossification center and the ossified humeral metaphysis. These findings were present in 24 (62%) of the pitchers and 14 (88%) of the catchers, but only 30 (53%) in Group III.

Among the players with roentgenological findings of separation of the medial epicondyle, episodes of soreness were reported by 14 out of 24 pitchers (58%), 4 out of 14 catchers (28%), and only 13 boys out of 30 other players (33%). In other words, only 46% of boys with X-ray findings of separation reported a history of soreness during their Little League careers.

Roentgenographic abnormalities of fragmentation of the medial epicondyle were found in 6 pitchers (15%), 9 catchers (56%), and 12 others (21%), with a total of 24%, or 27 boys (see Figure 1). Of those boys, episodes of soreness were reported by 3, 5, and 5 in each group, respectively.

These roentgenological findings appeared to have no correlation with the symptoms. Osteochondrosis or a vascular necrosis of the capitellum was found in one catcher (see Figure 2). Hang (1981) witnessed a tardy ulnar neuritis in a Little League baseball player 10 years ago; however, no player in this group presented this ailment.

Discussion

The present study indicated that the overwhelming majority of boys had hypertrophy of the humeral cortex in the dominant pitching arm and enlarged medial epicondyles. These findings probably represent a normal roentgenological response of a dominant arm in a preadolescent boy.

Although it was referred to as accelerated growth from repeated

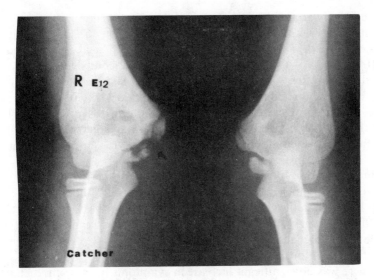

Figure 1 – A catcher with fragmentation of the medial epicondyle.

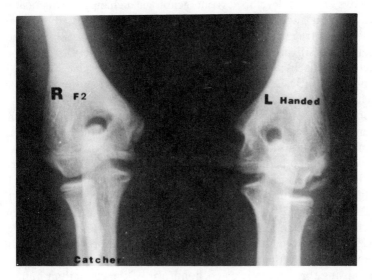

Figure 2 – A left-handed catcher demonstrating osteochondrosis of the capitellum humeri on the right hand side. He also had separation of the medial epicondylar apophysis.

stimulation of the epiphysis by Adams (1965) who in his follow-up X-ray studies observed premature closure of the epicondylar apophysis as compared with the opposite arm, this phenomenon was not observed in this study.

Of these youngsters, 61% presented some separation or widening of the epicondylar apophysis. Surprisingly this roentgenological picture was found in 88% of the catchers (14 out of 16) and only 62% of the pitchers (24 out of 39). The Houston survey (Gugenheim et al., 1976) reported this finding as avulsion fracture of the medial epicondyle, although none of the youngsters in this group experienced sudden pain or cracking noises in their pitching experiences. In addition, 45% of the boys with this roentgenological picture reported symptoms of soreness but none was sufficiently severe to keep him from playing a game. We are of the opinion that separation of the medial epicondyle is a result of chronic traction stress instead of avulsion fracture.

Fragmentation of the medial epicondylar apophysis was another common roentgenological finding. The incidence of this finding in this study was higher than the study conducted in Houston (9%) (Gugenheim et al., 1976) and Eugene (23%) (Larson et al., 1976), however Adams (1965) reported a higher incidence (50%) in the 9 to 14 year age group in Southern California.

There was no significant increase in symptoms in these children compared with those who had normal roentgenograms. Nearly as many individuals without symptoms (14 boys or 51%) as those with symptoms (13 boys or 49%) developed roentogenological change of fragmentation. The Houson survey (Gugenheim et al., 1976) suggested that medial epicondylar fragmentation in asymptomatic cases probably represented secondary ossification centers as described by Caffey (1972), but not true fracture. This hypothesis can hardly be accepted since such a large number of Little Leaguers have an accessory ossification center on the pitching arm only.

Compared with other studies conducted in the United States, the incidence of separation and fragmentation in this study is higher. The climatic conditions in Taiwan are quite similar to those in California and allow baseball to be played year-round. In addition, the overzealous attitude and expectations of coaches, parents, and local fans make Little League baseball one of the most competitive sports in Taiwan.

Osteochondrosis of the capitellum of the humerus was found in one catcher. Although this was a rare condition as described by Panner (1929), we believe that the repeated lateral compression force on the elbow peculiar to the pitching act should be considered a dominant factor causing the traumatic osteochondrosis of the capitellum humeri.

Another interesting phenomenon is that catchers presented a similar or even higher incidence of radiographical findings of separation and fragmentation (see Table 1). The number of pitches thrown by a catcher per inning and during practice probably is as great as that of a pitcher. In addition, whether throwing from a squatting or semi-squatting position causes a higher stress on the pitching elbow is a subject requiring further investigation.

Conclusion

In a clinical and radiographic study of 112 Little Leaguers, 64% reported a history of elbow symptoms, 4% valgus deformity and 11% flexion contracture. Radiographic changes of separation and fragmentation of the capitellum were found in one case. However, there was no significant correlation between radiographic findings and symptoms. The skeletal lesions appeared to be an adaptive reaction to valgus stress of the pitching motion. Catchers had a high incidence of radiographic change. We hope these youngsters will form the basis for a prospective study, and if they could be followed up to maturity a great deal of valuable information could be obtained.

Biomechanical Study

Through recent advances in modern technology, it is possible to analyze the forces acting at the human joint (Morrison, 1968). However, no published work has included the analysis of forces acting on the elbow joints during the throwing act.

It has been hypothesized that stress is considerably increased by throwing curves and other breaking pitches, which require more forceful flexion and pronation of the wrist (Adams, 1965; DeHaven and Evarts, 1973). There is also evidence indicating that by changing the style of delivery, one can improve the elbow ailment (Albright et al., 1978). This study was concerned with the means of recording forces acting on the elbow joint during the pitching motion and with correlating the elbow loading with the pitching style.

Theory

The experimental model was a linkage co-ordinate system consisting of the forearm and hand-ball segments (see Figure 3). By measurement of the acceleration of the center of mass of the two segments (Drillis, 1966), it is possible to calculate the resultant inertial force component along the forearm segment at the elbow joint by solving the dynamic equilibrium equation.

The elbow and wrist joints have forces and moments acting across them. These are the compressive forces acting on the bearing surfaces, and the moments generated by muscle forces acting across the joint. The differential equations of motion for the two segments can be written from a free-body diagram. In the forearm segment (see Figure 4), a force $(m_f \vec{a}_f)$ generated by the muscle to produce forearm motion exerted a joint reaction force at the elbow $(\vec{F}_e)$ and at the wrist $(\vec{F}_w)$, respectively. This force also produced a torque $(I_f \vec{\alpha}_f)$ which likewise exerted a moment $(\vec{M}_e)$

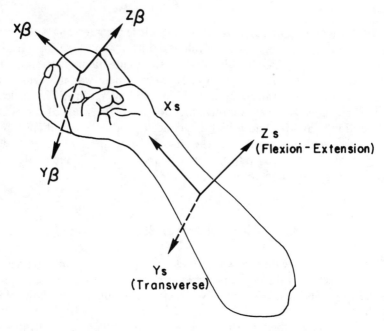

Figure 3—Axis systems of accelerometers on the forearm and hand.

at the elbow as well as a moment ($\vec{M}_w$) at the wrist. By the same token, the motion of the hand-ball segment (see Figure 5) was produced by a force ($m_h\vec{a}_h$) which exerted a joint reaction force, ($\vec{F}_w$) at the wrist together with a torque ($l_h\vec{\alpha}_h$) which exerted an angular moment across the wrist joint with a magnitude of $\vec{M}_w$.

From Newton's Second Law and the Angular Momentum Principle, the differential equations of motion of the two segments are related in such a way that:

$$m_f\vec{a}_f = \vec{F}_e + \vec{F}_w \tag{1}$$

$$l_f\vec{\alpha}_f = \vec{M}_e + \vec{M}_w \tag{2}$$

$$m_h\vec{\alpha}_h = \vec{F}_w \tag{3}$$

$$l_h\vec{\alpha}_h = \vec{M}_w \tag{4}$$

Substituting the value of $\vec{F}_w$ from equation (3) into equation (1) yields

$$\vec{F}_e = m_f\vec{a}_f - m_h\vec{a}_h \tag{5}$$

Substituting the value of $\vec{M}_w$ from equation (4) into equation (2) yields

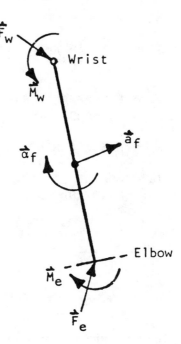

$\vec{\alpha}_f$ = ANGULAR ACCELERATION OF FOREARM

$\vec{a}_f$ = LINEAR ACCELERATION OF FOREARM

m_f = MASS OF FOREARM AND SPLINT

I_f = MOMENT OF INERTIA OF FOREARM AND SPLINT

(1) $m_f\vec{a}_f = \vec{F}_e + \vec{F}_w$

(2) $I_f\vec{\alpha}_f = \vec{M}_e + \vec{M}_w$

Figure 4—Free body diagram of forearm segment.

$$\vec{M}_e = I_f\vec{\alpha}_f - I_h\vec{\alpha}_h \qquad (6)$$

Value of normalized elbow force ($\vec{F}_e/m_f + m_h$) could thus be obtained.

Methods

The present study utilized accelerometers to obtain acceleration data of the moving forearm. A telemetered EMG system recorded muscular activity and stroboscopic photography was used to identify forearm rotation during the pitchng motion.

Because appreciable rotation occurred between the radio-ulnar and radio-humeral articulations, placement of the accelerometers on the forearm skin would be subject to rotational variation. In order to mount the accelerometers and maintain the sensitive axes with respect to the principal axes during the pitching act, we developed a special splint. This splint consisted of a plastic forearm plate attached to an upper arm cuff by a hinge joint aligned so that the plate maintained its longitudinal axis parallel to the forearm. The cuff was fastened to the upper arm and forearm in such a way that rotation of the forearm could occur beneath the plate while the splint maintained its coordinate axis with respect to the upper arm.

Initial studies were conducted with the forearm accelerometer package

$\vec{\alpha}_h$ = ANGULAR ACCELERATION
OF HAND

$\vec{a}_h$ = LINEAR ACCELERATION
OF HAND

m_h = MASS OF HAND AND BALL

I_h = MOMENT OF INERTIA OF
HAND AND BALL

$$(3) \quad m_h \vec{a}_h = \vec{F}_w$$

$$(4) \quad I_h \vec{\alpha}_h = \vec{M}_w$$

Figure 5—Free body diagram of hand-ball segment.

only. Two miniature piezoelectric accelerometers (Glennite A314-TMU) were attached to the splint at the estimated center of the forearm mass on a small aluminum block so that the sensitive axes were aligned with the longitudinal and tangential axes, i.e., varus-valgus plane, of the forearm. The accelerometer package weighed 60 g. The signal cables from the accelerometers were attached to connectors, passed along the arm over the shoulder and connected to a charge amplifier (Kistler 504) which was in turn coupled through a DC filter to remove low frequency DC drift that tended to occur with piezoelectric accelerometers. The signals were then fed to one of the terminals of a differential amplifier, and then to a dual beam oscilloscope.

The oscilloscope was set on a single sweep mode and the signal was triggered by a switch system activated by the pitcher's lead foot (opposite foot from the throwing arm) when stepping on it. The initiation of the signal was in close sequence to the beginning of the acceleration phase of the pitching act. The signal was terminated by a switch on the gripping surface of the thrown ball. At the moment of ball release the switch opened removing the DC offset voltage to prevent superimposition of the signals. The acceleration signals of axial and tangential direction between foot strike and ball release was recorded for analysis using a Polaroid camera.

To calculate the linear and angular acceleration components for each link, a minimum of six accelerometers were required. In practice, however, it was found that nine accelerometers were needed to compensate for cross-axis sensitivity of the accelerometers (Padgaonkar et al.,

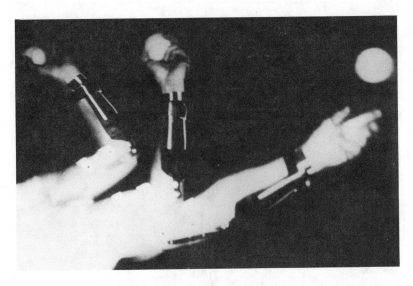

Figure 6—Forearm splint for placement of the accelerometers (see text). Figure also shows white marks on splint and marker on the wrist joint used to assess forearm rotation.

1975). Mounting nine accelerometers with associated cables, however, would present prohibitive problems to a pitcher. Rather than attempting to quantify all the forces and torques in the same experiment, the present study utilized a manageable number of accelerometers and only certain components of the forces were studied.

The acceleration of the pitching motion has a fundamental frequency of approximately 2HZ, and other phases of the action are even slower. Piezoelectric accelerometers lose accuracy below 2HZ because of charge leakage. Piezoresistive accelerometers are advantageous for this type of biomechanical study because their usable frequency range extends to a steady state as a lower limit.

Ten volunteer subjects participated in this study. All throws except for the instrumented ball studies were performed outdoors with a target at a distance of 60 ft. The instrumented ball studies were conducted indoors using a shock-absorbing mat placed at a distance of 25 ft. Recordings were taken only after the subjects were able to throw consistently both straight and curve pitches.

Since the extent and direction of forearm rotation has been a controversial subject, it was measured by placing two separated white markers on the splint and a corresponding marker along the ulnar border over the wrist joint. The relative motion between the forearm and the splint could be determined using strobophotographs (see Figure 6).

The muscle activities of the biceps, triceps and the wrist flexors were studied using a surface electrode EMG radio-telemetry system, and was

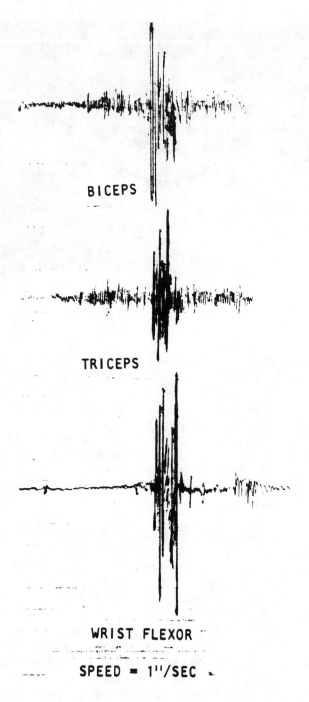

Figure 7—Telemeterized EMG indicates that simultaneous contraction of the biceps and triceps muscle occurs during pitching act.

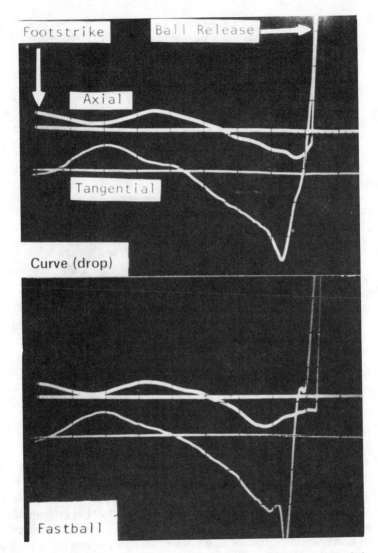

Figure 8 – Typical acceleration patterns of one pitcher for straight and curve pitches.

recorded by inking galvanometer recording tape. The speed of recording paper was adjusted at 1 inch/sec (see Figure 7).

Results

The acceleration analogs from the forearm splint and hand-ball segments revealed that the patterns were different for individual pitchers but quite consistent for each individual. Each tracing showed an abrupt increase in forearm tangential acceleration which initiated approximately 0.02 to

Table 2

Normalized Elbow Force $F_e/m_f + m_h$ (N/kg)

Pitcher	Straight	Curve
RA	89.9	88.1
BD	94.7	85.5
FL	72.7	91.4
DJ	65.6	82.5

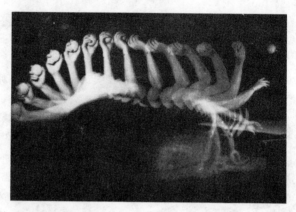

Figure 9—Single strobophotograph at 30 flashes/sec shows various phases of rotation of the forearm. The wrist is shown cocked in hyperextension approximately 0.13 sec before ball release.

0.04 sec before ball release (see Figure 8). The average time elapsed between foot step and ball release ranged from 0.14 to 0.28 sec.

The deceleration phase appeared to be more gradual without large deceleration peaks. No difference was found in the axial or tangential acceleration patterns for a given pitcher regardless of the type of pitches. A typical acceleration pattern of one pitcher for straight and curve pitches is shown in Figure 8. The values of normalized elbow forces for four different pitchers are given in Table 2.

Single strobophotographs showed that both pronation and supination may occur during various pitching styles. This rotation amounted to approximately 10 to 15°. Pronation was found with straight pitches and supination with curve-ball pitches (see Figure 9).

The telemeterized EMG showed that the muscular activity of the biceps, triceps, and wrist flexors occurred almost simultaneously. No discernible difference in EMG characteristics could be identified between the straight and curve pitch.

Discussion

This study demonstrated that consistent and reproducible acceleration patterns could be recorded for an individual. Although the patterns of acceleration signals varied, this problem was considered to be minor, since the subjects in this study served as their own controls. It was also found that there was no difference in the axial and tangential acceleration patern for an individual regardless of the type of pitch that was thrown.

Despite the apparent difference in technique of delivery between straight and breaking pitches, the values of the normalized elbow forces for four subjects were inconsistent. These normalized elbow force components, corresponding to the joint reaction force, displayed different magnitudes regardless of the pitching style. Therefore, it was not possible to correlate the calculated elbow stress as a function of the pitching style. These stresses could have been influenced by parameters such as ball velocity as well as muscular activity. This area requires future investigation.

Tullos and King (1973) observed only pronation of the forearm in the follow-through phase regardless of the type of pitch. We found that both pronation and supination of the forearm with respect to the splint occurred. This rotation amounted to approximately 10 to 15°. Pronation was associated with straight pitches and supination with curve-ball pitches.

Simultaneous activity of the biceps and triceps muscle during the pitching act was a phenomenon of interest (see Figure 7). The explosive triceps contraction generated a rapid elbow extension at the last part of the acceleration phase which appeared to be checked by the contraction of the biceps. We believe that this is one of the important factors in the prevention of hyperextension injuries of the elbow joint. No discernible EMG characteristics could be identified between straight and breaking pitches.

While wrist flexion has been identified by Collins (1960) as an important action contributing to ball velocity, the validity of this finding was questioned by Atwater (1970). From strobophotographic pictures at 30 flashes/sec, we found that the wrist joint was cocked in hyperextension approximately 0.13 sec before ball release (see Figure 9). We also observed that the wrist snap occurred approximately 0.05 sec before ball release (see Figure 9). A simultaneous abrupt increase in acceleration of the forearm and hand approximately 0.02 to 0.04 sec before release would seem to indicate that both the forearm acceleration and the wrist snap are the major controlling mechanisms culminating in ball velocity.

Conclusion

This preliminary study indicates that the accelerometer can be a useful tool in obtaining kinematic data of a moving extremity. The instrumentation designed and constructed is capable of recording the acceleration components in a consistent manner. Our study of EMG muscular activity, acceleration pattern and normalized elbow forces do not positively support the hypothesis that the breaking pitch increases the elbow stress compared to the standard straight pitch. We suggest that the main factor causing elbow injury is probably related to the amount of throwing rather than the type of pitching.

Acknowledgment

This study was supported in part by Research Grant TW-2151-01 International Fellowship from the National Institutes of Health, Public Health Service, U.S.A.

References

ADAMS, J.E. 1965. Injury to the throwing arm. A study of traumatic changes in the elbow joint of boy baseball players. California Med. 102:127-132.

ALBRIGHT, J.A., Jokl, P., Shaw, R., and Albright, J.P. 1978. Clinical study of baseball pitchers: Correlation of injury to the throwing arm with method of delivery. Am. J. Sports Med. 6:15-21.

ATWATER, A.E. 1970. Movement characteristics of the overarm throw: A kinematic analysis of men and women performers. Doctoral dissertation, University of Wisconsin, Madison, Wisconsin, USA (Univ. of Oregon Microcard, PE 1235).

BENNETT, G.E. 1947. Shoulder and elbow lesions distinctive of baseball players. Ann. Surg. 126:107-110.

BROGDON, B.G. 1960. Little league elbow. Am. J. Roentgenol. 83:671-675.

CAFFEY, J. 1972. Pediatric X-ray diagnosis. Sixth edition, Chicago, Year Book Medical Publishers.

COLLINS, P.A. 1960. Body mechanics of the overarm and sidearm throws. Master's Thesis, Univ. of Wisconsin, Madison, Wisconsin, USA (Univ. of Oregon Microcard. PE 508).

DEHAVEN, K.E., and Evarts, C.M. 1973. Throwing injuries of the elbow in athletes. Orthop. Clin. North Am. 4:801-808.

DOBBINS, D.A. 1970. Loss of triceps function on an overarm throw for speed. M.S. paper, University of Wisconsin, Madison.

DRILLIS, R. 1966. Body segment parameters. New York University. Distributed by Clearinghouse for Federal Scientific and Technical Information.

GUGENHEIM, J.J., Jr., Stanley, R.F., Woods, G.W., and Tullos, H.S. 1976. Little league survey: The Houston study. Am. J. Sports Med. 4(5):189-200.

HANG, Y.S. 1981. Tardy ulnar neuritis in a little league baseball player. Am. J. Sports Med. 9(4):244-246.

LARSON, R.L., Singer, K.M., Bergstrom, R., and Thomas, S. 1976. Little league survey: The Eugene study. Am. J. Sports Med. 4(5):201-209.

MORRISON, J.B. 1968. Bioengineering analysis of force actions transmitted by the knee joint. Biomed. Engineering, 3:164-170.

PADGAONKAR, A.J., Krieger, K.W., and King, A.I. 1975. Measurement of angular acceleration of a rigid body using linear accelerometers. ASMA Trans. 75-APMB-3.

PANNER, J.J. 1929. A peculiar affection of the capitellum humeri resembling Calve-Perthes disease of the hip. Acta Radiol. 10:234-238.

TORG, J.S., Pollack, H., and Sweterlitsch, P. 1972. The effect of competitive pitching on the shoulders and elbows of pre-adolescent baseball players. Pediatrics 49:267-272.

TULLOS, H.S., and King, J.W. 1972. Lesions of the pitching arm in adolescents. JAMA 220:264-271.

TULLOS, H.S., and King, J.W. 1973. Throwing mechanism in sports. Orthop. Clin. North Am. 43:709-720.

A.
Joint
Function

Mechanical Analysis of the Ankle Joint Replaced with an Alumina Ceramic Artificial Joint by the Two-Dimensional Finite Element Method

Hironobu Oonishi
Department of Orthopaedic Surgery, Osaka-Minami National Hospital, Osaka, Japan

Tatsuhiko Hasegawa
Century Research Center Co., Osaka, Japan

As the long-term clinical results of total hip replacement became accepted, the artificial ankle joint as well as the artificial knee joint were developed. We developed an artificial ankle joint, in which alumina ceramic and a combination of high density polyethylene (HDP) with alumina ceramic were employed, respectively, for the part contacting the bone and the sliding part. This ankle joint is a self-locking type without cement. In the sliding part, the radius of curvature of HDP differs from that of the ceramic in order to provide play between HDP and the ceramic. The component of the artificial ankle joint on the tibial side was designed to be loaded on the cortical bone at the anterior part of the tibia and on the cancellous and cortical bones at the posterior part.

Analysis by the FEM

Model Formation

Models were made of the lateral views principally for ankle joints. Positions standing on one leg vertically (V), inclined anteriorly (AI), and inclined posteriorly (PI) were considered for a) physiological conditions (P), and b) conditions after replacement with an artificial ceramic ankle joint, respectively (see Figure 1).

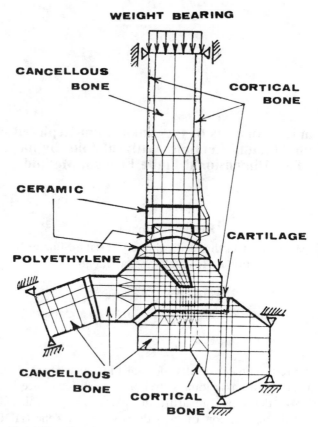

Figure 1—Model formation.

Boundary Conditions

First, the load was transferred only in the normal direction of the articular surface. Second, under conditions immediately after replacement (IAR), the ceramic and the bone were not fused. Third, when the ceramic and the bone were fused a few months after replacement, the areas between the ceramic and the bones were completely unified. Fourth, on weight bearing, the proximal end of the tibial model was restrained antero-posteriorly. Restriction was imposed on the up-and-down movement in the forefoot and the posterior part of the calcaneus.

Element and Mechanical Properties

The NASTRAN program was used in the analysis and the membrane element (TRMEM, QDMEM1) and the rod element were used for element divisions. The number of elements and nodes were about 440 and 430, respectively (see Table 1).

Table 1

	Young's modulus N/mm^2	Poisson's ratio
Cortical bone	15,000	0.3
Cancellous, bone	1,000	0.2
Cartilage	14	0.49
Al$_2$0$_3$ Ceramics	377,700	0.23
High density polyethylene	20.4	0.4

Weight Bearing Conditions

It was assumed that the resultant force of the body weight and the muscular force was applied to the joint vertically from the tibia in all the positions of standing. The joint was designed to incline at the maximum angle of 8.4° both anteriorly and posteriorly.

Results

Deformations

Results indicated that:

1. Deformations in the vertical position as compared with the physiological conditions showed the deformations around the joint were larger after replacement.

2. Deformations in the position of posterior inclination (P-PI, IAR-PI, AR-PI) indicated the tibia bends posteriorly and convexly. This deformation is largest under the physiological conditions (P). Large deformations of the cartilage and HDP were observed in the anterior part.

3. Deformations in the position of anterior inclination (P-AI, IAR-AI, AR-AI) indicated the tibia bends anteriorly and convexly. This deformation was larger after replacement with an artificial ankle joint (IAR, AR) than under the physiological conditions.

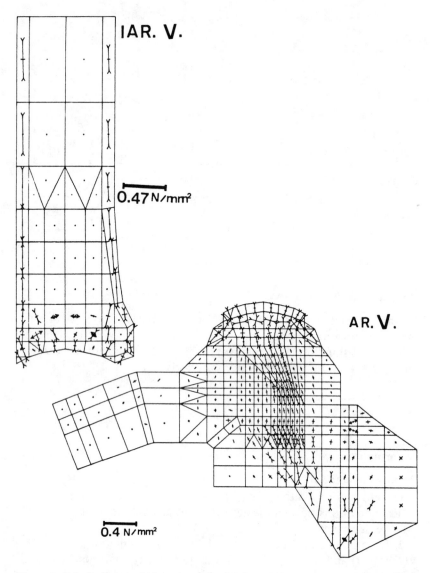

Figure 2 – Principal stress. IAR = immediately after replacement, AR = after replacement, V = vertical.

Principal Stresses

Under the Physiological Conditions (P). In the vertical position (P-V) the largest stress was distributed in the posterior cortical bone. These stresses were transferred to the talus through the joint cartilages and were dispersed relatively evenly over the entire surface of the talus.

In the position of posterior inclination (P-PI), the stress of the cortical bone was reduced in the posterior part of the tibia, and increased in the anterior part. The above stress reached the talus by transfer through the articular cartilages. The anterior half part of the talus received the larger stress, which gradually extended to the entire bone, finally reaching the inferior part of the talus.

In the position of anterior inclination (P.AI), the stresses distributed in the cortical bone in the posterior and anterior parts of the tibia increased to about 2.33 times as large as those observed in the vertical position (PV) and relatively large stress flowed to the posterior half of the talus.

After Replacement with an Artificial Ankle Joint. In the vertical position (V-IAR) the stress distribution of the tibial cortical bone immediately after replacement with an artificial ankle joint was almost the same as under the physiological condition. The stress of the anterior cortical bone was dispersed in the ceramic. About 2/3 of the stress of the posterior cortical bone moved to the ceramic, and the remaining stress (about 1/3) reached the posterior bulgy part. Most of the stresses flowed to the posterior half of the stem. A large stress flowed to the anterior part of the stem. These large stresses flowed to the tip of the stem and to areas around the cancellous bone posterior to the stem. In addition, large stress distribution was seen in the anterior part of the stem and in areas anterior in the stem. In the observation of the cancellous bone in areas of the talus where contact with the ceramic was present, compressive stress was distributed parallel to the surface of the stem both anteriorly and posteriorly. This compressive stress worked as a shear stress against the stem. At the area of contact with the horizontal surface of the ceramic, compressive stress was distributed perpendicularly to the ceramic surface. At the upper part of the anterior stem, compressive stress acts on the ceramics strongly at a right angle (see Figure 2).

In the vertical position after replacement (V-AR) when the fusion between the ceramic and the bone had occurred, the stress distribution hardly changed in the tibia. The stress distribution in the ceramic talus became almost uniform and larger stress flowed into the ceramic (see Figure 2).

In the position of posterior inclination (PI-IAR), the stress distribution in the cortical bone of the tibial diaphysis was almost identical with that of PI-P. At the ceramic of the tibial component, the stress was concentrated on the anterior part of the ceramic. The stress decreased markedly at the posterior part of the distal end of the tibia, namely, the bulgy part, as compared with PI-P. At HDP, the major stress was distributed on the anterior part perpendicularly to the articular surface. Compressive stress was transferred to the bone anterior to the stem, and in the cancellous bone in contact with the anterior and the upper part of the

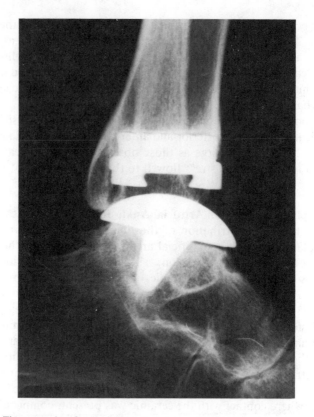

Figure 3—Three months after replacement.

stem, the compressive stress worked parellel to the stem surface. The latter compressive stress resulted in shear stress to the stem surface.

In the position of posterior inclination after replacement (PI-AR)—the state of being fused between the ceramics and the bone after replacement—the stress distributions in both the bone and the ceramics at the tibia were almost identical with those immediately after replacement (PI-IAR).

In the position of anterior inclination (AI) as compared with the posterior inclination (PI), a relatively large stress distribution was seen in the anterior cortical bone without extreme unevenness in distribution.

After replacement (AI-AR) in the fusion between ceramic and the bone, the stress distributions in both the bone and the ceramic at the tibia were almost the same as those immediately after replacement.

Relation with X-ray Pictures

The X-ray pictures a few months after replacement with an artificial ankle joint showed that the trabecular new bone formation appeared

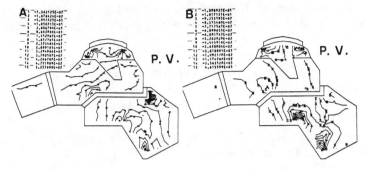

Figure 4—Contour lines of stress level. A = Major principal stress; B = Minor principal stress.

from the upper anterior part of the tibial component ceramic to the anterior cortical bone of the tibia, and from the upper posterior part of the ceramic to the posterior cortical bone of the tibia, and osteosclerosis was also seen. Between the ceramic and the bone in the talus, osteosclerosis was uniformly observed. These are quite in agreement with the analysis of results (see Figure 3).

Major and Minor Principal Stresses

The observation of the contour line of stress level of the major and minor principal stresses under the physiological condition indicate that the total stress lines, which are obtained by superimposing all the major and minor contour lines of stress level in the vertical condition (V) (see Figure 4), in the posterior inclination (PI), and in the anterior inclination (AI), coincided with the trabecular bone distribution pattern observed in the X-ray pictures. Therefore, the trabecular pattern after ankle replacement changed into an integrated state in all the diagrams after replacement.

Discussion

As the shear stress is applied to the anterior and posterior surfaces of the talus component's stem, if the stem is inserted to the bone loosely, it is desirable to carry out bone grafting to fill the gaps. To prevent the tibial component from sinking into the tibia, the tibial component was so designed as to be loaded mainly on the anterior cortical bone of the tibia and on the posterior cancellous and cortical bones. When the tibial and talus components are ideally inserted, the weight-bearing walk can be carried out immediately after replacement, as clarified in the analysis of results by the FEM. The effects on the neighboring joints are extremely small after replacement. If the talus component loosens, however, or if it is fixed with bone cement, the neighboring joints will be influenced.

References

BUCHHOLZ, H.W. 1973. Totale Sprunggelenksendoprothese Modell, St. George. Chirurg. 44:241.

FRANKEL, V.H., and Nordin, M. 1980. Basic Biomechanics of the Skeletal System. Lea and Febiger, Philadelphia.

GLUCK, T. 1890. Die Invaginationsmethode der Osteo-Und Arthroplastie. Berl. Klin. Wschr. 33:752.

Tribological Study Using Human Joint Models

T. Murakami, N. Ohtsuki, H. Chikama,
T. Toyonaga, H. Nishizaki, and A. Nishio
Kyushu University, Japan

The mechanism of human joint lubrication which is characterized by low friction under slow speed reciprocating motion has been investigated in detail for a long time and various explanations have been offered. It has been suggested by Dowson (1967) that the major lubrication mechanism would seem to be some form of elastohydrodynamic action determined by sliding or squeeze film action between porous surfaces with boundary lubrication providing the surface protection in cases of severe loading and little movement.

The purpose of this study was to investigate the tribological function of articular cartilage of human joints from the viewpoint of elastohydrodynamic lubrication by using simulation models. In this study, a two-dimensional photo-elastic technique was employed, and therefore both contact stress distribution and frictional behavior were simultaneously observed.

Procedure

In order to simulate the elastic deformation of articular cartilage, a model of cartilage was made of urethane rubber, and models of bones were made of epoxy resin. The Young's moduli of the simulated cartilage and bone are depicted in Table 1. Two kinds of two-dimensional joint models in conformal contact were prepared as shown in Figure 1. One of them was the composite model which was made by combining epoxy resin with urethane rubber. The other was the epoxy model which neglected the elastic deformation effect of cartilage. The size of the upper disc was 49.8 mm in its outer diameter and 10 mm in thickness. The thickness of the urethane layer was 3 mm. The size of lower common

Table 1

Comparison of Young's Moduli and Viscosity

Human joint		Model (24°C)	
Young's modulus (MPa)			
Cartilage	5 - 50	Urethane rubber	10
Cancellous bone	50 - 500	Epoxy resin	3000
Viscosity (Pa • s)			
Synovial fluid	0.01 - 1.0	P 150	0.056
		BS	0.95

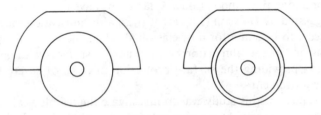

Figure 1—Composite model and epoxy model.

specimen was 50 mm in inner diameter and 15 mm in thickness. Accordingly, the radial clearance was 0.1 mm. The roughness of the contact surface of the upper and lower epoxy specimens was 0.38 μm and 1.2 μm r.m.s., respectively.

The friction measuring tests under the lubricated conditions were carried out by means of a pendulum method. Simultaneously, the dynamic variation of the contact stress distribution was visually observed with the aid of photoelastic method. The complete apparatus is shown in Figure 2.

For lubrication of the contact surfaces, paraffinic oils, i.e. P 150 and BS (bright stock) described in Table 1 were used. The contact surfaces were immersed in a transparent oil bath. The normal load per unit axial length was 10 kN/m. Immediately after loading, the pendulum including the upper specimen began to swing freely under the condition of an initial amplitude of 0.20 rad. After the pendulum stopped, the second run was carried out without unloading. The variation in amplitude of the

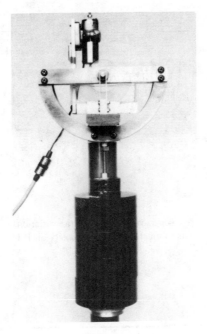

Figure 2 — Test apparatus.

swing was recorded with the aid of a transducer together with a signal indicating the instant of photographing.

The average coefficient of friction in the case of damping was estimated approximately with the following equation:

$$f = (l/r) (\theta_{n-1} - \theta_n)/4 \qquad (1)$$

where: f is average coefficient of friction, l and r are the distances from the center of rotation to the center of gravity of the pendulum (230 mm) and to the contact surface, respectively. θ_{n-1} and θ_n are the amplitudes of the n − 1 th and n th oscillation, respectively.

Results

Comparing the variation in amplitude in Figure 3a for the composite model with that in Figure 3b for the epoxy model in the experiments using low viscosity oil P 150, it suggests quite clearly that the elastohydrodynamic effect of the former predominates over the latter. Furthermore, in the case of the composite model, the damping was hastened in the second run without unloading. On the other hand, in the case of epoxy model the hydrodynamic film formation was not expected any more. The pendulum damped approximately linearly and the same frictional

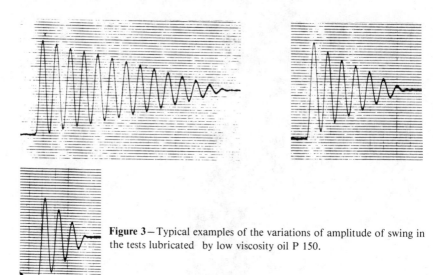

Figure 3 — Typical examples of the variations of amplitude of swing in the tests lubricated   by low viscosity oil P 150.

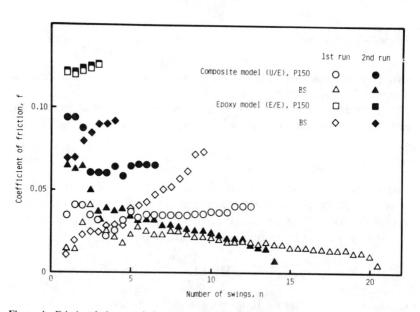

Figure 4 — Frictional characteristics by means of pendulum method.

behavior was observed in both the first and second runs. These findings show the effect of the squeeze film action due to higher compliance of composite model. After calculation of the coefficients of friction from Equation (1) they were plotted against the number of swings as shown in Figure 4.

Typical examples of the isochromatics are shown in Figures 5-8 together with the corresponding variations in amplitude. Figure 5 obtained for the composite model shows that the contact area was larger compared to the epoxy model and the upper specimen slides along the surface of lower specimen corresponding to the swing. On the other hand, Figure 7 obtained for the epoxy model shows that the contact region abruptly changes with each swing because of the breakdown of the hydrodynamic oil film.

Experimental results using high viscosity oil BS are shown in Figures 4, 6, and 8. In the case of the composite model, immediately after the start, the pendulum damped very slowly. After a few swings, the damping rate increased and fluctuated and then gradually decreased. For smaller amplitudes of swings, logarithmic damping was observed. These damping modes correspond to the tribological conditions. In the first region, the elastohydrodynamic condition was realized, because of thick oil film formation caused mainly by the squeeze film action. In the second run, this first mode was not observed. In the intermediate region, it is considered that the damping characteristics depend on the mutual relation between the oil film thickness and the amplitude of the swing. In the last region, the logarithmic damping perhaps corresponds to the oil entrapment by the urethane rubber or the internal damping of urethane rubber.

In the case of the epoxy model using high viscosity oil, BS, the hydrodynamic lubricating condition was realized. Therefore, the upper specimen slides along the surface of the lower one, corresponding to the swing as shown in Figure 8. It is further noted that the squeeze film effect was observed as shown in Figure 4. For smaller amplitudes, however, the damping hastened, which was caused by an asperity contact.

Discussion

In order to investigate the mechanism of human joint lubrication, the friction measuring tests were carried out by means of a pendulum method using two kinds of joint models, i.e., the composite model and the epoxy model. Simultaneously, the variations of contact stresses were observed.

In the case of the composite model which corresponded to the normal human joint with articular cartilage, the elastohydrodynamic oil film formation caused by mainly squeeze film action was observed. In this model, squeeze film action played a more important role in the oil film formation than the wedge action, because low friction for larger amplitudes of swing was observed only in the first run immediately after suddenly loading. In the second run without unloading, the damping increased and the first mode of damping was observed no longer. Unsworth et al. (1975) and Mabuchi et al. (1980) have pointed out the ef-

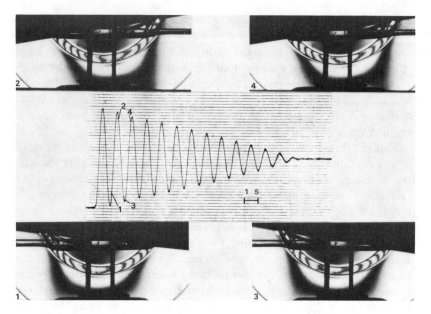

Figure 5 — Typical examples of isochromatics and the variation of amplitude of swing in the case of composite model lubricated by low viscosity oil P 150.

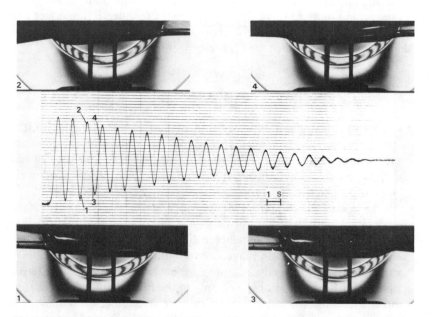

Figure 6 — Typical examples of isochromatics and the variation of amplitude of swing in the case of composite model lubricated by high viscosity oil BS.

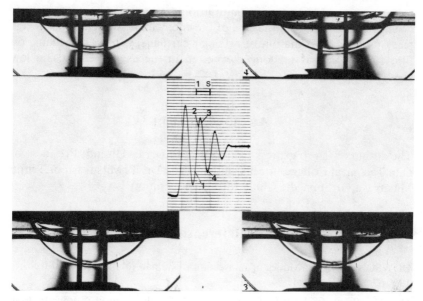

Figure 7 — Typical examples of isochromatics and the variation of amplitude of swing in the case of epoxy model lubricated by low viscosity oil P 150.

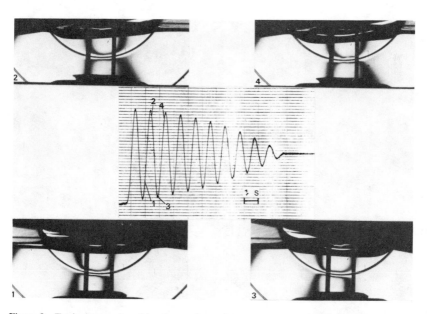

Figure 8 — Typical examples of isochromatics and the variation of amplitude of swing in the case of epoxy model lubricated by high viscosity oil BS.

fect of squeeze film action in natural joints.

On the other hand, in the case of the epoxy model which omits the effect of elastic deformation of articular cartilage, the hydrodynamic oil film formation was no longer expected in the experiments using low viscosity oil.

Acknowledgment

The authors wish to express their thanks to Dr. F. Hirano, President of Oita Technical College, for his guidance and Dr. T. Mitsuyasu of Tenjin Mitsuyasu Orthopaedic Clinic for his cooperation.

References

DOWSON, D. 1967. Modes of lubrication of human joints. Proc. I. Mech. E. 1966-67, 181, Part 3J:45-54.

MABUCHI, K. et al. 1980. The role of fluid film lubrication in joint friction. Physiology and Pathology of Bone and Cartilage Metabolism, 7:383-388. (in Japanese)

UNSWORTH, A., Dowson, D., and Wright V. 1975. The frictional behavior of human synovial joints. Part I: Natural joints. Trans. ASME, 97F(7):369-376.

Contact Pressure Distribution of the Hip Joint

Y. Miyanaga, T. Fukubayashi, H. Kurosawa,
T. Doi, T. Tateishi, and Y. Shirasaki
University of Tokyo, Tokyo, Japan

The magnitude and distribution of the contact pressure of the hip joint provides fundamental clinical information. The experimental results and the clinical approach for evaluation of the hip joint mechanics are presented here.

Experimental work was done on cadaveric hip joints to determine the mechanical behavior when subjected to compression load. Load-deformation curves and contact pattern were analyzed theoretically from the roentgenograms by rigid body-spring model (RBSM) (Kawai, 1980).

Materials

Four normal hip joints were taken from cadavers using routine autopsy procedures. The acetabulum was osteotomized including the iliac, pubic and ischial bones. The upper femur was cut about 10 cm below the minor trochanter. At the same time, preoperative and postoperative AP roentgenograms of osteoarthritic hip joints were prepared for determining the contact pressure.

Method

Preparation of the Specimen

The acetabulum was held at an angle of about 45° on its mounting plate by use of bone cement (polymethylmethacrylate). The femoral head was fixed to a special fixation device by means of an intramedullary rod which was inserted into a methacrylate-filled cavity. The joint component was aligned in the loading frame with the acetabulum side below, in the normal anatomical position.

Test Procedure

The Instron type testing machine (TOM-500) was used to examine the mechanical compression behavior at the various loads (50, 100, 150 kg) and cross head speeds (1, 3, 5 mm/min), and to obtain load-deformation curves. The temperature for the experiment was about 20°C. Throughout the test, the hip joint was kept moist by normal saline.

After each load-deformation test was performed, measurement of contact pressure was carried out at the same position, by use of Prescale (Fuji Film Co., Tokyo) which is a very thin, flexible, and pressure sensitive sheet made of two polyester films.

The contact area of the hip joint was measured by a casting method. Silicone rubber called 'Xantopren Blau' (Bayer, Co., Leverkusen) which is commonly used for dental purposes, was utilized in this procedure.

Roentgenogram Analysis

A two-dimensional analysis was performed. The magnitude and direction of the resultant forces were determined and drawn on the film. Kawai's model was used in order to analyze the magnitude and distribution of the contact pressure.

Results

Hip Behavior on Compressive Loading

The load-deformation curves were initially concave downwards and nonlinear. As the load increased gradually, the slope was steeper and the curve became more nearly linear.

Pressure Distribution

Figure 1 shows the pressure distribution under the various amount of loads with reference to the contact area. The outer line indicates the edge of cartilage over the acetabulum. Solid lines show the edge of contact area of the hip joint. The pattern of the pressure distribution was generally neither uniform nor symmetrical. Until a load of 100 kg (about two times body weight) was applied, pressure was distributed mainly over the anterior and posterior part of the acetabulum, and extremely low pressure was calculated at the zenith. Then peak pressure of about 20 kg/cm^2 was recorded also over the centers of the anterior and posterior

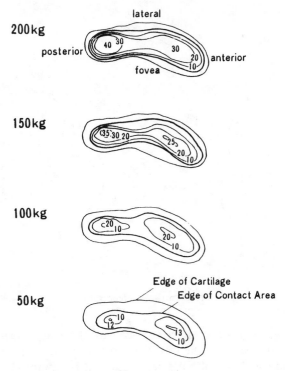

Figure 1 — Pressure distribution (kg/cm²) and contact area.

part. As the load was increased to 150 and 200 kg from 100 kg, the area of the pressure distribution enlarged all over the cartilage and at the same time the magnitude of peak pressure became larger. As for the regional differences, the pressure was distributed more over the anterior part than the posterior part and the amount of pressure was usually higher at the anterior part.

Contact Area

The amount of contact area was dependent on the load. With a 20 kg load (about 40% of body weight), the zenith of the acetabular cartilage surface did not make contact with the surface of the femoral head, and the anterior and posterior portions were the main contact areas. The contact maps show that with a load of over 50 kg, the total cartilage surface, including peripheral surfaces, was brought into contact. This contact area was also reproduced on the articular surface of the femoral head.

It was very characteristic that the overall contour of the edge of the contact areas with loads over 50 kg was kidney-shaped. The contact areas of the femoral head included the lateral side of the anterior, posterior aspects and superior aspects, while the area of the acetabulum ranged

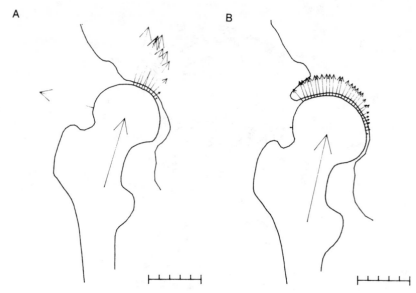

Figure 2—Female, 34 yr old, osteoarthritis. A: preoperative; B: postoperative (rotational acetabular osteotomy).

over most of the entire cartilage surface with the zenith narrowed. The contact area increased as the load increased, because both the articular cartilage and subchondral bone deformed well by compression and the joint become congruous. However, a band of articular cartilage on the perifoveal and inferior regions of the femoral head did not make contact, with increased loads at the neutral position. The maximum amount of contact area was about 14-16 cm^2.

Roentgenogram Analysis

The osteoarthritic hip showed the eccentric distribution of the pressure. The weight-bearing area had the highest pressure. The laterally directed forces were high. After various kinds of osteotomy, the pressure distribution became more uniform and the magnitude was lower (see Figure 2).

Discussion

There have been several reports on weight-bearing function by Greenwald and Haynes (1972), and on contact area and contact pressure by Harrison et al. (1953), Greenwald and O'Connor (1971), Bullough et al. (1973), and Day et al. (1975). However, test procedures used by these researchers were different and there has been no report in which weight-bearing function, contact area and contact pressure were analyzed at the same position, as in our technique.

The hip joint was found to behave reasonably in relation to the magnitude of load and strain rate. The contour of the contact area was fundamentally kidney-shaped and enlarged with increased loads. The pattern of the stress distribution was similar with that of the contact area. The peak pressure was recorded mainly at the anterior and posterior aspects of the acetabulum, and the magnitude was larger at the anterior part. There was one hip, however, in which peak pressure was found at the zenith. It is very interesting to note that the distribution of contact pressure and the size of the contact area are neither concentric nor symmetrical. This finding is quite different from the opinions of Kummer (1968). Greenwald and O'Conner (1971) classified the contact areas according to the range of loads and positions occurring in normal walking. Our experimental data were almost consistent with the results of Greenwald and Haynes (1972). The contact area was calculated to be 26.8 cm^2 by Greenwald and Haynes, and 14-16 cm^2 by us.

The theoretical pressure distribution analysis by Kawai's model showed not only the congruency of the hip joint but also the pressure distribution was smooth after femoral or acetabular osteotomies. This theoretical analysis is very simple and can be a useful solution for selection of operative procedures.

References

BULLOUGH, P., Goodfellow, J., and O'Conner, J. 1973. The relationship between degenerative changes and load-bearing in the human hip. J. Bone & Joint Surg. 55B:746-758.

DAY, W.H., Swanson, S.A.V., and Freeman, M.A.R. 1975. Contact pressures in the loaded human cadaver hip. J. Bone & Joint Surg. 57B:302-313.

GREENWALD, A.S., and O'Connor, J.J. 1971. The transmission of load through the human hip joint. J. Biomech. 4:507-528.

GREENWALD, A.S., and Haynes, D.W. 1972. Weight-bearing areas in the human hip joint. J. Bone & Joint Surg. 54B:157-163.

HARRISON, M.H.M., Schajowicz, F., and Trueta, J. 1953. Osteoarthritis of the hip: a study of the nature and evolution of the disease. J. Bone & Joint Surg. 35B:598-626.

KUMMER, B. 1968. Die Beanspruchung des menschlichen Hueftgelenks. I. Allgemeine Problematik. (The pressure of human hip joint I. general problem) A. Anat. Entwickl. -Gesch. 127:277-285.

KAWAI, T. 1980. Development of discontinuum mechanics (1): Seisan Kenkyu 32:267-272.

Photoelastic Study of Models Simulating
Human Knee Joint

H. Nishizaki, N. Ohtsuki, T. Murakami,
H. Chikama, T. Toyonaga, and A. Nishio
Kyushu University, Fukuoka-shi, Japan

The mechanical functions of the menisci of the human knee joint have been interpreted in terms of knee joint stability, shock absorption, congruity, lubrication (tribology), limits of extremes of flexion and extension, load transmission and reduction of contact stress. Although many investigators have pointed out the mutual relationship between meniscectomy and osteoarthritic changes, there are not enough data concerning the mechanical functions of the menisci.

In order to investigate the influence of meniscectomy on the stress distribution of the human knee joint, a three-dimensional stress freezing photoelastic technique was employed. The experiments were carried out using three types of models, i.e., the normal knee joint model, the partial meniscectomy model, and the total meniscectomy model.

Table 1

Comparison of Young's Moduli

Young's modulus		(MPa)	
Human joint		Model (180°C)	
Cartilage	5 - 50	Silicone rubber	1.5
Meniscus	17.5		
Cancellous bone	50 - 500	Epoxy resin	13.0

Figure 1 — Normal knee joint model, partial meniscectomy model and total meniscectomy model.

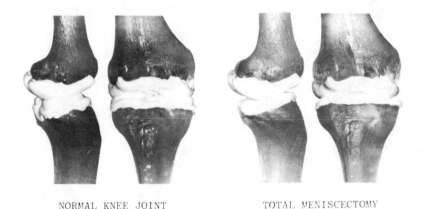

NORMAL KNEE JOINT   TOTAL MENISCECTOMY

Figure 2 — Normal knee joint model and total meniscectomy model.

Procedures

In this study, three-dimensional composite models were prepared to simulate the elastic modular ratio between the most important elements in the knee joint (Chand, Haug, and Rim, 1976; Prati and Freddi, 1979). The Young's moduli of the models at 130°C and the different human joints are shown in Table 1. The elastic modular ratio between subcondral bone and cartilage or menisci is regarded as approximately 5 to 10. Models of femoral and tibial bones were made of epoxy resin. On the other hand, models of articular cartilage and menisci were made of silicone rubber. The bone models were made by casting the skeletal bones in silicone rubber molds. The shape of the menisci model, including cartilage, was determined according to Kaku's (1957) anatomical studies. The partial meniscectomy model consisted of the entire peripheral rim of both menisci. Figures 1 and 2 show these models.

The procedure of stress freezing consisted of heating the models to 130°C followed by slow cooling under load to room temperature. The centered compressive load of 84 N was applied to the models at 0° flex-

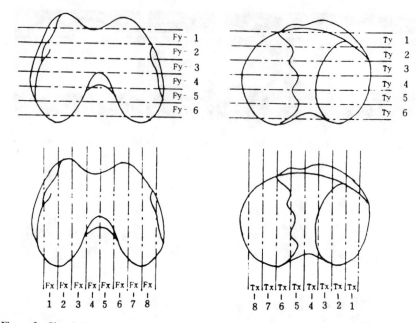

Figure 3—Sketch showing method of slicing femoral and tibial bones.

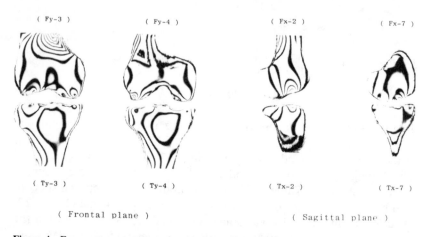

(Fy-3)   (Fy-4)   (Fx-2)   (Fx-7)

(Ty-3)   (Ty-4)   (Tx-2)   (Tx-7)

(Frontal plane)   (Sagittal plane)

Figure 4—Frozen stress patterns of normal knee joint model.

ion angle after reaching 130°C as shown in Figure 2.

Figure 3 shows the method of slicing the femoral and tibial frozen models. The frontal and sagittal slices of 5 mm thickness were cut from the two frozen models in each respective model.

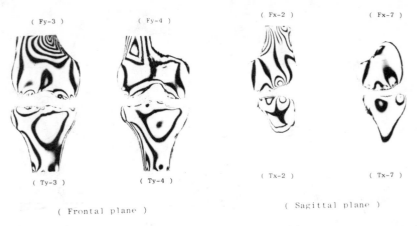

(Fy-3)   (Fy-4)   (Fx-2)   (Fx-7)

(Ty-3)   (Ty-4)   (Tx-2)   (Tx-7)

(Frontal plane)   (Sagittal plane)

Figure 5—Frozen stress patterns of partial meniscectomy model.

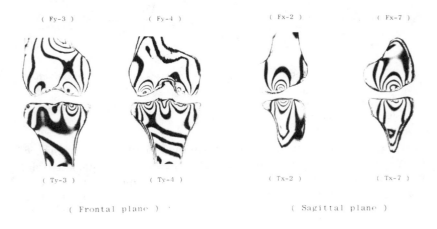

(Fy-3)   (Fy-4)   (Fx-2)   (Fx-7)

(Ty-3)   (Ty-4)   (Tx-2)   (Tx-7)

(Frontal plane)   (Sagittal plane)

Figure 6—Frozen stress patterns of total meniscectomy model.

Results

Photoelastic analysis was carried out only on the bones, i.e., the materials with higher modulus in the composite model.

Figures 4 to 6 show the typical examples of isochromatics. Figure 7 shows the distribution of principal stress difference in the frontal plane at the proximal interfaces of the tibial subchondral bones. These figures show that in the case of the total meniscectomy model the stress concentration appeared to be in the subchondral bone compared to the normal knee joint. The maximum principal stress difference at the proximal in-

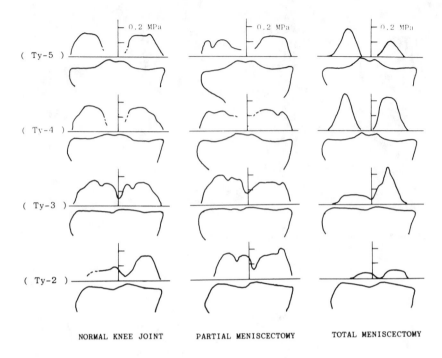

NORMAL KNEE JOINT   PARTIAL MENISCECTOMY   TOTAL MENISCECTOMY

Figure 7 — Distribution of principal stress difference: frontal plane at proximal interfaces of tibial subchondral bones.

terfaces of the tibial subchondral bones in the case of the total meniscectomy model was 1.5 times as high as that in the case of normal knee joint model. On the other hand, in the case of the partial meniscectomy model the stress concentration on the subchondral bone was considerably less.

Discussion

In this study, the mechanical functions of menisci in terms of load transmission and reduction of contact stress were investigated by menas of a three-dimensional stress freezing photoelastic method. We carried out the experiments using three types of models, i.e., the normal knee joint model, the partial meniscectomy model, and the total meniscectomy model.

In the case of total meniscectomy, the stress concentration appeared to be in the subchondral bone due to smaller load-bearing areas and lower conformity, compared to the normal knee joint model. The contact areas of the models were estimated from the measured results of contact pressure distribution by means of Prescale (Fuji Film Co., Tokyo), which is very thin, flexible, and pressure-sensitive sheet made of two polyester films. These measured contact areas corresponded to the measured areas concerning human knee joints with and without menisci by Walker and Erkman (1975) and Fukubayashi et al. (1978).

Furthermore, the result of the experiments using the partial meniscectomy model suggests that the application of the appropriate partial meniscectomy might prevent to a large extent the stress concentration on the subchondral bone.

Many investigators have pointed out that meniscectomy exerts a harmful influence on osteoarthritic changes. Himeno et al. (1980) reported on the multivariate analysis of results sometime after meniscectomy (the analysis of osteoarthritic changes in plain X-P). According to their investigation, a clinical evaluation and X-P score concerning appropriate partial meniscectomy are considered to be better than that concerning total meniscectomy. It might be expected, therefore, that our experimental results which showed the effect of reduction of stress in the case of a partial meniscectomy model correspond to results of clinical evaluations and osteoarthritic changes some tme after meniscectomies were performed.

Acknowledgments

The authors wish to express their thanks to Dr. T. Mitsuyasu of Tenjin Mitsuyasu Orthopaedic Clinic for his cooperation and to Dr. F. Hirano, President of Oita Technical College, for his guidance.

References

CHAND, R., Haug, E., and Rim, K. 1976. Stresses in the human knee joint. J. Biomechanics 9:417-422.

FUKUBAYASHI, T., Kurosawas, H., Murase, K., and Doi, T. 1978. The contact area of the knee joint. Proc. 5th Annual Meeting of Japanese Orthopaedic Biomechanics Research Society 5:142-150.

HIMENO, S., Mitsuyasu, T., Toyonage, T., and Chikama, H. 1980. Late results after meniscectomy—The analysis of osteoarthritic changes in plain X-P. Clin. Orthop. Surg. 15(12):1130-1139. (in Japanese)

KAKU, K. 1957. Anatomical studies on the human knee joint. Igaku Kenkyu (Acta Medica). 27(8):203-222. (in Japanese)

MAQUET, P.H.J. 1976. Biomechanics of the Knee. Springer-Verlag, Berlin.

PRATI, E., and Freddi, A. 1979. A critical evaluation of some current experimental models simulating the human knee joint. VII Int. Cong. Biomechanics.

WALKER, P.S., and Erkman, M.J. 1975. The role of the menisci in force transmission across the knee. Clin. Orthop. 109:184-192.

A Biomechanical Investigation of
Cricoarytenoid Joint Kinematics

Ione E. Sellars and Christopher L. Vaughan
University of Cape Town Medical School,
Cape Town, South Africa

An understanding of the mechanics of the human larynx, and especially the cricoarytenoid joint (CAJ) which controls vocal cord action, should be essential to those involved in speech therapy and related fields (Sellars, 1979). There is some confusion, however, as to how the CAJ functions, since most studies have tended to be very qualitative in nature.

Anatomically, the CAJ is found between a concave facet on the inferior aspect of the arytenoid, and an elongated convex facet on the upper aspect of the cricoid lamina. Sonesson (1959) described this joint as similar to a cylindrical joint with its long axis coinciding with the cricoid facet, and presented mean cadaver values for the orientation of this axis. Von Leden and Moore (1961) also used cadaver specimens and described the CAJ as a shallow elongated ball-and-socket joint. They identified three principal types of motion: 1) a rocking movement around the joint axis; 2) a linear glide parallel to this axis; and 3) a restricted rotary motion around the attachment of the cricoarytenoid ligament to the cricoid lamina. Frable (1961) performed a photographic study of CAJ motion using 14 cadaver specimens and gave a single equation, which, she claimed, represented the plane of rotation. Ardran and Kemp (1966) presented tomographic observations in living subjects and showed that the arytenoid cartilage rocked on the cricoid facet during movements of the vocal cords. Fink and Demarest (1978) described the arytenoids as sliding and rocking on the cricoid facets during inspiration. Recently Sellars and Keen (1978) dissected the CAJs of 45 larynges and found the anatomy to vary quite considerably. They found three types of movement of the arytenoid cartilages during adduction: inward rocking, sliding up the slope of the cricoid facet, and twisting to bring the vocal processes together. The movements were related to the shape and position of the cricoid facet.

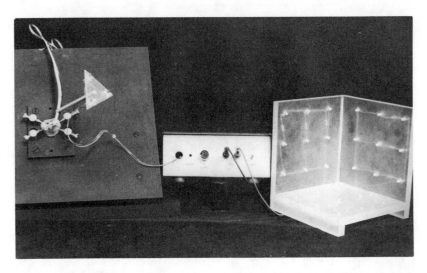

Figure 1—Cricoid jig, forceps clamped onto right arytenoid, pulsing LEDs, and object-space controller with LEDs 50 mm apart.

In summary, it would appear that a rigorous quantitative description of CAJ kinematics is still lacking. The purposes of the present investigation therefore were: 1) to develop a three-dimensional (3D) model of the CAJ and identify the mechanical parameters which characterize its kinematic behavior; 2) to develop an experimental apparatus and technique to collect the necessary data; and 3) to test the model with raw data from dissected larynges.

Methods

Five cadaver larynges were used in the study. They were dissected to the stage where the cricoid and arytenoids were free of the other laryngeal structures, the intrinsic musculature was removed, and only the ligaments and joint capsule were left intact. In each case only the right CAJ was studied.

The experimental apparatus consisted of: 1) a jug in which the cricoid was fixed, 2) a pair of forceps which clamped onto the arytenoid and to which was attached a triad of pulsing light emitting diodes (LEDs), 3) an object-space controller (OSC) of known 3D dimensions, and 4) two 35 mm cameras loaded with color slide film. Figure 1 illustrates these pieces of equipment. The triad of pulsing LEDs were similar to those described by Francis and Gabel (1976), while the OSC was based on the design of Karara (1974) except that LEDs were used to locate 3D coordinates.

Using the forceps, the arytenoid was moved through three ranges of

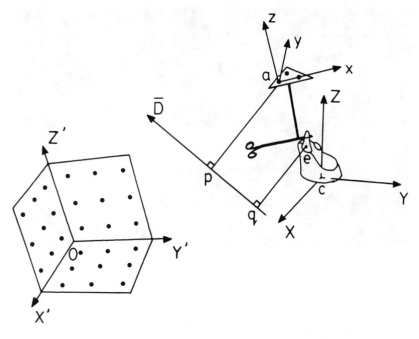

Figure 2 — Reference frames and three-dimensional vector methods used to locate screw displacement axis $\bar{D}$ (refer to text for more detail).

motion — rocking, sliding and twisting — in order to simulate intrinsic muscle action and the motion of the LED triad was recorded on the two slides. The slides were analyzed on a Houston HI-PAD digitizer and the 3D motion of the LED triad (and hence the arytenoid) relative to the cricoid was determined with the aid of a FORTRAN program.

Theory

For the purpose of defining the mechanical model, the cricoid and arytenoid cartilages were assumed to be rigid bodies. Consider the reference frames depicted in Figure 2. The 3D coordinates of the LED triad, obtained by the Direct Linear Transformation methods of photogrammetry (Abdel-Aziz and Karara, 1971), are in the terms of the OSC reference system (0:X′Y′X′). The reference system embedded in the cricoid cartilage (c:XYZ) is fixed with its origin c at the middle of the cricoid, its XY plane coinciding with the inferior margin of the cricoid, and its Y axis pointing anteriorly. The arytenoid reference system (a:xyz) is not fixed in space but moves as the arytenoid is moved. The unit vector triads $\hat{I}\hat{J}\hat{K}$ and $\hat{i}\hat{j}\hat{k}$ are parallel to XYZ and xyz, respectively.

First, it is necessary to express all points n in terms of the cricoid reference system by the following equation:

$$\bar{r}^C_{n/c} = [\alpha]\,(\bar{r}^F_{n/0} - \bar{r}^F_{c/0}) \tag{1}$$

where the superscripts C and F indicate the cricoid and OSC references systems, respectively, and $[\alpha]$ is a 3×3 transformation matrix whose elements are functions of the Euler angles orienting c:XYZ relative to 0:X′Y′Z′. The Euler angles are based on the convention of Goldstein (1965). In order to orient a:xyz relative to c:XYZ the following counterclockwise rotations take place: (1) $\phi\hat{K}$; (2) $\theta\hat{i}$; and (3) $\psi\hat{k}$. Once the Euler angles (θ, ϕ, ψ) have been calculated for each time interval, their first derivatives may be obtained — the quintic spline (Wood and Jennings, 1979) was used for this purpose — and an expression for the arytenoid's angular velocity $\bar{\omega}_A$ may be obtained:

$$\bar{\omega}_A = (\dot{\theta}\cos\phi + \dot{\psi}\sin\theta\sin\phi)\,\hat{I} + (\dot{\theta}\sin\phi - \dot{\psi}\sin\theta\cos\phi)\,\hat{J} + (\dot{\psi}\cos\theta + \dot{\phi})\,\hat{K} \tag{2}$$

As pointed out by Andrews and Youm (1979) in their study of wrist kinematics, $\bar{\omega}_A$ is parallel to the instantaneous screw displacement axis $\bar{D}$. The unit vector in that direction is thus

$$\omega = \omega_A / |\bar{\omega}_A| \tag{3}$$

$\bar{D}$ is located relative to a by the expression

$$\bar{r}_{p/a} = (\bar{\omega}_A \times \bar{v}_a)/(\bar{\omega}_A \cdot \bar{\omega}_A) \tag{4}$$

where the superscript C has been dropped for convenience and $\bar{v}_a$ is the linear velocity of a as seen in C. Furthermore,

$$\bar{r}_{p/c} = \bar{r}_{a/c} + \bar{r}_{p/a} \tag{5}$$

and since the location of $\bar{D}$ relative to the joint center e is more meaningful, the following expression for $\bar{r}_{q/c}$ can be achieved by simple vector algebra:

$$\bar{r}_{q/c} = \bar{r}_{p/c} + \omega\,(\omega \cdot [\bar{r}_{e/c} - \bar{r}_{p/c}]) \tag{6}$$

The screw displacement axis ($\bar{D}$) can thus be expressed as

$$\bar{D} = \bar{r}_{q/c} + u\omega \tag{7}$$

Table 1

Cricoarytenoid Joint Motion: Rocking (Refer to Text for More Detail)

Specimen	$r_{e/c}$ (mm)			$r_{q/c}$ (mm)			ω			β
	X	Y	Z	X	Y	Z	I	J	K	(degrees)
1	6.9	−5.7	20.2	9.7	−5.7	−6.6	.704	.706	.074	45.1
2	7.6	−5.2	22.0	2.6	−0.5	−7.9	.820	.571	−.049	53.0
3	10.8	−8.3	20.4	6.7	−2.2	−4.8	.827	.562	.001	47.6
4.	9.6	−8.9	24.9	3.0	−3.8	−0.4	.456	.875	.088	52.8
5.	7.3	−6.4	21.5	2.7	−7.0	−8.1	.411	.908	−.083	42.3
Mean	8.4	−6.9	21.8	4.9	−3.8	−5.6	.644	.724	.006	48.2
S.D.	1.5	1.5	1.7	2.8	2.3	2.8	.178	.146	.067	4.2

where u is an arbitrary scalar. This then gives the axis of rotation, and the range of motion (β) may be obtained by integration

$$\beta = \int_{t_i}^{t_f} |\bar{\omega}_A| \, dt \qquad (8)$$

where t_i and t_f are the initial and final times, respectively.

Results and Discussion

The results for rocking, sliding and twisting are presented in Tables 1 and 2. The vector (D), exactly 100 mm long and centered at $r_{q/c}$, is seen from three orthogonal views in Figure 3. Each of the three principal axes of motion — rocking, sliding and twisting — are illustrated for specimen number 1.

From the tables it is clear that the values for $r_{p/c}$ and ω (representing the axis of rotation) and β (range of motion) are similar for each of the three types of motion, although there is some variability among the specimens. This was expected however, since the anatomy and the geometry of CAJs vary considerably (Sellars and Keen, 1978).

A number of interesting observations may be made. First, the rocking axis does not coincide with the cricoid facet — as has been suggested by some authors — but actually lies inferior to the cricoid. Second, the sliding motion is not linear in nature but rather is angular and takes place about an axis also inferior to the cricoid. Third, the near-vertical axis about which twisting takes place does not pass through the CAJ but somewhat posterior to it in agreement with the qualitative suggestion of

Table 2

Cricoarytenoid Joint Motion: Sliding and Twisting (Refer to Text for More Detail)

Specimen	Sliding $r_{q/c}$ (mm)			ω			β (degrees)	Twisting $r_{q/c}$ (mm)			ω			β (degrees)
	X	Y	Z	I	J	K		X	Y	Z	I	J	K	
1	4.8	-4.9	-8.4	-.601	.797	.066	35.3	10.6	-21.4	22.2	.069	.142	.987	62.7
2	12.8	-2.0	-12.0	-.565	.825	-.008	37.9	23.2	-13.6	22.4	.010	.057	.998	74.0
3	7.9	-7.2	-6.5	-.715	.695	.100	38.2	17.9	-27.9	20.6	.219	.089	.972	60.3
4	13.2	-10.9	-20.1	-.594	.800	-.083	35.4	14.8	-24.2	26.2	.048	-.160	.986	60.5
5	7.4	-4.5	-7.2	-.560	.827	.051	35.1	13.7	-21.4	24.7	.043	.202	.979	65.9
Mean	9.2	-5.9	-10.8	-.607	.789	.025	36.4	14.0	-21.7	23.2	.078	.066	.984	64.7
S.D.	3.3	3.0	5.0	.056	.048	.064	1.4	5.2	4.7	2.0	.073	.123	.009	5.1

(a) ANTERIOR VIEW

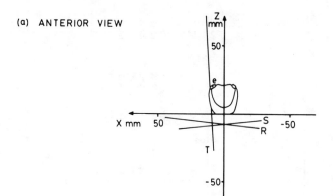

(b) SUPERIOR VIEW

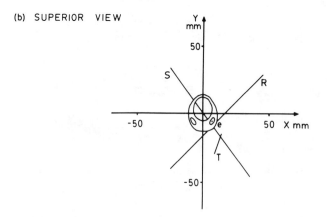

(c) LATERAL VIEW

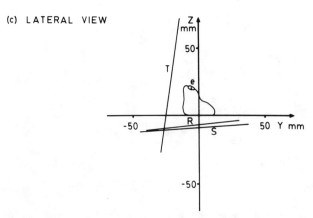

Figure 3 — Orientation of axes of rotation for specimen 1. R = rocking, S = sliding, T = twisting. (a) frontal plane, looking along -Y axis; (b) horizontal plane, -Z axis; (c) sagittal plane, -X axis.

von Leden and Moore (1961). When studying the large range of motion about this axis, it should be realized that the motion was constrained only by the joint capsule and associated ligaments. For the in vivo situation it is perhaps unlikely that the intrinsic musculature could undergo sufficient length changes to accommodate such large ranges of motion.

Since the data presented in this paper are based on cadaver specimens, they should be interpreted with this fact in mind. They do, however, give some indication of the mechanisms by which the CAJs—and hence the vocal cords—move during phonation and respiration.

Conclusions

In this study:

1. The motion of the right CAJ in five cadaver specimens was studied and found to consist of three principal axes of rotation.

2. The orientation of these axes in 3D space was determined in addition to the range of motion about the axes.

3. A quantitative description of the rotational kinematics of the CAJ —hitherto lacking in the literature—has been presented.

4. The results provide some indication of how the CAJ behaves in vivo.

Acknowledgments

The authors would like to acknowledge the financial aid given by the South African Medical Research Council, the experimental assistance rendered by Mr. Neville Smidt and Mrs. Elize Fuller, and the construction of the OSC and LED driver by Mr. John Ireland and Mr. Keith Willenberg.

References

ABDEL-AZIZ, Y.I., and Karara, H.M. 1971. Direct linear transformation from comparator coordinates into object-space coordinates in close-range photogrammetry. Proc. ASP Symp. Close-Range Photogrammetry, Urbana, Illinois.

ANDREWS, J.G., and Youm, Y. 1979. A biomechanical investigation of wrist kinematics. J. Biomech. 12:83-93.

ARDRAN, G.M., and Kemp, F.H. 1966. The mechanism of larynx. Brit. J. Radiol. 39:641-654.

FINK, B.R., and Demarest, R.J. 1978. Laryngeal Biomechanics. Harvard University Press, Cambridge, Massachusetts.

FRABLE, M.A. 1961. Computation of motion at the cricoarytenoid joint. Arch. Otolylarynology, 73:551-556.

FRANCIS, P.R., and Gabel, R. 1976. LED drivers: useful tools in biomechanics. In: P.V. Komi (ed.), Biomechanics V-B, pp. 456-463. University Park Press, Baltimore.

GOLDSTEIN, H. 1965. Classical Mechanics. Addison-Wesley Publishing Company, Reading, Massachusetts.

KARARA, H.M. 1974. Aortic heart valve geometry. Photogrammetric Engineering. 40:1393-1402.

SELLARS, I.E. 1979. A reassessment of the control of vocal cord movement. S. Afr. J. Communication Disorders. 26:120-126.

SELLARS, I.E., and Keen, E.N. 1978. The anatomy and movements of the cricoarytenoid joint. The Laryngoscope. 88(4):667-674.

SONESSON, B. 1959. Die funktionelle Anatomie des Cricoarytaenoidgelenkes. Zeitschrift fur Anatomie und Entwicklungsgeschichte. 121:292-303.

VON LEDEN, H., and Moore, P. 1961. The mechanics of the cricoarytenoid joint. Arch. Otolaryngolgy. 73:541-550.

WOOD, G.A., and Jennings, L.S. 1979. On the use of spline functions for data smoothing. J. Biomech. 12:477-479.

Response of the Human Shoulder
to External Forces

Ali Erkan Engin

Ohio State University, Columbus, Ohio, U.S.A.

During the last decade and a half, about a dozen total-human-body models have appeared in the literature. These models have gained increasing attention, in both vehicle crash victim studies and aerospace-related applications, in view of the high cost of the experiments with human cadavers and/or anthropometric dummies. The most sophisticated versions of the total-human-body models are articulated and multi-segmented to simulate all the major articulating joints and segments of the human body in three-dimensional space. A brief review of the mathematical models of the human body was provided in a paper by Engin (1979) and an extensive treatment of the same subject appeared in a review article by King and Chou (1976).

Short-time response of the multi-segmented models to predict accurately live human response requires proper characterization of the passive resistive force and moment properties in articulating joints. Furthermore, in these models the shoulder complex has been the most challenging for the biomechanist because of the lack of appropriate data and the complicated anatomical nature of the shoulder complex. The passive resistive force and moment data in the shoulder complex have been already presented in a recent article (Engin, 1980b).

Simulation of biodynamic events lasting more than a fraction of a second require the incorporation of the active resistive force and moments of muscles into the multi-segmented models and constitute long-time response of the model. The next generation models of the human body will most likely have contributions of various active muscles in determining the motion of one body segment with respect to the adjacent body segment, in addition to the passive resistance of soft-tissue elements. This article is concerned with a brief description of a research program which was developed to collect active resistive muscle force and moment

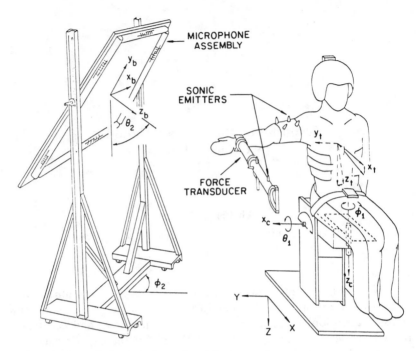

Figure 1 — Schematic drawing of experimental setup.

data at the human shoulder complex when the arm is subjected to various external forces. The numerical results are presented from experiments conducted on six subjects.

Procedure

The major components of the specially designed and built experimental apparatus for this research are a subject restraint system, a force-applicating device which employs three sonic emitters, and an upper arm cuff which holds four sonic emitters as shown in Figure 1. The three-dimensional force application direction and the orientation of the upper arm with respect to the torso were determined by monitoring these seven sonic emitters by a sonic digitizer* having four linear microphones as sensors for the sound impulses generated at the tips of the emitters. The operation of the sonic digitizer is based on the amounts of time required for the sound waves to travel from the sonic emitters to the microphones and calculation of the shortest (the slant range) distances to the four microphones. The slant range distances are subsequently used in a type

---

*Commercially available by Science Accessories Corporation under model name GP-6-3D Graf/Pen.

of triangulation analysis to calculate the x, y, and z coordinates of the tips of the emitters. In the three-dimensional coordinate determination mode of operation which was utilized in this research the sonic emitters had to be within an effective working volume defined by the 75 × 150 cm rectangular area of the microphone assembly and a fixed distance of 180 cm from this rectangular area.

The force applicating device, which is terminated by a force transducer capable of measuring all three components of the force and moment vectors, was designed to perform two tasks: 1) to determine the direction of the force application, and 2) to determine the location of the force application on the arm with respect to the torso coordinate system (x_t, y_t, z_t) or laboratory coordinate system (X, Y, Z). These tasks are accomplished by monitoring three strategically located sonic emitters on the force applicator. The knowledge of the locations of the sonic emitters on the force applicator and their sequential firing order provides sufficient information for a vector analysis to determine both the position and direction of the force application. The necessary computer software for a PDP 11/34A was developed to collect data simultaneously from the sonic digitizer and the force transducer.

Six healthy subjects (three male, three female) between the ages of 20 and 22 years and with no special training in athletics, were tested. Selected anthropometric measurements of these subjects were taken (Engin, 1980a) according to the definitions given by Hertzberg (1972). During anthropometric measurements and the subsequent testing phase, male subjects wore swimming trunks, female subjects wore gymnastic leotards. Active muscle resistance of subjects against externally applied loads were determined for various positions of the arm with respect to the torso. The isometric force applications by the subjects lasted for approximately 4 sec and the tests were repeated twice in a test session for each subject.

Results and Discussion

The orientation of the upper arm with respect to the torso was defined by means of the spherical coordinate angles, $\ominus$ and Φ. The $\ominus$ angle refers to the angle between the z-axis of the torso and the long-bone axis of the upper arm (this angle also defines the shoulder flexion-extension in the sagittal plane). The Φ angle refers to the angle between the projection of the long-bone axis of the upper arm on the xy-plane and the x-axis (the positive and negative values of this angle also define the shoulder abduction and adduction, respectively). Note that for the orientation of the arm shown in Figure 1, the $\ominus$ and Φ values are 90°. Since numerous combinations of $\ominus$ and Φ angles can be considered, the results are quite extensive. Due to space limitations only the plots of the maximum

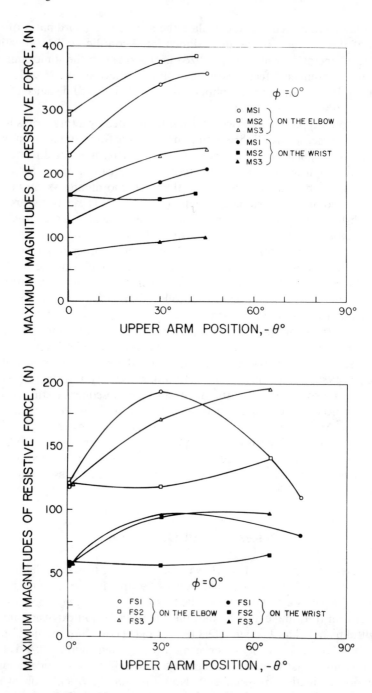

Figure 2 — Maximal values of resistive muscle force by the male (MS1, MS2, MS3) and the female (FS1, FS2, FS3) subjects against the external force application on the elbow and the wrist.

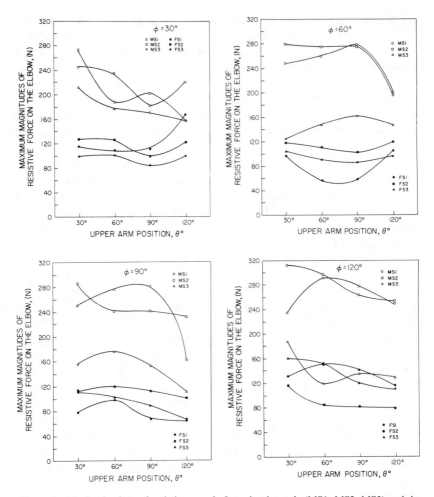

Figure 3—Maximal values of resistive muscle force by the male (MS1, MS2, MS3) and the female (FS1, FS2, FS3) subjects against the external force application on the elbow for various orientations of arm.

magnitudes of the resistive forces are presented in Figures 2 and 3. It should be emphasized that because of the static equilibrium, the magnitude of the force vector obtained from the force transducer acting on the arm also represents the magnitude of the resistive force vector in the shoulder complex.

Although the resistive moment results are not provided in this article, they were also computed about a shoulder joint point which was determined in the following manner by using the coordinate information obtained from the sonic emitters on the upper arm cuff. The position of the cuff was first defined with respect to the bony landmarks such as the medial and lateral epicondyles of humerus; then, the long-bone axis of humerus was determined by using coordinate data of the sonic emitters. The shoulder joint point about which the moments were calculated was assumed to be at a prescribed distance on this analytically determined long-bone axis. Note that definition of shoulder joint is somewhat arbitrary since one can choose an infinite number of points for the moment computation in the "shoulder complex" space. This arbitrariness can be attributed to the presence of four independent articulations among the bones of the shoulder complex, i.e., the clavicle, scapula, humerus and the thorax.

The average of the maximal magnitudes of the resistive forces from two tests are plotted in Figures 2 and 3. In particular, Figure 2 shows the results for the external force application on the elbow and the wrist. In this figure the angle $-\ominus$ along with $\Phi = 0°$ defines shoulder extension positions in the sagittal plane. The most striking feature of the results of Figure 2 is that the resistive muscle forces of subjects against the force application on the elbow are approximately twice as high as those of the force application on the wrist. Because of the difference between the moment arms, however, the magnitudes of the corresponding resistive moments at the shoulder complex are comparable. Finally, in Figure 3 the maximum values of resistive muscle forces against the external force application on the elbow are displayed for 16 combinations of $\ominus$ and Φ.

It is reasonable to expect that the values of the active muscle forces of the human arm and their corresponding moments at the shoulder depend not only on sex as indicated in this article but also on anthropometry, age, physical fitness, and, particularly, degree of training for muscular strength. Although there are both intra- and inter-subject variations for the maximum values of these muscle forces and their moments, incorporation of their magnitudes and behavior into the multi-segmented mathematical models of the human body should improve the biodynamic response capabilities of these models.

Acknowledgment

This research was supported by the U.S. Air Force Office of Scientific Research, Bolling Air Force Base, D.C. 20332, USA.

References

ENGIN, A.E. 1979. Passive resistive torques about long bone axes of major human joints. Aviat. Space Environ. Med. 50(10):1052-1057.

ENGIN, A.E. 1980a. Long Bone and Joint Response to Mechanical Loading. The Ohio State University Research Foundation Report RF Project No. 761590/711899.

ENGIN, A.E. 1980b. On the biomechanics of the shoulder complex. J. Biomech. 13(7):575-590.

HERTZBERG, H.T.E. 1972. Engineering anthropology. In: H.P. Van Cott and R.G. Kinkade (eds.), Human Engineering Guide to Equipment Design, U.S. Government Printing Office, Washington, D.C.

KING, A.I., and Chou, C.C. 1976. Mathematical modeling, simulation and experimental testing of biomechanical crash response. J. Biomech. 9:301-317.

Instability of the Hip Joint and Its Contact Pressure

Shinkichi Himeno
Fukuoka Children's Hospital, Fukuoka, Japan

Tadahiko Kawai and Norio Takeuchi
Tokyo University, Tokyo, Japan

Accurate evaluation of instability and contact pressure is indispensable for optimum planning to treat a patient with a disordered hip joint. Surgeries for the hip joint are aimed at altering the mechanical conditions around the hip and restoring the stable and painless joint. The mechanical non-linearities of the joint contact surfaces, however, make it quite difficult to assess pre-operatively the instability of the joint by conventional methods, such as the Finite Element Method (FEM). This is one of the major reasons why the results of treatment are unstable and unsatisfactory.

A new method has been developed to estimate the instability of the hip joint and its contact pressure from an ordinary X-ray film. This method is based on a 'Rigid Body Spring Model' (RBSM) proposed by Kawai et al. (1977). An analysis of a hip joint requires just 2 to 3 min in a micro-computer system, a convenience that will lead to wide applications in the practice of medicine.

Methods

In this article, the mechanical non-linearities of the 'contact problems' are discussed and a means of coping with these difficulties.

Non-linearities of the Joint

No stress but compression can be transmitted through joint contact surfaces. Animal joints including human joints are known for their low friction. The friction coefficient is usually 0.01 to 0.001 in a normal joint.

When shear stress acts on the contact surfaces, two surfaces easily slide over each other. When tensile stress occurs, the two surfaces come apart easily. Thus, neither tension nor shear stress is transmitted through the joint contact surfaces.

At the beginning of an analysis, one knows only the potential contact area. In some parts of the area, tensile stress may occur and become a non-contact area. Interactive calculations are required for this reason.

Limitation of FEM

Recent developments in mechanical analysis are remarkable. The most important one is the use of the Finite Element Method (FEM). This is a powerful method to analyze linear systems such as buildings, ships, etc. where strain is considered to be a linear function of loads. However, this kind of linearity is rather rare in natural phenomena. The destruction of material, non-Newtonian fluids, contact problems, and the like are highly non-linear problems. In these situations, FEM requires too extensive calculations to solve these problems practically.

RBSM

It is felt that the difficulties come from the model of FEM itself, namely, isotropy and continuity of strain in the element. Another way to cope with the non-linear problems is to establish a new model for discrete analysis.

RBSM was proposed for this purpose. This model consists of non-deformable and arbitrary-shaped rigid elements and interconnecting springs. Deformation occurs *inter*-elementarily (between elements) in this model, while in conventional FEM deformation occurs *intra*-elementarily (inside of elements). This property enables one to easily formulate the 'sliding phenomenon.'

Because an element is rigid, displacement of any arbitrary point on the element can be calculated after one knows the displacement of one point on the element (the center of gravity, for example). This means that the shape of an element can be complex without increasing the degrees of freedom. The total degrees of freedom (unknown variables) are dramatically reduced in some instances. Usually the amount of calculation (or CPU time) is proportional to the square of the number of unknown variables. A decrease in degrees of freedom is thus effective for research turn-over.

Formulation of Contact Problems

The RBSM is considered to be best suited for contact problems. When friction is negligible, spring coefficients of all shear springs on contact

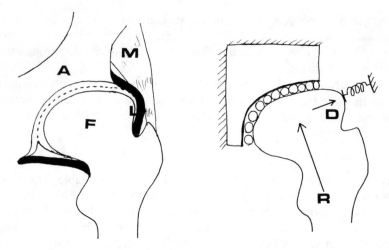

Figure 1—Equivalent model for the hip joint: A = acetabulum, F = femoral head, M = abductor muscles, L = capsular ligament, R = resultant force, and D = dislocative force.

surfaces are assumed to be zero and contact surfaces can slide freely on each other.

At the beginning of an analysis it is impossible to know if some potential contact area is really in contact or not. It is assumed at first that all of the contact surfaces are in contact. Calculations are carried out, knowing the stresses on normal springs at the contact surfaces. When tensile stresses appear, the spring coefficient of the normal spring with maximum tension is reset to zero. This small area is considered to be a non-contact area. Global stiffness matrix is reset and calculation continues. Interactive calculations are continued until no tension is detected.

Equivalent Model for the Hip Joint

Figure 1 shows the equivalent model for the hip joint used in this analysis. There are three elements: acetabular, femoral, and stopper. The joint contact surface between the acetabular and the femoral element is approximated by the polygonal sides. Every side has a normal spring and a shear spring whose coefficients are always set to zero. Frictionless rollers at the joint space represent these pairs of springs. The stopper element, which represents the capsular ligament in the human hip joint, prevents the femoral element from dislocating when the joint is unstable.

When a human stands on a leg, the force of gravity which acts on the center of the body rotates the body medially around the center of the femoral head. Abductor muscles, especially M. Gluteus medius, contract against this rotation. The muscle force is calculated from lever arm ratio. Thus, the resultant force (R in Figure 1) of muscle force and body weight acts on the center of the femoral head.

Figure 2—A microcomputer system used in this analysis.

When the resultant force is applied, the rollers at the joint space bear this load. If it fails, however, the femoral element moves outward to push the stopper spring. The magnitude of compression of the stopper spring is defined as the 'dislocative force' and is an indicator of the instability of the hip joint. Distribution of contact pressure is known by calculating compressive stresses of the normal springs at the joint space.

Microcomputer System

Figure 2 shows the microcomputer system used in this analysis. Geometrical data are input as a series of XY coordinates from a digitizer by moving a cursor along the shadow of a hip on a plain X-ray film (right side of the photo). Data are input to a microcomputer (Hewlett Packard 9845T) for the calculations. After the computation, results are displayed on the graphic display.

Calculating time required for an analysis is approximately 2 to 3 min. The entire procedure including input and output time does not exceed 5 min. If a large-scale computer is available, calculating time may be reduced to several seconds. The computer program is about 1000 lines long in enhanced BASIC language.

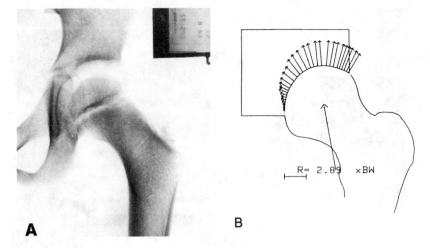

Figure 3A & 3B — Normal hip joint and its analysis.

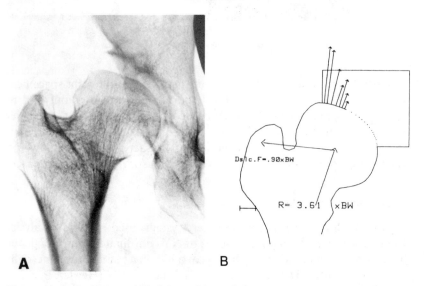

Figure 4A & 4B — Deformed hip joint and its analysis.

Results

Some results are shown for normal and deformed hip joints for purposes of illustration.

Normal Hip

Figure 3A shows a plain X-ray film of a normal hip joint of a child. Normal femoral head is spheric and well covered with acetabulum.

Figure 3B shows a result of the analysis. Dislocative forces do not exist, namely, joint cartilage bears the load completely and the joint is stable. Contact pressure is distributed evenly.

Deformed Hip

Figure 4A shows a deformed hip after congenital dislocation of the hip. The patient is a 39-year-old female who has complained of hip pain and limping for 10 years. The femoral head is elliptical and the joint space inclines steeply.

Figure 4B shows a result. A dislocative force of up to 90% of body weight indicates severe instability of this joint. The dotted line signifies a non-contact area widely spread medially. Consequently, distribution of contact pressure is concentrated on the marginal areas of the acetabulum. This is the major cause of the pain.

Discussion

Non-linear problems are quite complicated and require much time and effort to solve but are essential for understanding natural phenomena. One way to cope with these difficulties is to use a large-scale, high-speed digital computer. Another way is to construct a model expressly to illustrate the phenomenon. An example of the second method of solving contact problems has been demonstrated. Usually in FEM, a mesh-division is inevitable and a considerable number of nodes inside of the bone are created. Most of CPU time is consumed by these nodes. However, information gained from these nodes is of little importance clinically. This waste is avoided by the introduction of rigid body assumptions and the effort is concentrated on the marginal form of the joint. Fortunately, this assumption is not completely disassociated from reality, for X-ray images of loaded and unloaded hip joints are usually identical.

RBSM is effective not only for contact problems but also for other non-linear problems where discontinuity is essential. This method will now be extended to generalized joint dynamics, biotribology, soft tissue mechanics, etc.

References

HIMENO, S., et al. 1981. The stability of the deformed hip joints after C.D.H. and Perthes' disease. Rinsho Seikeigeka. (in press)

KAWAI, T., et al. 1977. A new element in discrete analysis of plain strain problems. Seisan Kenkyu. 29:204-207.

B.
Trauma

A Biomechanical and Clinical Study of
Skier's Ankle and Boot Top Fracture

Seturo Kuriyama, Etsuo Fujimaki,
Shokichi Uemura, and Yoshihisa Tashiro
Showa University, Hatanodai, Shinagawa-ku, Tokyo, Japan

The records for 25 years, from 1956 to 1981, showed 44,091 persons or 46,173 cases of injury due to skiing. When divided by sex, these figures involved 31,160 males (71%), and 12,931 females (29%).

Statistical Results of Ski Injuries

These cases of injuries were classified into the following types of trauma: 19,512 cases (42.3%) were sprains, 12,311 cases (26.7%) were fractures, 10,063 cases (21.8%) were lacerations, 2,982 cases (6.4%) were contusions, 814 cases (1.7%) were dislocations, and 491 cases (1.0%) were others.

When classified by region of the body, the highest frequency — 34,632 cases (75.0%) — involved trauma to the lower extremities. Specifically, 15,554 (33.7%) injuries to the ankle joint, 9,939 (21.5%) to the knee, and 8,063 (17.5%) to the lower legs represented the highest incidences. Additional injuries were 5,766 (12.6%) to the arms, 4,550 (9.9%) to the head and face, and 1,225 (2.7%) to the trunk.

When the type of trauma and the regions were cross-tabulated, the following figures were obtained: Of 19,512 sprains, the greatest number of cases appeared in the ankle joint (9,160 or 46.9%) and in the knee (8,984 or 46.0%). Fractures (12,311) appeared in the malleolus of the ankle joint (6,172 or 50.1%), and in the bone shaft of the lower leg (4,705 or 38.2%). Lacerations (10,063) appeared to the head and face (3,893 or 38.7%) and to the arms (2,767 or 27.5%). In contrast to the cases of sprain and fracture which appeared predominantly in the legs, laceration wounds appeared more frequently to the head, face and arms. Contusions appeared in all parts of the body, while dislocations occurred

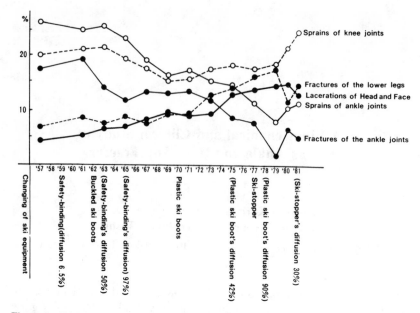

Figure 1 — Ski injuries and changing of ski equipment.

almost entirely to the shoulder joint.

But, the percentage of these types of trauma changed from year to year, especially as the development and changes in ski equipment, safety bindings and ski boots occurred (see Figure 1). In 1960, the safety bindings or the release bindings were developed and used generally; their use increased year by year. As expected, sprains and fractures of the ankle joint decreased, but sprains of the knee and fractures of the lower legs increased. In 1970, the plastic high-backed ski boots were developed and were generally used. These plastic boots were stiffer and had higher backs than the leather ski boots. Now, a new type of trauma is prevalent, this trauma is 'the boot top fracture.' This fracture occurs when bending a leg at the top with the stiff ski boot acting as a fulcrum.

These boot top fractures can be classified into three types: the transverse fracture, the oblique fracture, and the fibular fracture. From 1956 to 1965, the boot top fractures were very rare (0.7% of all fractures), but they have increased and, from 1975 to 1979, they involved 7.0% of the fractures, namely, they increased ten times in the last 20 years (see Figure 2).

When considering the types of boot top fractures and the length of the distal fragment, from 1966 to 1971, the average length of the distal fragment was 2.84 cm, but from 1975 to 1979, it increased to 4.0 cm. The oblique fracture decreased, and the other two types increased.

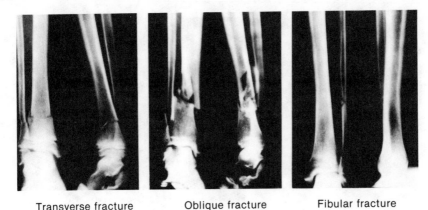

Transverse fracture   Oblique fracture   Fibular fracture

Figure 2 — Boot top level fractures.

Photoelastic Experiments

In order to understand the mechanism of occurrence of fractures at the ankle joint and the boot top fractures, two-dimensional photoelastic experiments were conducted with a model of the skeleton of the lower extremity prepared with D.A.P. resin so that the distribution of stresses under variously applied loads could be determined. Under loads of adduction at the ankle joint, tensile bending stresses act upon the fibula, while compressive stresses act upon the outer and inner margins of the fibula, but the maximum stress is at the level of the joint line. Compressive stresses act upon the inner aspect of the inferior articular surface of the tibia near the talus. The lines of maximum shearing stress act at about the point nearly coinciding with the direction of fracture lines of the X-ray film (see Figure 3).

Under loads of abduction of the ankle joint, tensil stresses act upon the inner margin of tibia, and abruptly increase at the tibiomalleolar part. Compressive bending stresses act upon the fibula, while the tensile stress and compressive stress act, respectively, upon the inner and out margins of the fibula.

The maximum stress is located at the level of the joint line on the inner margin, while for the outer margin it is located at a slightly higher level. In the cases of a tear of the medial ligament of the ankle joint, and loads in abduction, the compressive bending stress upon the fibula increases even further. The maximum stress on the outer margin of the fibula increases and, in this connection, it is characteristic of the large compressive stress focused upon the outer surface of the inferior articular surface of the tibia opposite the talus. Furthermore, if a tear of the inferior tibiofibular syndesmosis joint takes place, the maximum compressive bending stress of the fibula increases approximately 25 percent

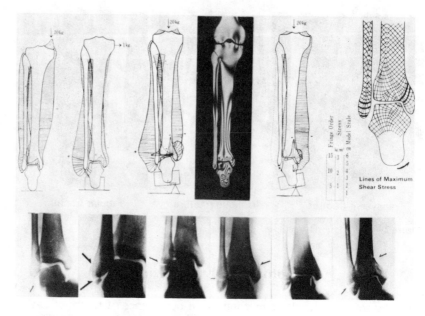

Figure 3—Stresses under loads of adduction.

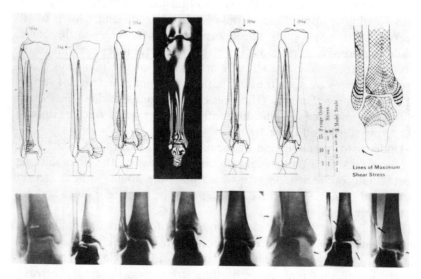

Figure 4—Stresses under loads of abduction.

at the periphery of the fibula. The lines of maximum shearing stress at this point coincide very well with the run of fracture lines on the X-ray film (see Figure 4).

In the lateral projection experiments, if there is dorsal flexion of the

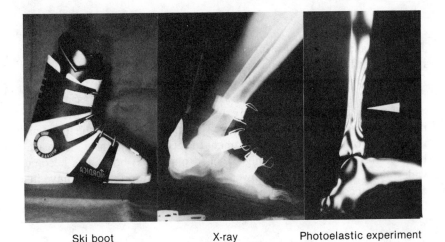

Ski boot   X-ray   Photoelastic experiment

Figure 5—Photoelastic experiment.

ankle, great compressive stress acts upon the level 1/5 of the way downward along the frontal margin of the tibia and at the front of the inferior articular surface of the tibia, while, in the cases of plantar flexion of the ankle, large compressive stresses act upon the dorsal part of the inferior articular surface of the tibia. The lines of maximum shearing stress coincide with the fracture lines.

Wearing a high-backed ski boot, and underloading the dorsal flexion in the ankle joint, the compressive stress acts on the frontal margin of the tibia and the tensile stress similarly acts on the posterior of the tibia. But, the maximum stress acts at the top end of the ski boot (see Figure 5).

Hardness and Thickness of the Bone

The hardness and thickness of the bone was studied using a fresh amputated leg. The hardness was measured by a micro-Vickers hardness meter. Measuring the tibia, the hardest point is slightly distal from the center of the tibia, and the hardness decreases distally or proximally from this point. So at the condyle and the malleolus the hardness is only about ⅔ of that at the center of the tibia.

The thickness of the tibia shows the same tendency as the hardness. When under loading in the neutral position of the leg, the peripheral stress distribution is in accord with the thickness and the hardness. Thus, the bone was made for its purpose.

The condyles and malleolus of the tibia are thin and relatively soft, and the trabeculae construction of the bone supplements these weak parts. This trabeculae construction is in accord with the line of principle stress trajectories.

Conclusions

This study showed 44,091 injuries with ratio of males to females of 7 to 3 in 44,327 affected locations, covering the period of 25 years, from December of 1956 to April of 1981. Injuries to the ankle have decreased because of the plastic ski boots and the safety bindings. Especially, the 'ski fractures' or the abducted and external rotated fractures of the lateral malleolus have decreased remarkably. The knee sprains, or, the ligamentous strains of medial colateral ligament of the knee joint have increased, because the use of stiff and high backed plastic ski boots has become widespread. Fractures of the lower legs and boot top fractures have increased in accordance with the development of the ski boots, from the leather boots to buckled boots to plastic boots.

References

FUJIMAKI, E. 1970. Photoelastic study on fractures of the ankle, Journal of the Showa Medical Association (Japan), 30(3):158-185.

FUJIMAKI, E., and Uemura, S. 1972. Ankle Injuries in Skiing, International Congress of winter sports medicine, Sapporo, pp. 47-55.

FUJIMAKI, E., Kuriyama, S., and Tashiro, Y. 1980. Boot top fracture in ski injuries, Orthop. Traum. Surg. (Japan), 23(13):1641-1651.

KURIYAMA, S., Fujimaki, E., and Uemura, S. 1980. Current trends in ski injuries and boot top fracture, J. Physical Fitness (Japan), 29:177-187.

SEKI, H., Fujimaki, E., Uemura, S., and Kuriyama, S. 1979. Ski injuries with special reference to dislocation of the shoulder joint. Proceedings of International Symposium on Science of Skiing, Zao, Japan, pp. 126-138.

On-Line Measurement of Moving Objects Using a Special-Purpose Microprocessor and a TV Camera

Yoshikuni Okawa
Gifu University, Gifu, Japan

Recently, much interest has been shown in the field of special purpose processors for image data processing. Two types of such processors have been proposed: (1) multiple arithmetic processors (e.g., ILLIAC III [McCormick, 1963] or Flexible Processor [Lillestrand, 1972]) and (2) pattern matching executors (GLOPR) (Preston, 1971) or PPM (Kruse, 1973). They all were designed for fast execution of image processing.

The subject of this article is the design of an image processing device of the second type. The main concern was the real-time processing. The term 'real-time' is meant for processing data taken from the real world as it is the case with the human eyes. Its ultimate goal is to produce numbers which describe properties of the interested objects in the scene. It is called the Dynamic Image Processor (DIP).

The voltage output of a TV camera is coded into two distinct values, 0 or 1, and fed into a data-arranging logic which in turn generates a locally rearranged image data stream. A pair consisting of a template and a mask operates on the data stream to signal the processor to start or close the programmed operations.

These experiments were carried out to calculate the position and size of a white disc moving freely on a black background. The results show that the machine can be a powerful tool for research in the field of real-time image processing.

Hardware Implementation

The Dynamic Image Processor consists of the following two principal parts: a microprogrammable processor (MPP) and a Template Matching Logic (TML) as shown in Figure 1. The microprogrammable processor is

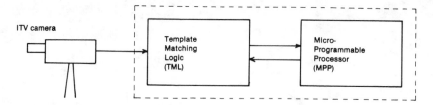

Figure 1 — Two principal parts of the Dynamic Image Processor.

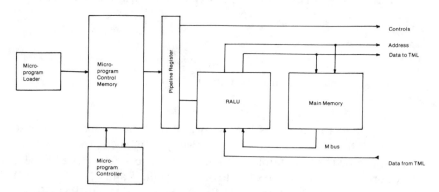

Figure 2 — The configuration of the microprogrammable processor.

constructed using custom bipolar LSIs.

Figure 2 shows the configuration of the microprogrammable processor. The horizontal width of one picture element in the image plane limits the basic operation time to 164 nano sec. Both MPP and TML are clocked simultaneously by this unique clock source. The main parts of the processor are the microprogram control memory, a pipeline register, a microprogram control unit, RALU (16-bit data and address buses), a main memory, and a hard-wired microprogram memory loader.

The processor is dynamically microprogrammable. The DIP has the characteristics of a general purpose computer to be able to perform a large variety of tasks, and fast execution ability of instructions. All of the flexibility of the system originates from this dynamically microprogrammable capability.

The outline of the Template Matching Logic is shown in Figure 3. The TML is connected to and controlled by the MPP. The TML consists of the following parts: a commercial TV camera, voltage handler, data arranging logic, a locally arranged image data stream, MOS memory, and a pair consisting of a template and a mask.

The main features of this system are the locally rearranged image data stream and ordered multiple templates. The sequential characteristic of the image processing algorithm is handled by this multi-template match-

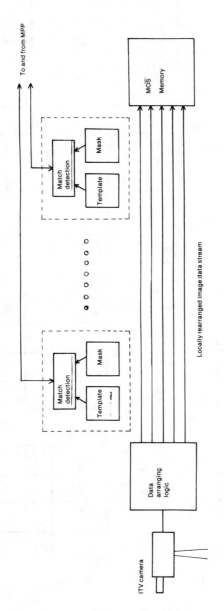

Figure 3 – The block diagram of the template matching logic.

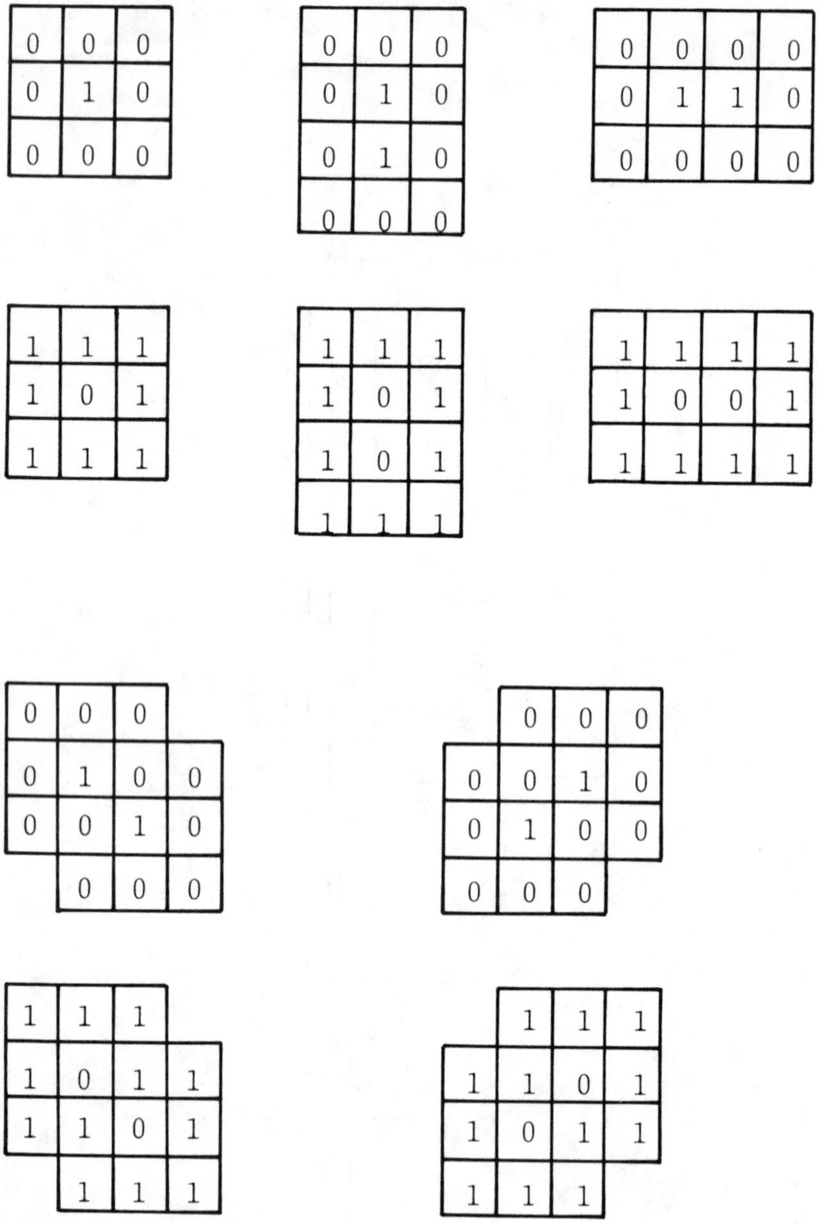

0	0	0
0	1	0
0	0	0

0	0	0
0	1	0
0	1	0
0	0	0

0	0	0	0
0	1	1	0
0	0	0	0

1	1	1
1	0	1
1	1	1

1	1	1
1	0	1
1	0	1
1	1	1

1	1	1	1
1	0	0	1
1	1	1	1

0	0	0	
0	1	0	0
0	0	1	0
	0	0	0

	0	0	0
0	0	1	0
0	1	0	0
0	0	0	

1	1	1	
1	0	1	1
1	1	0	1
	1	1	1

	1	1	1
1	1	0	1
1	0	1	1
1	1	1	

Figure 4—Examples of the fixed templates.

63	56 55 54 53 52	48 47 46 45 44	40 39	24 23 22	16 15 12 11	8 7 6	0

	P_3	EPE	B	P_2	EPE	A	E	P_1	F	FC	BT	P_0	AC

NAME	BIT	CONTENT
AC	0-6	Address control field
BT	8-11	Branch and test control field
FC	12-15	Flag control field
F	16-22	Function field
E	24-39	Emit field
A	40-44	General pulse A field
B	48-52	General pulse B field
EPE	45, 46 53, 54	Enable pulse in emit field
P_0	7	Parity check bit for 0-15
P_1	23	Parity check bit for 16-31
P_2	47	Parity check bit for 32-47
P_3	55	Parity check bit for 48-63
	56-63	(currently not used)

Figure 5 — The field specification of a microinstruction.

ing process. The maximum width of a template and mask is 5×5 in the present system, but may be easily extended to any desired dimension.

When the template matches the image data, the match detection logics emit signals to MPP and the other part of the TML according to the pre-assigned mode. The signal may toggle the center element of the data stream that is seen by the template. It may increment or decrement the specified counters or extract the coordinate information to the coordinate registers.

There are two types of templates: a fixed template and a variable template. Diode AND-OR matrices are wired for fixed templates whose content can not be altered easily. Some of the fixed templates now in use for noise detection and cleaning are illustrated in Figure 4. If some of these fixed templates are activated by the program, and if any one of them matches the present data stream, the center element(s) of the data stream is (are) toggled. The variable templates and their masks consist of flip-flops and can be loaded with any pattern by the MPP.

Instruction Examples

The size of the microprogram control memory is 56 bits $\times$ 1024 words. The format and definition of its subfields are listed in Figure 5. Eighteen bits of AC, FC, and F fields originate from the specification of the LSI maker. The emit field is for the emission of a constant of 16 bits or less, but a trick is deviced to generate some pulses controlled by the content of the EPE field. The detailed description of the content of each field is listed in Table 1.

The macro-instructions are unusual as compared with those of ordinary computers. A few of them are described in some detail in the following illustrations.

Table 1

The Definition of the Microcommands Currently Being Used

	general pulse B 48,49,50,51,52	general pulse A 40,41,42,43,44	Emit field 36,37,38,39	Emit field 32,33,34,35	Emit field 28,29,30,31	Emit field 24,25,26,27
00000	NOP	NOP	CSTA clear status	ENMB enable M bus	SACS set AC status	IHCK inhibit CPE clock
00001	SAIB set address 1 bus	SCAR set carry	CLCT clear counter	DSMB disable M bus	SSTA set stat A	LDXB load X bus
00010	RSTK reset K-bus	SETK set K-bus	LADL load ADRS lump	LDDL load data lump	RSTA reset stat A	MADC M address clear
00011	RLKB reset lower K-bus	RUKB reset upper K-bus	SSTD set stat D	SSTC set stat C	SSTB set stat B	MWRI M write status
00100	SLKB set lower K-bus	SUKB set upper K-bus	RSTD reset stat D	RSTC reset stat C	RSTB reset stat B	MRED M read status
00101	EFLK emit field to lower K-bus	EFUK emit field to upper K-bus	ARCX ADRS rst. ch. and X bus	ENCN enable conversion	SMBS set M busy status	MSCY M start cycle
00110	EFKB emit field to K-bus		MSK1 set mask 1	DSCV disable conversion	STHL set horizontal latch	MLOP M load and operate
00111	HCLK halt clock		MSK2 set mask 2	SCBS set conversion busy status	STVL set vertical latch	WSTK write stack
01000			TMP1 set template 1	WOPH horizontal window open	WOPV vertical window open	
01001			TMP2 set template 2	WCLH horizontal window close	WCLV vertical window close	
01010			NMK1 set noise mask 1	MWRI M write status	MADC M address clear	
01011			NMK2 set noise mask 2	MRED M read status	MOPR M operate	MIHA M inhibit address load
01100			NTP1 set noise template 1	SIPS set image proc status	MCPU M CPU mode	
01101			NTP2 set noise template 2	SVLT set threshold voltage	MAUT M automatic mode	
01110			XSCV start conv.			
01111		LDCT load counter				

The image field of the ITV camera

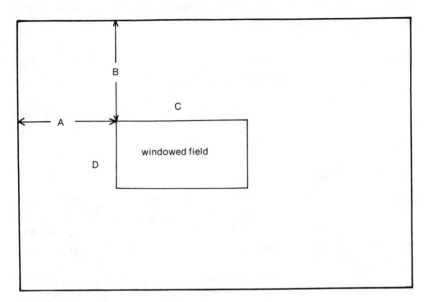

Figure 6 — Setting of a windowed field.

1. WINDOW A,B,C,D

This instruction restricts the image field to a rectangular field of interest as shown in Figure 6. The data stream of TML stops outside the window. The pattern matching operations are executed only within the window.

2. LOAD N,A

Data stored in the address A to A + 3 in the main memory are sent to the nth template and mask of the TML.

3. EXECUTE A,B

This instruction passes the control to the address A (or microprogram control jumps to the address A) when the scanning beam of the TV camera reaches a horizontal window end and the address B when the scanning beam reaches the horizontal and vertical window end. If WINDOW A,B,C,D are executed before this instruction, the program A is initiated C times and the program B is initiated one time in a frame of the TV picture.

4. SAMPLE A

The gray scale data of an image in the window is converted into 8-bit digital format and stored in the MOS memory from address A. Because the conversion time of ADC is 2 msec, if $C \geq 30$, the execution time of this instruction is about 0.50 sec.

Experimental Results

Various experiments have been carried out. Two of them are described in the following.

Object Following

Since the environmental conditions are artificial by nature, a simple mark pursuit problem is solved using the DIP. A white disc is attached on a movable string and is designed to be transferred to any desired position on a black background plane. The position of the disc is measured by two potentiometers.

The DIP calculates the position of the disc by the moment method. The templates detect the rising and falling edges of the disc. The X-Y coordinates are extracted into the coordinate registers of the TML at the moment when one of the patterns is matched to the data stream. At each right end of the window the DIP checks the coordinate registers of the TML. If the registers are loaded with a new X-Y information in that scanning time, the DIP calculates the partial sum of the following quantities.

$$A = \Sigma \, (HC_f - HC_r + 1) \qquad (1)$$

$$Y = \Sigma \, (HC_f - HC_r + 1) \, VC \qquad (2)$$

$$X = \Sigma \, (HC_f - HC_r + 1) \, (HC_f - HC_r) \qquad (3)$$

HC_f = horizontal coordinate when the template is matched, that is, the falling edge of the object.

HC_r = horizontal coordinate when the template is matched, that is, the rising edge of the object.

VC = vertical coordinate of the scanning line.

Finally, at the lowest right corner of the window, the DIP calculates the position of the disc according to the following equation.

$$X_G = \frac{\Sigma(HC_f-HC_r+1)\,(HC_f+HC_r)}{2\Sigma(HC_f-HC_r+1)} = \frac{X}{2A} \tag{4}$$

$$Y_G = \frac{\Sigma(HC_f-HC_r+1)\,VC}{\Sigma(HC_f-HC_r+1)} = \frac{Y}{A} \tag{5}$$

Thus DIP computes the position of the disc 60 times/sec. An EXECUTE A,B type of macro-instruction is applicable. One example of the sampled data is shown in Figure 7. The positions computed by the DIP are seen to be delayed from those measured by the potentiometers. If the scene is not as simple as this, more complicated algorithms are required.

Perimeter Measurement

In order to classify objects according to their size, the perimeter is an interesting number (Bacus and Gose, 1972). The 2 × 2 templates are matched with the image data stream. Whenever one of the templates of Group A matches the image, CTR 1 is incremented by 1. A template of type B counts up CTR 2. Finally, the DIP calculates

$$Z = CTR1 + p \times CTR2 \tag{6}$$

where $\frac{1}{2} < p \leq \sqrt{2}/2$. p must be pre-adjusted properly considering the geometric figure of the handled objects. The moving circular discs,

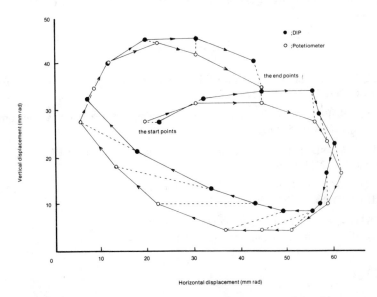

Figure 7 — Comparison of observed and computed displacement of the object.

triangles and various rectangles are shown to the DIP through the TV camera and it computes their perimeters which are compared with their true values. The errors are within 5%. But if a geometric figure of the objects is restricted to only one of them (e.g., a circular disc), the error is expected to decrease.

Conclusion

The design and some experimental results of a real-time image processor are discussed. The image data from a TV camera was locally rearranged to flow through the processor. Multiple templates were equipped to take matches with the image data. Counting the number of matches, extracting coordinates information into the registers, rewriting the elements in the data stream, etc., were carried out to extract information about the real world.

Both object-following and perimeter-measuring experiments were carried out to show some possible applications in the field of measuring moving objects in 3D space.

The basic source of difficulties which were met with during the experiments was not the problem of circuit construction, but that of program writing. The lack of the supporting tools for microprograms puts heavy burdens on programmers.

The described experimental results, although they are not such complicated ones, show practical usefulness of applying the Dynamic Image Processor to the pattern classification problem in the real world.

References

BACUS, J.W., and Gose, E.E. 1972. Leukocyte pattern recognition. IEEE Trans. on SMC. SMC-2(4):513-526.

KRUSE, K. 1973. A parallel picture processing machine. IEEE Trans. on Computers. C-22(12):1075-1087.

LILLESTRAND, R.L. 1972. Techniques for change detection. IEEE Trans. on Computer. C-21(7):654-659.

MCCORMICK, B.H. 1963. The Illinois Pattern Recognition Computer ILLIAC III. IEEE Trans. on EC. EC-12(5):791-813.

PRESTON, K., Jr. 1971. Feature extraction by Golay hexagonal pattern transforms. IEEE Trans. on Computer. C-20(9):1007-1014.

Avulsion Fractures of the Tibial Tuberosity as a Result of Violent Muscle Contraction

Osamu Kameyama, Hideo Oka, Fujio Hashimoto,
Kiyone Nakamura, Izumi Hatano, and Tsutomu Okamoto
Kansai Medical University, Osaka, Japan

Minayori Kumamoto
Kyoto University, Kyoto, Japan

Recently, athletic injuries have been increasing with the greater participation in various kinds of sports. Possibly the athlete's own muscle power as well as indirect force may be the cause of some of these injuries, especially in the injuries of the extensor apparatus of the knee joint, i.e., fractures of the patella, ruptures of the patellar tendon ligamentum, ruptures of the quadriceps tendon, and fractures of the tibial tuberosity. Clinical findings strongly suggest that indirect force often plays an important role in the occurrence of such injuries.

In the avulsion fractures of the tibial tuberosity, Watson-Jones (1976) mentioned that this affliction occurred when the knee was bent by great force against the maximally-contracted quadriceps tendon. In 100 reported cases within our experience, however, we found that 72% of the injuries with accurate history occurred at the take-off of the jump and 28% at landing. Furthermore, 65% occurred in the vertical jump (high jump, 43%; shooting jump shot in basketball, 10%; somersault, 8%; volleyball spike, 4%), while only 6% happened in the broad jump and 29% in other accidents.

This study was an attempt to elucidate what mechanism might be responsible for fractures most commonly occurring at the take-off of the vertical jump in terms of serial double joint action (Kumamoto and Takagi, 1980).

Method

In the present experiments, four healthy young adult male subjects were tested. Two of the subjects were well-trained in jumping. Electromyograms (EMGs) were recorded from the tibialis anterior (Ta), the lateral gastrocnemius (Gl), the vastus medialis (Vm), the vastus lateralis (Vl), the rectus femoris (Rf), the biceps femoris (Bf), the semimembranosus (Sm) and the gluteus maximus (Gm) utilizing a multichannel electroencephalograph (60 mm/sec) and an electromagnetic oscillograph (200 mm/sec) with surface electrodes. Foot switches were used to record the following moments: 'heel-off' (the heel lifted up) and 'toe-off' (the foot completely off the ground). Video tape from the side of the body, electrogoniograms of the ankle, knee, and hip joints, and vertical and horizontal curves of the ground reaction forces were simultaneously recorded with the EMGs while the subjects were performing vertical and standing broad jumps with various upper body postures.

Results

Vertical Jump

The representative EMGs of the well-trained subjects during the vertical jump with the upper body erect is shown in Figure 1. The angle of the ankle joint changed from dorsiflexion to plantarflexion at about 100 msec prior to toe-off. The main discharge of the Gl appeared before heel-off and ceased before toe-off, thus producing strong plantar flexion in order to push off the ground. The sequentially strong discharge of the Ta might account for the regulation of the ankle joint to prevent hyperplantar flexion. The strong discharge of the Gm occurred throughout the entire extension of the hip ending about 50 msec before toe-off, whereas in the initial stage of hip extension, the Sm and Bf varied from a fairly large discharge to no discharge depending on the upper body posture. The more erect the upper body, the less the activity in the Sm and Bf. The gradually increasing activity of the Vm functioned for regulation of the knee joint movements, passive knee flexion and active knee extension, bearing the upper body load. The Rf acted in the same fashion as the Vm except for a period where the discharge of the Rf seemed to be depressed by the discharge of the Sm and Bf. It was noted that the discharges of the Sm and Bf were suddenly depressed at the time the Rf increased its activity which corresponded to the period of most rapid extension of the knee.

Broad Jump

Figure 2 shows the representative EMGs of a well-trained subject during

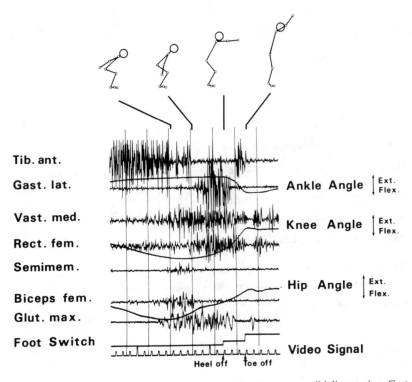

Tib. ant.

Gast. lat.   Ankle Angle | Ext. ↓Flex.

Vast. med.   Knee Angle | Ext. ↓Flex.

Rect. fem.

Semimem.

Biceps fem.   Hip Angle | Ext. ↓Flex.

Glut. max.

Foot Switch

Video Signal

Heel off   Toe off

Figure 1 — EMGs of the take-off in the vertical jump. Tib. ant. = tibialis anterior, Gast. lat. = lateral gastrocnemius, Vast. med. = vastus medialis, Rect. fem. = rectus femoris, Semimem. = semimembranosus, Biceps fem. = biceps femoris, Glut. max. = gluteus maximus. Time scale = 100 msec, and Video signal = one every two frames.

the broad jump. There was no significant difference in the ankle joint movement in the vertical and broad jumps.

The discharge of the Gl of all subjects showed double peak patterns. The first burst appeared from 100 msec prior to heel-off and ceased at heel-off, where the heel-off is a preparatory motion for the broad jump. The second burst of the Gl appeared from 200 msec prior to toe-off and ceased about 100 msec before toe-off. Subsequently, a marked discharge of the Ta appeared which might account for the control of the ankle joint in order to prevent hyperplantar flexion, as occurred in the vertical jump. The hip joint movement changed from flexion to extension at about the time of heel-off. The knee joint began to extend at 150 msec before toe-off. Hip extension could be fully attributed to the strong discharge of the Gm continuing until just before the toe-off.

In the initial stage of hip extension when the knee joint was still flexed (probably passively flexed), the activities of the Sm and Bf, the bifunctional muscles, were about the same as in the vertical jump with the upper body flexed. The gradually increasing discharge of the Vm func-

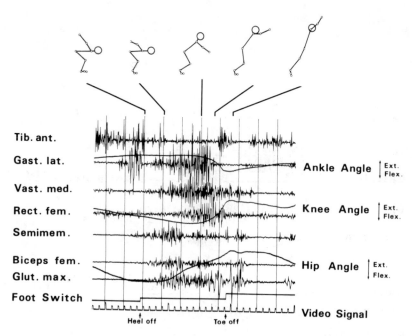

Figure 2—EMGs of the take-off in the broad jump. The abbreviations, time scale, and video signal are the same as in Figure 1.

tioned to regulate the knee joint which was passively flexed and actively extended. The Rf functioned in the same fashion as the Vm except for the period where the discharge of the Rf seemed to be depressed by the activities of the Sm and Bf, as was also noted in the vertical jump. The definite tendency for decreased discharges of the Sm and Bf was observed when the Rf showed the large burst for about 200 msec before the toe-off.

In the case of the untrained subjects, a clear reciprocal discharge pattern of the bifunctional muscles was seldom observed during the take-off. Not only in the broad jump but also in the vertical jump, the co-contraction between the Sm, Bf and Rf was usually recognized. The marked discharge of the hamstring muscles with the Rf prior to the toe-off might have generated the great force to pull the lower extremity backward.

Discussion

At take-off in the vertical and broad jumps, the knee and hip joints extend simultaneously. In such simultaneous knee and hip extension movements, it has been pointed out that the bifunctional leg muscles play an important role in reflecting the dynamic features of the double joint movements in both the static (Yamashita, 1979) and dynamic (Kumamoto and Takagi, 1980) conditions. According to Kumamoto and Takagi (1980), in the double joint movements, the resultant force output was not the sum of the force output of all the joints but was limited by the weaker joint. When the resultant force output was limited by the knee joint, the facilitation in the Rf and the inhibition in the Sm and Bf were observed during simultaneous extension of the knee and hip joints. On the other hand, when the limiting factor was the hip joint, the facilitation in the Sm and the Bf, and the inhibition in the Rf were observed.

Therefore, the reversal of the discharge patterns in the bifunctional muscles usually observed in the broad jump indicated that the limiting factor of the resultant force output changed from the hip joint to the knee joint during the jump; in other words, the joint which was predominant in bearing the total external load changed from the hip to the knee. In the vertical jump, however, the more erect the upper body posture, the longer the knee was the limiting factor. In the extreme cases, there was no discharge in the Sm and Bf but only in the Rf of the bifunctional muscles. Thus, the knee joint was predominant in bearing the load in the vertical jump.

In addition to this, the discharge pattern of the bifunctional Bf and Sm in the untrained subjects sometimes showed co-contraction with the Rf which might result in a contraction countering the knee extension during the vertical jump. This might account for 65% of the avulsion fractures of the tibial tuberosity occurring in the vertical jump but only 6%, in the broad jump.

In 91 of the cases reviewed, the avulsion fractures occurred in males between the ages 11 to 20 years, but mainly in the 13 to 18-year-old group, and recently there has been a tendency for such fractures to occur in younger people. This injury most commonly occurs before the age of 18, the age when the epiphysis of the tibial tuberosity and the head of the tibia fuse to the shaft of the bone. It has been classified into six types according to the condition of the epiphysis (Ogden, Tross, and Murphy, 1980).

The results obtained here suggest that possible poor coordination among the bifunctional muscles of the leg might cause the great stress at the inserted portion of the patellar tendon ligamentum and induce the separation of the tibial tuberosity from the head of the tibial shaft which is only loosely connected at the growth plate of the tuberosity.

References

KUMAMOTO, M., and Takagi, K. 1980. Dynamic and neurophysiological features in serial multijoint movements. In: K. Takagi and M. Kumamoto (eds.),

Regulation of Physical Activity, pp. 207-229, Kyorin Shoin Ltd., Tokyo.

OGDEN, J.A., Tross, R.B., and Murphy, M.J. 1980. Fractures of tibial tuberosity in adolescent. J. Bone and Joint Surg. 62-A:205-214.

WATSON-JONES, J. 1976. Fractures and Joint Injuries. 5th ed. Churchill of Livingstone, Edinburgh, London, p. 1046.

YAMASHITA, N. 1975. The mechanism of generation and transmission of forces in leg extension. J. Human Ergol. 4:43-52.

C.
Miscellaneous

Stress Fractures of Athletes in Japan:
A Clinical-Roentgenologic and Biomechanical Study

Yasuo Sugiura
Division of Orthopaedic Surgery,
Nishio Municipal Hospital, Aichi, Japan

Yoshiteru Mutoh
Division of Orthopaedic Surgery,
Tokyo Kosei-Nenkin Hospital, Japan

Etsuo Fujimaki
Department of Orthopaedic Surgery,
Showa University School of Medicine, Tokyo, Japan

A stress fracture, or fatigue fracture, is almost the same phenomenon as a fatigue break in metals. The bone is broken by a stress fracture by repeated minor trauma to the bone during walking, running, jumping or swinging and/or by repeated contraction of the muscles without any major impact or trauma.

Clinical-Roentgenologic Study on Stress Fractures of Athletes in Japan

One hundred sixty-two cases of stress fractures of Japanese athletes were observed roentgenologically from 1960 to 1980. There were 111 males and 51 females. Most of the subjects ranged in age from 15 to 19 years. The frequency of occurrence of stress fractures was as follows: 1) tibia-88 cases, 2) metatarsals-25 cases, 3) fibula-21 cases, 4) ribs-14 cases, and 5) others-14 cases.

Stress fractures of the metatarsals were caused by the start in the sprint, full-speed running short distances, and/or high hurdling. Most of the cases of stress fracture of the fibula were caused by repeated use of so-called 'rabbit jump' which seemed to be a unique type of training for athletes in Japan. Most of the cases of stress fracture of the ribs were caused by repeated shots with a driver in golf.

Stress fractures of the tibia were classified into two types. The first one (81 cases) was a running type of fracture which was observed most frequently at the upper 1/4 to 1/3 of the tibia. Most of them were caused by repeated long distance running, especially on roads in the winter. The second one (7 cases) was a jumping type of fracture which was observed as an Umbauzone (transverse-fracture-like-line) at the anterior cortex of the middle of the tibia. All of them were caused by either high jumping and/or hurdling.

Photoelastic Experiment on Running Type of Stress Fracture of the Tibia

In order to investigate the differences in sites of the running-type stress fractures of the tibia between the Caucasians and Japanese from a biomechanical point of view, a two-dimensional photoelastic experiment was performed. A model of the skeleton of the lower leg was made from Epoxy resin.

The results obtained are summarized as follows (see Figures 1a, 1b, 2a, 2b, 3a, and 3b):

The loading of the inner side of the crus varum (forced crus varum position) caused compression stress concentrated in the upper 1/3 region of the tibia, while loading of the outside of the crus valgum (forced crus valgum position) created a compression stress concentrated between the middle and the lower 2/5 of the tibia.

This experiment indicated that site differences of running-type stress fractures of the tibia between the Caucasians and Japanese was caused by a difference in shape of the lower legs (crus varum in the Japanese, crus valgum in the Caucasians) and loading (inside loading in the Japanese, outside loading in the Caucasians.)

Summary

There were 162 cases of stress fractures in Japanese athletes observed from 1960 to 1980 surveyed roentgenologically. A photoelastic experiment on the running type of stress fractures of the tibia revealed that the site differences between the Caucasians and the Japanese were caused by a difference in the shapes of the lower legs and the loading.

Crus varum

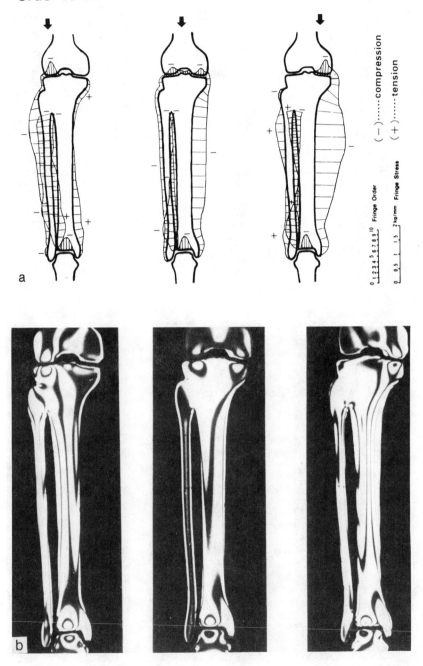

Figure 1—Photoelastic experiment on the crus varum. a: envirous stress distribution, b: photoelastic stripes.

Crus rectum

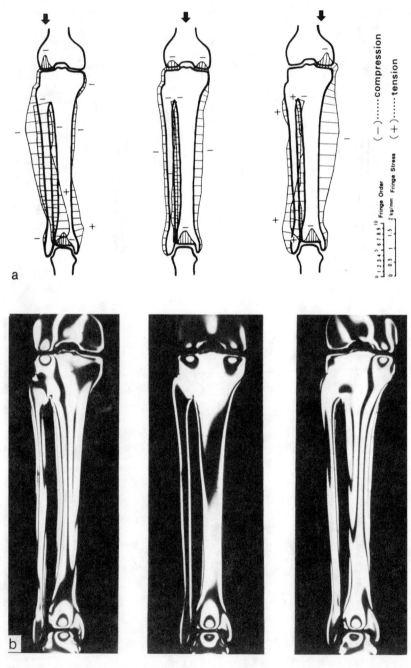

a

b

Figure 2—Photoelastic experiment on the crus rectum. a: envirous stress distribution, b: photoelastic stripes.

Crus valgum

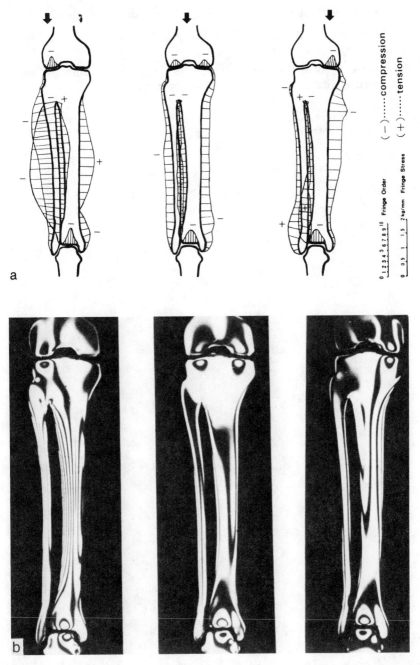

Figure 3—Photoelastic experiment on the crus valgum. a: envirous stress distribution, b: photoelastic stripes.

References

DEVAS, M. 1975. Stress Fractures. Churchill Livingstone, Edinburgh.

MORRIS, J.M., and Blickenstaff, L.D. 1967. Fatigue Fractures: A Clinical Study. Charles C. Thomas, Springfield.

SUGIURA, Y. 1979. Überlastungsschäden (Stress fracture in athletes). Seikeigeka (Orthopaedic Surgery), 30:675-682. (in Japanese)

Recent Experimental Results in the Sheffield Heart-Simulator of a New Mono-Cusp Artificial Heart Valve Model

R. Brouwer, A. Cardon, and W. Welch

Free University of Brussels, Brussels, Belgium

Clinical and experimental data obtained with various currently employed prosthetic heart valves show that in most cases the power loss caused by the presence of the opening valves is sufficiently low. The main problems remaining today are hemolysis, due to unfavorable systolic and diastolic turbulence patterns and thrombogenecity, caused by blood-material interaction and regions of stasis. To reduce these two artifacts a smooth design was chosen without the presence of any rigid superstructure (as is the case in ball-type valves) or hinges (present in some disc-type valves), since there is evidence that they might promote the forming of thrombus, eventually interfering with valve motion.

The form of the valve (see Figure 1) is obtained by removing one part of a sphere at a radius of about 80% of the spherical radius. Along this circumference a reinforcement-ring is attached to obtain sufficient lateral elasticity. One side of the valve is reinforced by gradually increasing the thickness of the cusp. At present the valve is made of a very flexible rubberlike material with an E-modulus of 1.18 N/mm². The valve is attached at the most rigid part on the aortic annulus over half of its circumference (see Figure 2).

During systole a buckling process takes place by which the uppermost flexible part moves downward towards the aorta and is located into the fixed, more rigid part of the cusp, thus enabling forward flow. At the onset of diastole the inverse phenomenon occurs.

Procedure

The new design was compared with several clinically-used valves of the same diameter (23 mm): a Björk-Shiley standard disc valve, a Björk-

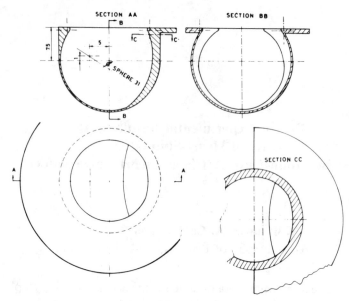

Figure 1 – Dimensions and form of the valve.

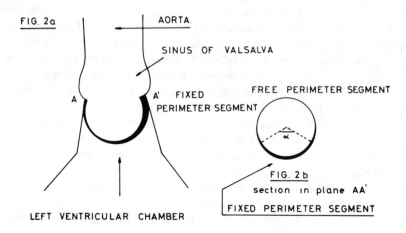

Figure 2 – Positioning on the aortic annulus.

Shiley convex-concave disc valve, an Omniscience disc valve, a Smeloff-Cutter ball valve, and a Hancock glutaraldehyde-preserved porcine xenograft. Tests were carried out in a pulse duplicator, furnished by the University of Sheffield (United Kingdom), previously described in a paper by Martin et al (1978). See Figure 3.

This system can be used both in pulsatile and steady flow conditions, despite a few modifications. The model circulation was designed in such

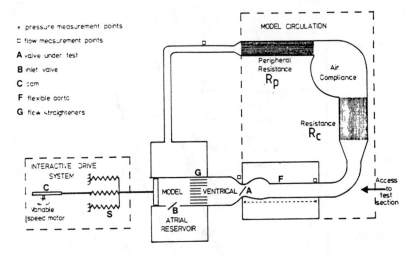

Figure 3—A schematic of the pulse duplicator.

a way that the aortic impedance was carefully simulated. Together with the use of flexible anatomically-shaped aortic tubes (F) made of transparent silicone rubber, this provided quasi-physiological pressure and flow-curves in the test-section (A) under pulsatile flow conditions. Water at room temperature was used as test fluid.

Pressure recordings were obtained just before the tested valve in the model ventrical and at 7 cm behind the valve in the aorta. Ventrical pressure was measured using a Druck PDCR 75 pressure transducer, aortic pressure by a Gaeltec 16 CF catheter tip pressure transducer, with a frequency response of at least 40 Hz. Flow rate was monitored with a SKALAR TRANSFLOW 601 electromagnetic flowmeter system. Pressure and flowcurves were recorded on an ultraviolet recorder at a paper speed of 0.5 m/sec. After digitalization of the signals, mean systolic pressure drops were obtained by integrating the pressure drop-signal over the period of positive flow. Amounts of regurgitation were obtained by integrating the flow signal over the period of negative flow. Velocity profiles and turbulence intensities were measured using a hot-wire anemometry system (DISA).

Results

Pressure Measurements

Steady flow pressure drops were measured at nine different flow rates of physiological interest, ranging from 115 ml/sec to 370 ml/sec. Figure 4 shows that the spherical valve induces about the same pressure drop as

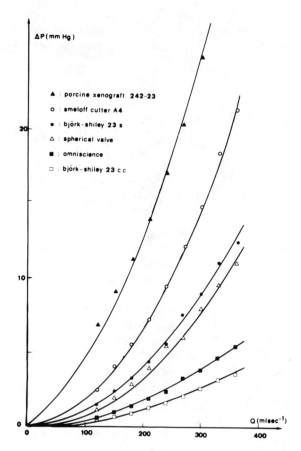

Figure 4 — Steady flow, pressure drop vs flow.

the Björk-Shiley standard disc valve, which is sufficiently low. Two disc-type valves — Omniscience and Björk-Shiley convex-concave — give the lowest pressure drops. For all valves it is observed that the pressure drop is related to the flow rate in a power-law fashion:

$$\Delta p = C Q^2 \tag{1}$$

where C is the overall valve drag coefficient. This is obvious in a log-log plot of Δp vs. Q (see Figure 5), which gives straight lines of slope ~ 2 (worst correlation coefficient r = 0.95). This indicates that all valves create turbulent flow.

Systolic pressure drops vs. systolic ejection flow rate are given in Figure 6 for the different valves tested. Again, the spherical valve and Björk-Shiley standard valve fall within the same range.

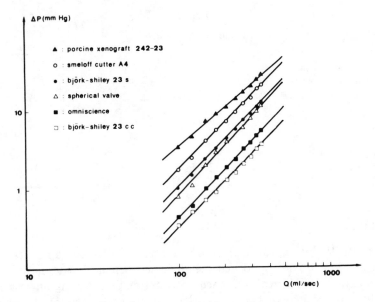

Figure 5 – Steady flow LOG-LOG plot, pressure drop vs flow.

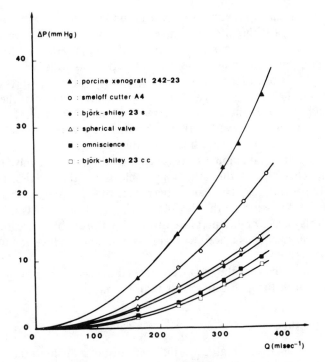

Figure 6 – Pulsatile flow, systolic pressure gradient vs systolic ejection flow rate.

Table 1

Amounts of Regurgitation

Amount of regurgitation (% of forward volume)	On closure	Total
Björk-Shiley S	6	8
Björk-Shiley cc	5	6
Smeloff-Cutter	4	7
Porcine Xenograft	3	3
Spherical value	5	13

Backflow

To be sufficient, a valve prosthesis must show limited amounts of regurgitation. It was found that the angle and height of fixation on the aortic annulus had great influence on backflow. Decreasing angle and height strongly increased the insufficiency of the valves. As this decrease lowered induced pressure drops, an optimum valve had to be found. This turned out to be an angle of fixation of 180° (half of the circumference) and a fixed height of 5 mm.

Table 1 compares amounts of regurgitation, given in % of forward pulsed volume, for the different valves, at a flow rate of 70 beats/min. The aortic pressure was maintained at $\sim$ 125/75 mm Hg.

Regurgitation on closing the spherical valve was satisfactory (5%). This is probably due to the fact that the beginning of closure starts before reversed flow has occurred as a result of the lateral elasticity of the cusp. At the end of systole the closure is already about 80% of total closure, the rest of it caused by reversed flow. This was clearly seen on high speed 16 mm cinematographic recordings (60 frames/sec).

Regarding the total amount of regurgitation the spherical valve is least sufficient. This is due to the rather uneven closure of the soft tissue of the cusp on the aortic ring. This can be overcome by increasing the height of the reinforcement-ring on the flexible moving part of the valve. Unfortunately this causes higher pressure drops and a probably increased risk for hemolysis problems since the contact-surface area between cusp and ring is augmented.

It must be noted that the tests were carried out with water, which will more easily leak through tiny spaces than blood.

Velocity Profiles

Velocity profiles were obtained at different distances behind the valves, ranging from 3 cm to 10 cm. All valves were placed in the valve test

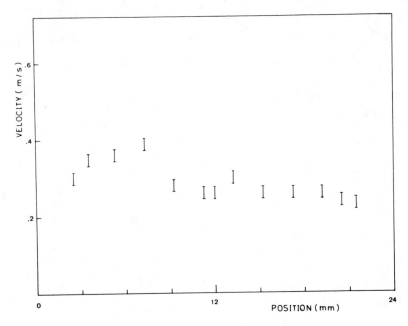

Figure 7—Velocity profile, spherical valve, 40 mm downstream.

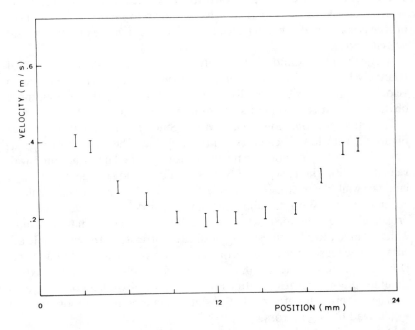

Figure 8—Velocity profile, Björk-Shiley valve, 40 mm downstream.

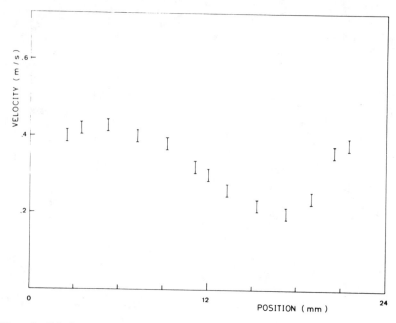

Figure 9 — Velocity profile, Smeloff-Cutter valve, 40 mm downstream.

chamber symmetrically with regard to the vertical axis. An illustrative velocity profile of the spherical valve is shown in Figure 7. Though the profile is somewhat asymmetrical, this valve gives a more or less constant velocity profile.

A Björk-Shiley standard disc valve (see Figure 8) shows a zone of relatively low velocity just behind the opened disc, when compared to velocity at the two sides. At distances nearer the valve this is even more obvious, with a real zone of stagnation in the vicinity of the disc.

The same phenomenon occurs with a Smeloff-Cutter ball valve (see Figure 9), with lower velocities a little eccentric. These zones of stagnation could be sites of possible thrombus formation. The rather high wall velocities with these types of valves could be advantageous to high laminar wall shear stresses $T_{rx} = \mu \frac{dV}{dr}$, though it is very difficult to estimate them. With this hot-wire anemometry system we could approach the flexible aorta walls at a distance of approximately 3 mm, which is much too far to estimate wall shear stresses with reasonable accuracy. The reason for the use of this hot-wire system instead of a Laser-Doppler system was the fact that flexible deformable aortas were used, to simulate physiological conditions as closely as possible. Focusing of a laserbeam through deformable walls at a single point is impossible and would lead to large errors.

In opposition to disc-type and ball-type valves, no regions of stagnation could be found with the spherical valve. From the turbulence-intensity measurements made it was obvious that all valves created very turbulent flow, with intensities up to 50%. It was, however, very difficult to make any comparison between the different valves, because—depending on the position of the hot-wire probe—all valves showed turbulence intensities of the same size.

Conclusion

Though regurgitation still causes some problems, the spherical valve shows promising features: a smooth closure, no rigid flow obstructing superstructure, no regions of stasis and a sufficiently low pressure drop.

References

BROUWER, R., Cardon, A., Hiel, C., and Welch, W. 1980. Some results about a new mono-cusp aortic valve prosthesis. Inter. Congr. on Rheology, Naples. 8:1-6.

GHISTA, D.N., and Reul, H. Optimal prosthetic aortic leaflet valve. J. Biomechanics. 10:313-324.

MARTIN, T.R.P., Palmer, J.A., and Black, M.M. 1978. A new apparatus for the in vitro study of aortic valve mechanics. I Mech. E. 7(4).

TILLMAN, W., Runge, J., and Reul, H. 1977. In vitro measurement of wall shear stress as hydrodynamical criterion for artificial heart valves. ESAO IV, London, Nov., 28-30.

YOGANATHAN, P., and Corcoran, W.H. 1978. In vitro velocity measurements in the vicinity of aortic prosthesis. J. Biomechanics. 12:135-152.

The Relation Between Sit-Up Exercises and the Occurrence of Low Back Pain

Yoshiteru Mutoh and Takemi Mori
Tokyo Kosei-Nenkin Hospital, Tokyo, Japan

Yoshio Nakamura and Mitsumasa Miyashita
University of Tokyo, Tokyo, Japan

A 43-year-old male patient with lumbar disc herniation was operated on in 1978. On admission he stated that repeated sit-up exercises with the knees extended caused pain in his lower back. Since then, this question has been asked of all patients with low back pain. Similar answers were obtained from a large number of these patients.

The purpose of this study was to investigate the relationship between sit-up exercises and the occurrence of low back pain and to determine the most appropriate abdominal strengthening exercises.

Procedure

Clinical Study

The roentgenograms and records of 29 patients with low back disorders related to sit-up exercises with the knees extended seen at Tokyo Kosei-Nenkin Hospital between April 1, 1980, and March 31, 1981, were reviewed with particular attention to the mechanism of the development of this type of pain. The total number of new outpatients during this period was 7,274.

Biomechanical Study

During the performance of a series of sit-up exercises electromyographic records were obtained of the rectus abdominis and the rectus femoris muscles of six healthy male physical education majors ranging in age from 19 to 25 years. The following four types of sit-up exercises were studied: Exercise I: with knees extended, ankles unsupported, Exercise II: with knees extended, ankles supported, Exercise III: with knees flexed, ankles unsupported, Exercise IV: with knees flexed, ankles supported. Surface EMG electrodes were secured to the skin over the muscles studied. Two electrodes, 3 cm apart, were placed over the mid-portion of the left rectus abdominis following an imaginary line 3 cm lateral to the umbilicus and over the mid-portion of the left rectus femoris, respectively.

Results

Clinical Study

The summary of 29 patients is shown in Table 1. There were 21 male and 8 female patients. The ages of the patients ranged from 12 to 51 years with an average of 21.2 years. All the patients did sit-up exercises with the knees extended and ankles supported for the purpose of strengthening the abdominal musculature.

These patients were divided into three groups according to the following situations:

Sit-up exercises with knees extended and ankles supported, which were done as:
a) part of an athletic training program, giving rise to low back pain, b) part of an athletic training program, aggravating the preexisting low back pain, and c) a therapeutic exercise, aggravating the preexisting low back pain.

Group A comprised 18 patients (62.1%), Group B, 7 patients (24.1%), and Group C, 4 patients (13.8%). The average ages were 17.4 years in Group A, 21.0 years in Group B, and 39.0 years in Group C.

In this group of patients, 25 were athletes. The types of sport activity were baseball, volleyball, basketball, track and field, swimming, jogging, etc. For the purpose of strengthening the abdominal muscles, sit-up exercises with knees extended and ankles supported were adopted in a variety of sports.

The diagnoses of 29 cases consisted of lumbosacral strain (22 cases, 75.9%), lumbar disc lesion (4 cases, 13.8%), spondylolysis (2 cases, 6.9%), and spondylosis deformans (1 case, 3.4%). The clinical symptoms, in almost every case, subsided in one to six weeks with conservative treatments.

Table 1

Summary of Cases

Case	Age (yrs)	Sex	Diagnosis	Type of Sport	Situation*
1	15	M	Lumbosacral strain	Swimming	
2	12	M	Spondylolysis	Baseball	
3	15	F	Lumbosacral strain	Gymnastics	
4	17	M	Lumbosacral strain	Baseball	
5	24	M	Lumbosacral strain	Baseball	
6	18	M	Lumbar disc lesion	—	
7	18	M	Lumbosacral strain	Volleyball	
8	16	F	Lumbosacral strain	Basketball	
9	18	M	Lumbosacral strain	Sumo-wrestling	A
10	19	M	Lumbosacral strain	—	
11	13	M	Lumbosacral strain	Basketball	
12	22	F	Lumbosacral strain	Ballet	
13	18	F	Lumbosacral strain	Volleyball	
14	29	M	Lumbosacral strain	Volleyball	
15	15	M	Lumbosacral strain	Volleyball	
16	14	F	Lumbosacral strain	Track and field	
17	13	M	Lumbosacral strain	Swimming	
18	17	F	Lumbosacral strain	Track and field	
19	20	M	Lumbar disc lesion	Judo	
20	16	M	Lumbosacral strain	Baseball	
21	15	M	Lumbosacral strain	Basketball	
22	20	F	Lumbosacral strain	Climbing mountains	B
23	17	M	Lumbosacral strain	Track and field	
24	37	M	Lumbar disc lesion	Baseball	
25	22	M	Lumbar disc lesion	Baseball	
26	37	M	Lumbosacral strain	Jogging	
27	33	M	Lumbosacral strain	—	C
28	51	F	Spondylosis deformans	—	
29	35	M	Spondylolysis	Jogging	

*explained in text

Biomechanical Study

The electrical activity of the rectus abdominis (RA) formed two main peaks during the phases of trunk flexion and trunk extension, and decreased or disappeared during the middle phase of the sit-up exercises (see Figure 1).

During the phase of trunk flexion, the electrical activities of RA showed the maximum magnitude in the first half of trunk flexion and then disappeared before the completion of trunk flexion. On the other hand, during the trunk extension phase, the electrical activities of RA

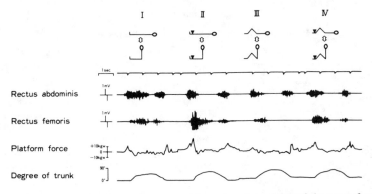

Figure 1 — Representative electromyograms of the rectus abdominis and the rectus femoris during the four types of sit-up exercises.

Table 2

Integrated Electrical Activities of the Rectus Abdominis

Subject	Exercise I	Exercise II	Exercise III	Exercise IV
1	100[a]	92.3	98.3	101.0
2	100	71.0	81.6	70.0
3	100	77.1	103.0	88.8
4	100	92.8	104.5	106.7
5	100	106.7	81.1	101.0
6	100	48.6	96.2	51.4

[a]Values expressed are percentages.

showed the maximum magnitude at the last half of trunk extension and disappeared before completion of trunk extension. The patterns of activities for RA were almost the same whether the knees were flexed or extended, or the ankles supported or not.

The electrical activities of the rectus femoris (RF) also showed the same pattern. During the trunk flexion phase, the electrical activities of RF showed the maximum magnitude in the last half. The maximum electrical activities of RF during trunk extension were seen in the first half.

Integrated electrical activities of RA and RF were compared with the four types of sit-up exercises. Within each subject, integrated figures were expressed in percentages. Integrated electrical activities of RA were not affected by the position of the knees. The integrated electrical activities of RA when the ankles were unsupported tended to be higher than when the ankles were suported (see Table 2). The integrated electrical activities of RF when the knees were extended tended to be higher than when the knees were flexed, and when the ankles were supported tended to be higher than when the ankles were unsupported (see Table 3).

Table 3

Integrated Electrical Activities of the Rectus Femoris

Subject	Exercise I	Exercise II	Exercise III	Exercise IV
1	43.8	100[a]	17.7	85.0
2	34.5	100	13.1	74.2
3	69.4	100	8.6	36.7
4	69.2	100	84.6	99.5
5	43.9	100	29.0	59.8
6	68.7	100	57.3	106.0

[a]Values expressed are percentages.

Discussion

As shown in these cases, abdominal strengthening exercises in Japan often take the form of sit-up exercises with the knees extended and the ankles supported.

Electromyographic studies already have revealed that, during sit-up exercises, the first 45° of trunk flexion as well as the last 45° of trunk extension are the responsibility of the abdominal muscles, while the middle phase of the exercises shows a definite decrease in muscle activities (Flint, 1965a). Similar results were obtained from this study. During the middle phase, the hip flexors such as the psoas muscle are primarily involved (Flint, 1965b). This muscle is known to act during the sit-up exercise only after the first 30° from the long-lying positions and through the entire range in the hook-lying position (Laban et al., 1965).

According to this study, the integrated electrical activities of the rectus femoris tended to be lower with the knees flexed than when the knees were extended. Though both the psoas muscle and the rectus femoris are hip flexors, the former is a one-joint muscle and the latter is a two-joint muscle. For this reason, with the knees flexed, activity of the rectus femoris is decreased. Therefore, for the purpose of strengthening abdominal muscles, sit-up exercises should be executed with the knees flexed.

As pointed out by Kendall (1965), those individuals with a weakness of the abdominal muscles can do sit-up exercises more easily when the ankles are supported owing to contraction of the hip flexors. Since the hip flexors also act as extensors of the lumbar spine, lumbar lordosis is increased. This study also demonstrated that the integrated electrical activities of the rectus abdominis tended to be lower and that of the rectus femoris tended to be higher when the ankles were supported than when the ankles were unsupported. Therefore, sit-up exercises with the ankles unsupported leads to strengthening of the abdominal muscles more effectively without causing an increased lumbar lordosis.

Nachemson's data (1976) on loads on the third lumbar disc during various positions and activities indicated that the loads of sit-up exercises are the same as that of bending forward 20° with 10 kg in each hand. Radin et al. (1979) described that repeated flexion of the lumbosacral spine, from a mechanical point of view, is contraindicated for patients who have spondylosis or spondylolisthesis at L_5, S_1. They also stated that abdominal strengthening exercises should be done in the recumbent position with the lumbosacral spine splinted by the floor or a mat.

In these cases, low back disorders probably occurred due to repeated sit-up exercises with the knees extended and ankles supported. This brought about increased mechanical loads in the lumbosacral region and increased lumbar lordosis causing hip flexor contraction. Therefore, it is necessary to educate people that sit-up exercises are not appropriate for the purpose of strengthening abdominal muscles, and that sit-up exercises with the knees extended and the ankles supported should be avoided. Especially, in those persons with weakness of the abdominal muscles and/or preexisting low back disorders, this point should be stressed. It may be concluded that, as an exercise to strengthen abdominal muscles effectively without injuries, partial sit-up or curl-up exercises with the knees flexed and ankles unsupported are the best.

References

FLINT, M.M. 1965a. Abdominal muscle involvement during the performance of various forms of sit-up exercise. Am. J. Phys. Med. 44(5):224-234.

FLINT, M.M. 1965b. An electromyographic comparison of the function of the iliacus and the rectus abdominis muscles. Am. J. Phys. Ther. Asso. 45(3):248-253.

KENDALL, F.P. 1965. A criticism of current tests and exercises for physical fitness. Am. J. Phys. Ther. Asso. 45(3):187-197.

LABAN, M.M., Raptou, A.D., and Johnson, E.W. 1965. Electromyographic study of function of iliopsoas muscle. Arch. Phys. Med. Rehab. 46:676-679.

NACHEMSON, A.L. 1976. The lumbar spine. An orthopaedic challenge. Spine 1(1):59-71.

RADIN, E.U., Simon, S.R., Rose, R.M., and Paul, I.U. 1979. Practical Biomechanics for the Orthopedic Surgeon. John Wiley & Sons, New York.

II.
NEURO-
MUSCULAR
CONTROL

Keynote Lecture

The Stretch Reflex as Motor Control

Saburo Homma
Chiba University, Chiba, Japan

Skeletal muscle, when stretched, tends to contract. This response, called the segmental stretch reflex, operates in a closed loop. Impulses conducted through the stretch reflex arc were recorded from Ia afferent fibers and α motoneurons. During stretching of the muscle, the decoding process from Ia impulses to motoneuronal spikes was studied from the view point of motor control.

The Stretch Reflex Activated by Brief Muscle Stretch

The gastrocnemius muscle was stretched briefly with the tapping of the tendon, and the activity in the reflex arc was recorded from various places (Homma et al., 1962).

The muscle stretch is recorded in trace 4 in Figure 1 and a sample of Ia discharge in a filament separated from the dorsal root is shown in trace 1. Trace 3 is an intracellular record from an α motoneuron supplying the stretched muscle; it shows the excitatory postsynaptic potentials (EPSPs). When the EPSPs reached a critical level for firing, the α motoneuron initiated a spike which was conducted along the efferent nerve. Trace 5 shows a unit spike recorded from the ventral root. Trace 2 shows the muscle action potential recorded with a needle electrode, and trace 6 gives the reflex tension developed by the stretched muscle. The right column of records in Figure 1 was obtained when the muscle was stretched more rapidly. Trace 1 shows an increased Ia discharge rate, and the two EPSP ripples are summed earlier. Therefore, the summed EPSPs rise more steeply and make the α motoneuron fire earlier. The ventral root spike shown in trace 5 also appears earlier. The gastrocnemius action potential increases in amplitude as shown in trace 2, and the reflex tension is greater.

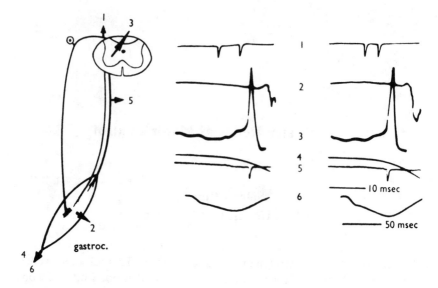

Figure 1 – Cat, gastrocnemius. The muscle was stretched by a twitch contraction of the gastrocnemius of the other leg. The six traces are records from the positions shown in the diagram, and are described in the text. The stretch for the right-hand set of records was applied more rapidly than for those on the left (Homma et al., 1962).

The EPSPs produced by Ia impulses were summed temporally. The rising slope of the summed EPSPs depends on the temporal pattern of the successive Ia impulses and this in turn can be varied by changing the rate of muscle stretch. The steeper the extension, the more the increase of the reflex muscle tension.

Accommodation of Firing Level for Slow Rising EPSPs

The Achilles tendon was stretched at different rates and the responses of the gastrocnemius motoneuron was monitored by an intracellular electrode (Homma, 1966).

Three responses with the associated EPSPs are shown in Figure 2.

The spike with minimum latency was excited by the fastest stretch applied to the muscle. The critical threshold level for firing is clearly distinguished as an upward inflexion on the rising phase of the EPSPs. Spikes appear almost at the top of the steep EPSPs. The second spike, induced by a slower stretch, was generated after summation of two successive EPSPs, and the third spike only after summation of three EPSPs. The more rapid the muscle stretch, the earlier the spike response developed. It is seen clearly that the critical threshold levels at the onsets

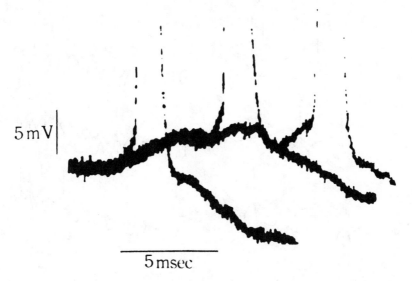

Figure 2 — Three responses of an extensor motoneuron recorded via an intracellular electrode. The spike with minimum latency (to the left) was excited by the fastest stretch applied to the muscle. The spike appears at the top of a steep EPSP. The second spike, induced by a slower stretch, was generated after summation of two successive EPSPs, and the third spike only after summation of three EPSPs (Homma, 1966).

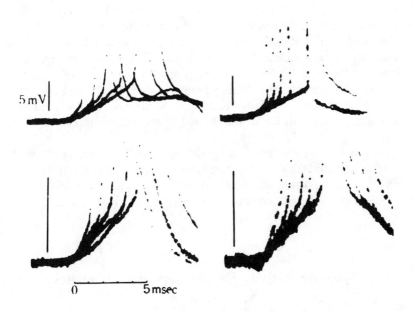

Figure 3 — Succession of depolarizations in four motoneurons produced by increasing rates of muscle stretch and associated changes in critical threshold for firing (Homma, 1966).

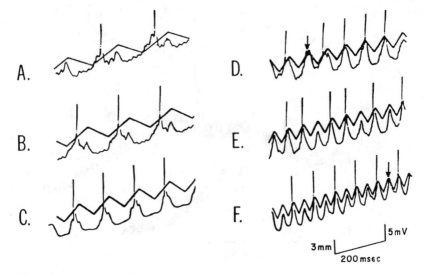

Figure 4—A further example of synaptic activity during repetitive stretch as frequency was increased beyond that at which an action potential always was activated reflexly by each individual stretch. Arrows (D and F traces) indicate those individual stretches which did not trigger action potentials even though the rate and magnitude of accompanying EPSP summation would appear sufficient for firing to have occurred (Homma et al., 1970).

of the three spikes are not identical. The second spike is generated at the highest level of EPSP. That is, the steeper the rising slope of the EPSPs, the lower the critical depolarization level for firing.

Succession of EPSPs in four motoneurons produced by increasing rates of muscle stretch and associated changes in critical threshold for firing are shown in Figure 3. Each motoneuron showed a higher threshold level at the more slowly rising depolarization. It has been suggested that the higher threshold is due to accommodation of the motoneuron to the rising slope of the EPSP (Araki and Otani, 1959; Sasaki and Otani, 1961).

Thus, a slower EPSP and higher critical threshold prolonged the latency from the initiation of the EPSP to the spike. The rising slope of EPSPs is an important determination of latency and firing level.

EPSPs of the fast slope produced by quick muscle stretch is effective in activating the stretch reflex.

Synaptic Activity During Repetitive Stretches

Relations between muscle stretch and EPSP slope is obtained by repetitive stretches of variable frequencies. Synaptic activity during repetitive

'5m V

——100msec'100μ

Figure 5—Cat gastrocnemius motoneuron. The upper trace shows EPSP ripples and motoneuronal spikes during 100 Hz vibration (Homma and Kanda, 1974).

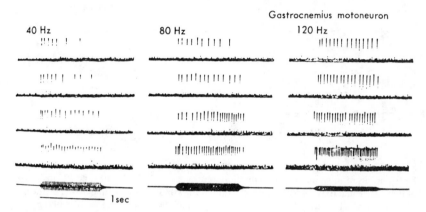

Gastrocnemius motoneuron

40 Hz   80 Hz   120 Hz

1sec

Figure 6—Motoneuronal unitary firing of the gastrocnemius at various vibration frequencies. Twenty-five repeated stimulations of 1 sec duration were applied to the gastrocnemius muscle. Four sets of records during 40, 80 and 120 Hz vibration were selected arbitrarily at a different stage of the series of repeated vibration (Homma et al., 1971).

stretches was also observed by intracellular recording of the α motoneuron (see Figure 4).

The activity of the α motoneuron innervating the gastrocnemius was intracellularly recorded while the Achilles tendon was longitudinally and triangularly stretched (Homma et al., 1970). EPSP ripples, corresponding to each triangular wave, were observed. When the EPSP ripples reached the critical firing level they caused the excitation of α motoneurons. Spikes always fired during the rising phase of individual EPSP ripples. EPSP responses progressively increased in slope and decreased in size and duration as stretch frequency was increased.

The Achilles tendon was sinusoidally stretched at higher frequencies (Homma and Kanda, 1974; Westbury, 1972). EPSP ripples still corresponded to each stretch, and they were temporally summed. When these summed potentials reached the critical firing level, they initiated spikes (see Figure 5). Spikes always fired during the rising phase of individual EPSP ripples, and inter-spike intervals were integral multiples of the sinusoidal cyclic time. The relation between the frequency of spikes, referred to as M_f, and the sinusoidal stretch frequency of Ia af-

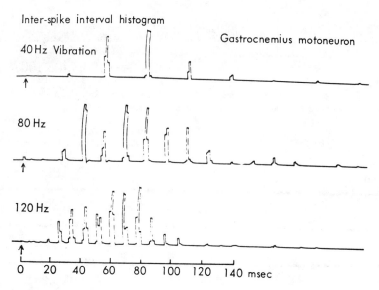

Figure 7—Inter-spike interval histogram of gastrocnemius motoneuron discharge. Distributions of intervals at 40 Hz vibration were around 25, 50, 70, 100 and 125 msec. All of them are equal to the integer multiplication of the vibratory cyclic time of 40 Hz (the uppermost trace). The situation was the same at 80 and 120 Hz (Homma et al., 1971).

ferent impulses which were equal to sinusoidal cycle, referred to as Ia_f, showed $M_f = \frac{1}{n} Ia_f$.

In this relation $\frac{1}{n}$ is called the decoding ratio. n is an integer, $1, 2, 3, \ldots$ n (Homma et al., 1972).

Motoneuronal spikes descend through the ventral root and elicit motor unit spikes in muscle. A motor unit discharge was recorded from a single functional fiber separated from the central cut end of the ventral root.

Figure 6 shows motoneuronal unitary firing of the gastrocnemius at various frequencies of sinusoidal stretches (Homma et al., 1971a).

Inter-spike intervals were measured and their histograms are shown in Figure 7. All of intervals were equal to the integer multiplication of the cyclic times of sinusoidal stretches. Instead of ventral root recordings of cats, motor unit spikes were recorded as EMG for human subjects.

A cylinder-type motor vibrator was fixed onto the patellar tendon by a rigid rubber band. Forced vibration of the quadriceps femoris muscle elicited gradually increasing involuntary muscle contraction. These phenomena have been referred to as tonic vibration reflex (TVR) (Hagbarth and Eklund, 1966). Motor unit spikes were recorded with a needle electrode inserted into the quadriceps femoris muscle (Hirayama et al., 1974).

Figure 8 shows EMG spikes of the quadriceps femoris muscle during a vibration of 55 Hz to the patellar tendon. Figure 8d shows a non-sequential inter-spike interval histogram processed from the EMG data.

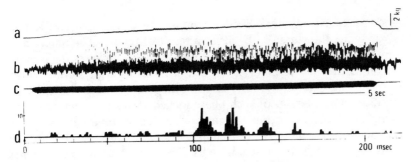

Figure 8 — Motor unit spikes during TVR of the human quadriceps femoris muscle. a, muscle tension; b, EMG spikes; c, monitored vibration is indicated by the thick part of the line. All three parameters were recorded simultaneously. Vibratory frequency 55 Hz. d, non-sequential inter-spike interval histogram processed from the data shown in b. Record a, c, and d are independently calibrated as shown in the illustration (Hirayama et al., 1974).

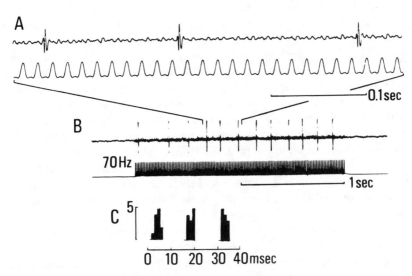

Figure 9 — A and B show single motor unit spikes and 70 Hz triangular taps. A is illustrated at a more expanded time scale than B. C is a cross-correlogram of the motor unit spikes with the triangular taps (Homma and Nakajima, 1979a).

Approximately four dominant peaks can be recognized, all being separated by the vibratory cyclic time of 55 Hz. It is possible to conclude that the quadriceps motoneuron fired preferentially at integer multiple intervals of the vibrating cyclic time. The above integral ratio, i.e., $M_f = \frac{1}{n}$ Ia_f, also held for human subjects.

Cross-correlation Between Vibration and Motor Unit Spikes

The successive Ia impulses generated by vibration of muscle monosynaptically produced EPSP ripples on the α motoneurons. These EPSP ripples were called vibratory EPSP (Homma, 1976; Homma and Kanda, 1974). When the vibratory EPSP as a result of temporal summation attains the critical firing level of the α motoneuron, it causes the α motoneuron to fire. Since spikes of the α motoneurons always occur during the rising phase of the vibratory EPSP, the spikes are phase-locked to the vibration, and intervals of such phase-locked spikes are equal to integer multiples of the cyclic time vibratory frequency. In order to analyze statistically the relations between phase-locked spikes and vibration, triangular stretches of muscle were used instead of sinusoidal stretches. Triangular stretches are equally effective as vibratory stimulation in stimulating the primary endings of the muscle spindle. It seems that even in humans, taps of a tendon with a triangular stimulus of small amplitude might excite the primary endings.

A human Achilles tendon was repeatedly tapped with triangular waves, with rising and falling times of 4 msec (Homma and Nakajima, 1979b). Such rapid taps are supposed to be the most sensitive stimuli for a primary spindle ending. Motor unit spikes were recorded from the soleus muscles. Figure 9B shows that motor unit spikes appeared immediately after triangular tapping at 70 Hz, and continued to fire until the tapping ceased. Figure 9A illustrates this phenomenon on a more expanded time scale than B. Intervals of the motor unit spikes were integer multiples of the cyclic time of triangular waves. A cross-correlogram of the motor unit spikes and triangular waves is illustrated in Figure 9C, and shows three correlation peaks. The interval between the peaks is the same as the cyclic time of the triangular waves. These peaks indicate that the motor unit spikes are phase-locked to the triangular taps. Cross-correlograms between motor unit spike and triangular tapping at various frequencies are shown in Figure 10.

In each correlogram in Figure 10, several peaks are observed and the intervals between the peaks are equal to the cyclic time of the tapping frequency as indicated on the left of each ordinate. A summation of the results of cross-correlograms of the motor unit spikes elicited by taps of 10 to 100 Hz is shown in the lowest part of Figure 10. In this summation an obvious peak appeared, related to the motor unit spikes which directly corresponded to the triangular waves. The time lag of the peak, i.e., the time lapse from a tap to the resultant motor unit spike, is the response time. The mean value of the response time in 55 soleus motor units was 33.7 ± 1.8 msec, which is a little longer than that corresponding to H-waves obtained by electrical stimulation of the tibial nerve. These motor unit spikes always occur during the rising phase of the vibratory EPSP elicited on the motoneuron. Consequently, the width of the peak indicates a probability density function of motor unit spikes which appear during the rising phase of the vibratory EPSP. Therefore, the width corresponds to the time-to-peak of the EPSP elicited by the triangular tap.

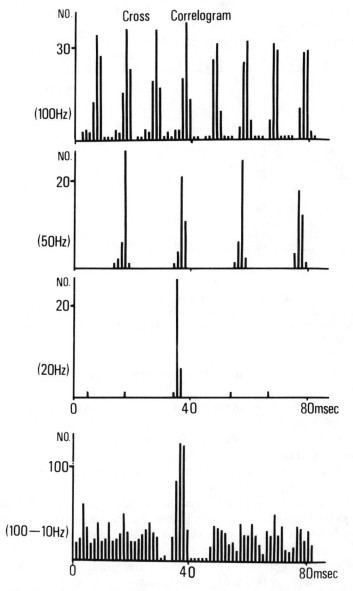

Figure 10—From the top to the third row, cross-correlograms of motor unit spikes with triangular taps whose frequencies were 100, 50 and 20 Hz, respectively. The lowest row shows a summation of the above three correlograms and ones with 90, 80, 70, 60, 40, 30 and 10 Hz triangular taps. The exaggerated peak is called a correlation peak (Homma and Nakajima, 1979b).

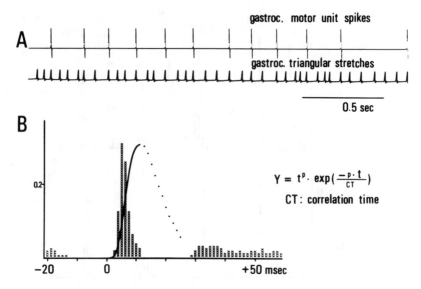

Figure 11 — A: Motor unit spikes (upper trace) of the gastrocnemius motoneuron-activated by random triangular stretches (lower trace) of the gastrocnemius muscle. B: Cross-correlogram of the motor unit spikes and the stretches. The solid line was obtained by integrating the correlation peak. Ordinate and abscissa indicate probability of spike occurrence and recurrence time, respectively (Homma, 1976).

Random Triangular Taps

The correlogram shown in the lowest part of Figure 10 was a summation of several cross-correlograms of the motor unit spikes elicited by taps of various frequencies. Such a correlogram is obtained by taps whose intervals alternate randomly: in these experiments, the minimum and maximum intervals were 20 msec and 80 msec, respectively (Homma and Nakajima, 1979a).

The upper trace of Figure 11A shows motor unit spikes reflexively elicited by random triangular stretches as shown in the lower trace. Figure 11B shows the cross-correlogram of the motor unit spikes and the onset of the random triangular waves. The prominent peak was seen in the cross-correlogram as in the summed correlogram in the lowest part of Figure 10. The peak clearly corresponds to the probability density distribution of the motor unit spikes which responded to the triangular waves with a suitable conduction time and synaptic delay. The time lag of the peak indicates the time from the onset of a stretch to a resultant motor unit spike, the so-called response time. The mean value of the minimum response time was 3.2 ± 0.6 msec for the 28 gastrocnemius motor unit spikes. It is apparent that the mean value indicates that the motor unit spikes are elicited by a monosynaptic-transmission mechanism in the stretch reflex. On the other hand, the distribution width of

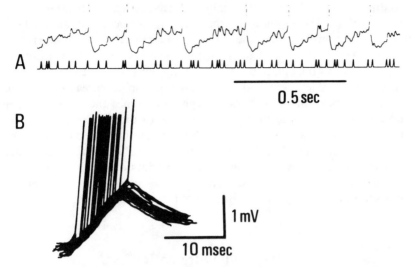

Figure 12 — A: Membrane potential change (upper trace recorded intracellularly from the gastrocnemius motoneuron during random triangular stretches (lower trace) of the gastrocnemius muscle. B: Superimposed spike potentials of A. Only components which deflected toward the overshoot potential were superimposed. Timing was taken at the beginning of the stretches (Homma, 1976).

the peak is assumed to indicate a probability density function of motor unit spikes occurring during the rising phase of EPSPs. These experimental results strongly support the theory that the width of the peak is related to the derivatives of postsynaptic potentials. The peak was integrated as shown in Figure 11B. The line rises slowly after the onset, then becomes very steep, and slows down again near the summit. Therefore, the line probably illustrates the time course of the rising phase of an EPSP. The falling phase of the EPSP is shown by means of dots because it is only based on calculations from the equation:

$$Y = t^p \exp\left(\frac{-p \cdot t}{CT}\right) \tag{1}$$

Thus, the peak makes it possible to calculate the time course and the time-to-peak of an EPSP. The former was initially slow, then steep and finally slow. The mean value of the latter in the 28 gastrocnemius motor units was 4.7 ± 1.1 msec.

The question of whether it is possible to calculate the time course of EPSPs with these statistical analyses, was investigated by means of intracellular recordings from an α motoneuron during random triangular stretches of the muscle.

As shown in Figure 12A, ripples of EPSP corresponding to the triangular stretches were intracellularly recorded from an α motoneuron.

Continuous application of the triangular stretches caused temporal summation of the EPSP ripples and when the summed potentials attained the critical firing level, the motoneuron fired. Figure 12B shows a superposition of the EPSP ripples which elicited spike potentials. Obviously the spike potentials take place during the rising phase of the EPSPs. Therefore, it is possible to conclude that motor unit spikes 'break out' within the time-to-peak of the EPSPs. Furthermore, since these spikes occur most frequently on the steepest rising slope of the EPSPs and less frequently on the slower slopes both at the start and near the summit of the EPSPs, one can calculate the time course of an EPSP from a probability density distribution of the spikes. The distribution of the spikes corresponds to the peak obtained from the cross-correlogram of the motor unit spikes and the onsets of the random triangular waves.

Polysynaptic Stretch Reflex

Forced vibration of the tendon in humans elicits the TVR phenomenon. Inter-spike intervals of the motor unit spikes were integral multiples of cyclic time of such vibratory frequencies as shown in Figure 8. But sustained vibration gradually increases involuntary muscle contraction and decreases the inter-spike intervals. Therefore, the decoding ratio increases.

These statements were confirmed by intracellular recording of the cat motoneuron. Recordings are shown very schematically in Figure 13.

The uppermost trace shows locked spikes triggered during the rising phases of the EPSP ripples. EPSP ripples were elicited by vibrations of regular frequency. The critical firing level of the motoneuronal membrane is shown by means of a broken horizontal line. The motoneuronal membrane was slowly and incrementally depolarized during sustained vibration (Homma and Kanda, 1974). As shown in the middle trace, the potential level was enhanced. The summed EPSP ripples reached the critical level for firing earlier; thus, inter-spike intervals were shortened, but spikes continued to be elicited during the rising phases of the EPSP ripples. When the membrane was further depolarized as shown in the lowest trace, the depolarization was over the critical level and caused intermittent firings which led to the recruitment of the spikes. Timing of the firing of the spikes was independent of the rising phases of the EPSP ripples. Therefore, TVR was shown to be elicited by activation of motor unit spikes in the following two categories: (1) spikes fired during the rising slope of the EPSP ripples and, (2) spikes fired independently of the EPSP ripples.

The former were called 'locked-spikes' and the latter 'unlocked-spikes' for the vibration (Homma, 1976).

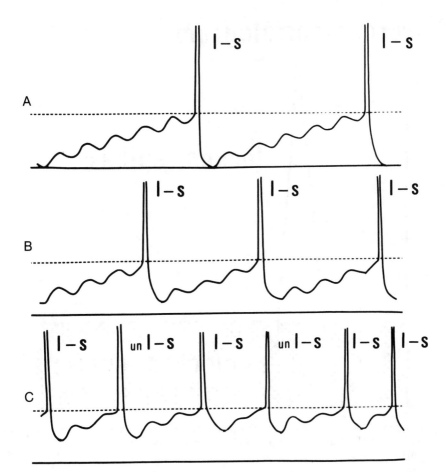

Figure 13 — Intracellular recording of motoneurons during muscle vibration. A: EPSP ripples produced by muscle vibration are called the vibratory EPSP. The vibratory EPSP, when as a result of temporal summation attains the critical level drawn by the broken horizontal line, makes the motoneuron fire. Those spikes which occur during the rising phases of the vibratory EPSPs are called locked-spikes, l-s. Intervals of locked spikes are integer multiple intervals of the vibrating cyclic time. B: The membrane was slowly depolarized. The summed EPSPs reached the critical level at an earlier point. Inter-spike intervals were shortened. Spikes continued to be locked for the vibration. C: The membrane was further depolarized. The depolarization was over the critical level, so some spikes fired independently of the rising phases of the vibratory EPSPs. These spikes are called unlocked-spikes, un l-s.

Both the locked and unlocked spikes are identified by a cross-correlation of vibration and motor unit spikes. The prominent peak as seen in the cross-correlogram is composed of the locked spikes, and the remaining part is made up of the unlocked spikes as shown in Figure 14.

cross-correlogram

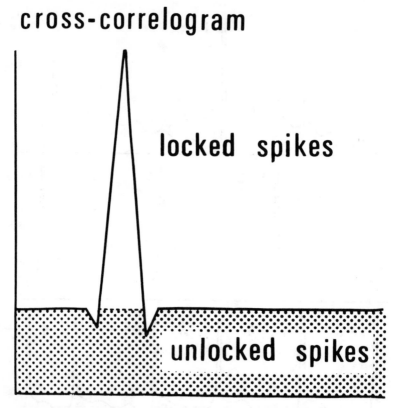

locked spikes

unlocked spikes

Figure 14 — Cross-correlation of vibration and motor unit spikes. Spikes which belong to the prominent peak are locked-spikes, and the others are unlocked-spikes.

Summary

It has been known that the tendon jerk reflex was caused by a monosynaptic transmission from the fastest conducting muscle afferents to homonymous motoneurons, and the receptor origin consisted of the spindle primary endings, and the monosynaptic connection from the Ia fibers was the exclusive afferent source (Lundberg and Winsbury, 1960; Homma, 1963). Recently, the existence of polysynaptic Ia pathways parallel to the monosynaptic pathways was reported with reliable results (Granit et al., 1957; Tsukahara and Ohye, 1964; Kanda, 1972; Homma, 1976). In these experiments, most of the motor unit spikes elicited by random triangular stretches were phase-locked to the stretch waves with a latency compatible with a monosynaptic connection. But a few spikes were not locked, and during repetitive stimulation unlocked-spikes increased gradually. These unlocked spikes were elicited through polysynaptic pathways.

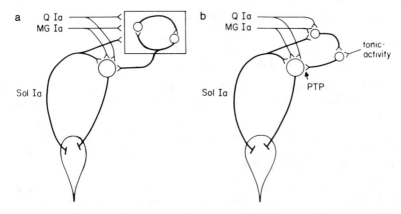

Figure 15 — Two fundamentally different, possible explanations for the long-lasting increase in motoneuronal excitability. a: The excitability increase is maintained by a feedback system (a loop), drawn in the figure as a box with two interconnected neurons but which can be anything from a loop within the neurons to diffuse interconnected networks of one or several neuron populations. b: The excitability increase is caused by PTP at synapses from interneurons mediating polysynaptic Ia excitation to the motoneurons. Increased excitation is obtained since the interneurons would also receive an inflow of tonic activity (Hultborn and Wigström, 1980).

With activities of the polysynaptic reflex, unlocked-spikes were built up very slowly and motoneuronal activities were maintained during sustained Ia discharges. The former is considered as an integration and the latter as a maintenance of the activity. Two fundamentally different, possible explanations for both integration and maintenance in motoneuronal activity have been suggested by Hultborn and Wigstrom (1980).

The left diagram in Figure 15 shows that the excitability increase is maintained by a feedback system, drawn as a box with two interconnected neurons. The right diagram in Figure 15 shows that the excitability increase is caused by post-tetanic potentiation (Granit et al., 1956; Homma et al., 1962), (PTP) at synapses from inter-neurons mediating polysynaptic Ia excitation to the motoneurons. Increased excitation is obtained since the interneurons would also receive an inflow of tonic activity.

Such increased excitation was not observed in the cat anesthetized with Nembutal but easily elicited in the decerebrate cat. It is likely to provide an important contribution to the enhanced stretch reflexes observed in that preparation. Though the increased excitation in the decerebrate cats and the augmenting phenomenon observed in the TVR in humans is not identified, comparisons of both observations are very fruitful. Transmission through this complex neuronal circuit is an important condition for the occurrence of tonic stretch reflexes.

References

ARAKI, T., and Otani, T. 1959. Accommodation and local response in motoneurons of toad's spinal cord. Jap. J. Physiol. 9:69-83.

GRANIT, R., Henatsch, H.D., and Steg, G. 1956. Tonic and phasic ventral horn cells differentiated by post-tetanic potentiation in cat extensors. Acta Physiol. Scand. 37:114-126.

GRANIT, R., Phillips, C.G., Skoglund, S., and Steg, G. 1957. Differentiation of tonic from phasic alpha ventral horn cells by stretch, pinna and crossed extensor reflexes. J. Neurophysiol. 20:470-481.

HAGBARTH, K.-E., and Eklund, G. 1966. Motor effects of vibratory muscle stimuli in man. In: R. Granit (ed.), Muscular Afferents and Motor Control, Nobel Symposium I, pp. 177-186. Almquist and Wiksell, Stockholm.

HIRAYAMA, K., Homma, S., Mizote, M., Nakajima, Y., and Watanabe, S. 1974. Separation of the contribution of voluntary and vibratory activation of motor units in many by cross-correlograms. Jap. J. Physiol. 24:293-304.

HOMMA, S. 1963. Phasic stretch of muscle and afferent impulse transmission in tonic and phasic motoneurons. Jap. J. Physiol. 13:351-365.

HOMMA, S. 1966. Firing of the cat motoneuron and summation of the excitatory postsynaptic potential. In: R. Granit (ed.), Muscular Afferents and Motor Control, Nobel Symposium I, pp. 235-244. Almquist and Wiksells, Stockholm.

HOMMA, S. 1976. Frequency characteristics of the impulse decoding ratio between the spinal afferents and efferents in the stretch reflex. In: S. Homma (ed.), Symposium on Understanding the Stretch Reflex, pp. 15-30. Elsevier, Amsterdam.

HOMMA, S., Ishikawa, K., and Stuart, D.G. 1970. Motoneuron responses to linearly rising muscle stretch. Am. J. Physical Med. 49:290-306.

HOMMA, S., and Kanda, K. 1974. Impulse decoding process in stretch reflex. In: A.A. Gydikov, N.T. Tankov and D.S. Kosarov (eds.), Motor Control, pp. 45-64. Plenum Press, New York.

HOMMA, S., Kanda, K., and Watanabe, S. 1971. Monosynaptic coding of group Ia afferent discharges during vibratory stimulation of muscles. Jap. J. Physiol. 21:405-417.

HOMMA, S., Kanda, K., and Watanabe, S. 1972. Preferred spike intervals in the vibration reflex. Jap. J. Physiol. 22:421-432.

HOMMA, S., Kano, M., and Takano, K. 1962. On phasic stretch of the annulo spiral endings. Symposium on Muscle Recepters, pp. 125-131. Hong Kong Univ. Press. Hong Kong.

HOMMA, S., and Nakajima, Y. 1979a. Coding process in human stretch reflex analyzed by phase-locked spikes. Neuroscience Letters. 11:19-22.

HOMMA, S., and Nakajima, Y. 1979b. Input-output relationship in spinal motoneurons in the stretch reflex. In: Granit and Pompeiano (eds.), Reflex Control of Posture and Movement. Progress in Brain Research. 50:37-43. Elsevier, Amsterdam.

HULTBORN, H., and Wigström, H. 1980. Motor response with long latency and maintained durations evoked by activity in Ia afferents. In: J.E. Desmedt (ed.), Spinal and Supraspinal Mechanisms of Voluntary Motor Control and Locomotion. 8:99-116. Prog. Clin. Neurophysiol.

KANDA, K. 1972. Contribution of polysynaptic pathways to tonic vibration reflex. Jap. J. Physiol. 22:367-377.

LUNDBERG, A., and Winsbury, G. 1960. Selective adequate activation of large afferents from muscle spindles and Golgi tendon organs. Acta Physiol. Scand. 49:155-164.

SASAKI, K., and Otani, T. 1961. Accommodation in spinal motoneurons of the cat. Jap. J. Physiol. 11:443-456.

TSUKAHARA, N., and Ohye, C. 1964. Polysynaptic activation of extensor motoneurons from Group Ia fibers in the cat spinal cord. Experientia 20:628-629.

WESTBURY, D.R. 1972. A study of stretch and vibration reflexes of the cat by intracellular recording from motoneurons. J. Physiol. 226:37-56.

A.
Muscular
Function

Muscle Activity around the Ankle Joint as Correlated with the Center of Foot Pressure in an Upright Stance

Morihiko Okada and Katsuo Fujiwara

The University of Tsukuba, Ibaraki, Japan

While the uniqueness of the human foot structure, its longitudinal arch for instance, has long been discussed by numerous authors from the functional-morphological view-points (e.g., Morton, 1935), few of them seem to have dealt with the human foot during its actual activity. On the other hand, extensive measurements in recent times of the foot force and EMGs during physical activities like the posture and gait are seldom related to foot morphology.

For an understanding of the functional significance of the human foot structure, two attempts were made in this study: 1) to examine at which part of the foot the center of foot pressure (CFP) remains during a quiet standing in natural balance, and 2) to examine with EMGs the attitudes of muscles around the ankle joint when the CFP is located in various parts of the foot. For the sake of simplicity, we treated exclusively the CFP in relation to the longitudinal axis of the foot.

Subjects and Methods

Ten male students of physical education between 18 and 24 years of age served as experimental subjects. They were asked to stand on a force platform (Gravicorder 2301, Anima Co.) in the Romberg's position, i.e., with the feet closed and placed in parallel, and to make each of the following performances with the eyes open:

1. to hold a quiet standing in natural balance for 20 sec, looking at a target displayed 2 m in front of the subject.

2. to lean forward very slowly from the above position, eventually to reach the forward limit of leaning within 20 sec.

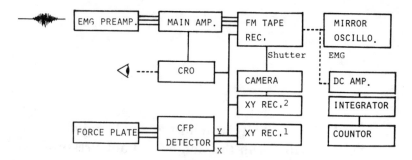

Figure 1 — Block diagram of the experimental set-up.

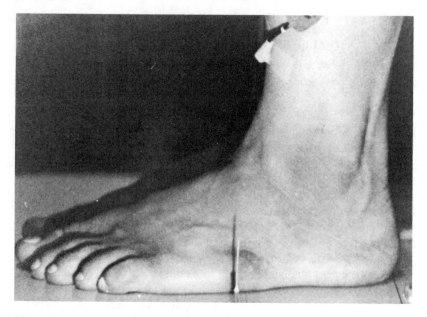

Figure 2 — The center of foot pressure projected on the foot surface.

3. to lean backward in the same manner as above, but to reach the backward limit within 15 sec. While performance 1 was repeated 20 times, performances 2 and 3 were repeated three times, by each subject in a random order.

Figure 1 shows the block diagram of the experimental set-up. An amplifier-calculator (Gravicorder 4301, Anima Co.) detected the longitudinal-transverse coordinate of CFP on the force plate in an actual dimension, which was displayed with an X-Y recorder. The longitudinal coordinate of the CFP was optically projected on the lateral surface of the left foot (see Figure 2). Thus, while the behavior of the CFP in each of the performances was being directly observed, the foot projected with the CFP was telephotographed transversely at intervals of 1 sec.

For the sake of inter-individual comparison, the CFP was shown in relative distance (%) from the hindmost point of the heel (pternion), regarding the foot length as 100%. The foot length was defined as a distance between the pternion and the foremost part of the toes along the inner margin of the foot. The element of the foot accommodating the CFP was identified by collating the picture of the performing foot mentioned above with an X-ray picture of the same foot taken transversely.

During each of the performances, bipolar electromyograms were picked up from the tibialis anterior, lateral and medial gastrocnemius, soleus, and abductor hallucis muscles of the left side, with Beckman surface electrodes of 3 mm in diameter of the contact area. Fine-wire intramuscular EMGs were also obtained for three of the ten subjects for screening of the contamination of activities inherent to surface EMGs. As a means of comparing the level of muscle activity during the performances, EMG was registered in a voluntary maximum contraction for 3 sec for each of the above muscles. EMGs and the longitudinal coordinates of the CFP were FM-tape-recorded together with the signals of photographic exposure. After playing back, they were recorded with a direct-writing mirror oscillograph. The longitudinal mean coordinate of the CFP during performance 1 was calculated through the time integral of the CFP fluctuation divided by 20 sec. EMG signals were full-wave rectified and integrated with a reset-level integrator (RFJ-5, Nihonkohden Co.), and the relative magnitude of muscle activity was calculated according to the methods described previously (Okada, 1972). Because of the limited number of tape-recorder channels, it was necessary to hold two experimental sessions for each subject.

Results

In Table 1 are shown the CFP position (mean ± SD for 20 trials) of quiet standing in natural balance, along with the CFP (an average of 3 trials) at the forward and backward limit of leaning, respectively. It can be seen that the longitudinal mean CFP in the quiet standing ranges from 44% to 53% of foot length from the pternion. Consulting the X-ray picture from subject to subject, this area proved to extend in the foot from the anterior-half of the navicular to the central portion of the medial cuneiform bone. In the extreme forward leaning, the CFP moved as far forward as the vicinity of the first proximal phalanx, while in the extreme backward leaning the CFP approached the lower end or posterior margin of the fibula, i.e., around the rear end of the talar trochlea.

Successful EMGs were obtained from the subject muscles except for two cases in the soleus and abductor hallucis, and one case in the medial gastrocnemius. In the quiet standing mentioned above, the soleus in-

Table 1

CFP (% Foot Length) of Upright Stance in Natural Balance,
and at the Forward and Backward Limit of Leaning

Subj.	Natural (%)	Forward (%)	Backward (%)
H.T	52.2 ± 3.71	94	8
H.S	50.9 ± 2.21	87	16
N.M	47.4 ± 4.54	90	12
T.K	52.2 ± 2.98	92	11
M.N	53.0 ± 4.63	86	16
M.S	44.8 ± 2.87	90	12
A.H	48.7 ± 3.18	82	13
Y.K	44.1 ± 2.54	90	13
K.O	47.9 ± 2.42	92	12
W.R	50.2 ± 2.96	89	17

variably exhibited a stable activity of the magnitude of 2 to 3% maximum. In the case where the CFP was located relatively forward as to enter the medial cuneiform, the medial gastrocnemius participated with a slight activity.

On leaning backward from the natural balance position, thus shifting the CFP gradually from the 40% to the 30% area, activity of the soleus became reduced, and eventually disappeared (see Figure 3). The CFP position at which the soleus ceased to discharge corresponded, though with a variation between subjects, to an area in the vicinity of the navicular tuberosity or the transverse-tarsal joint (Chopart's joint). Almost synchronously with the disappearance of EMG in the soleus, the tibialis anterior started to fire and drastically increased its activity with enhancement of the backward leaning. The activity of this muscle in the extreme backward leaning occasionally exceeded 40% maximum.

On the other hand, when leaning forward from the natural balance, the soleus and medial gastrocnemius gradually augmented their activities in the initial stage. Then, on shifting the CFP further forward to an area between 50% and 60%, namely, around the tarso-metatarsal joint or proximal portion of the metatarsals, the abductor hallucis became abruptly active, and enhanced its activity sharply with the leaning. Activity of this muscle in the extreme forward leaning often attained 60% maximum. The medial gastrocnemius and soleus also exhibited in this position remarkable activities above 30% and 20% maximum, respectively. Activity of the lateral gastrocnemius was generally weak under the present experimental conditions, thus attaining only a 5 to 6% maximum level even in the extreme forward leaning. It was noticed in half of the subjects that a considerable activity occurred abruptly in the tibialis

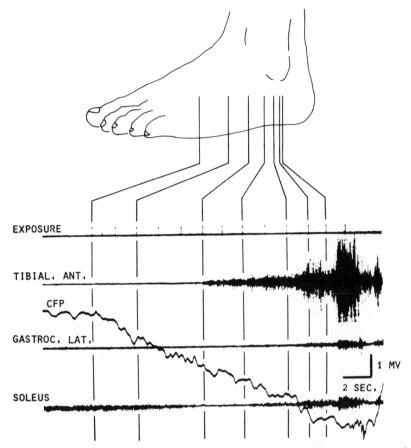

Figure 3—Modification of EMGs as the CFP was shifted backward from the natural balance position. Vertical lines indicate the CFP position and its counterpart on the chart, in which time passes from left to right. Muscle activities observed in the lateral gastrocnemius and soleus in the extreme backward leaning are derived from the vigorous one occurring in the tibialis anterior.

anterior in a position near the forward limit of leaning, namely, when the CFP entered the first metatarso-phalangeal joint, or proximal phalanx.

Table 2 summarizes the CFP positions where each of the muscles commenced or ceased its activity.

Discussion

Since Steindler (1935) reported that the perpendicular from the center of gravity in a standing human subject lies about 4 cm in front of the ankle joint, a number of investigators have reported on the CFP position in

Table 2

CFP Position (% Foot Length)
Where Each of the Muscles Commenced or Ceased Its Activity

Subj.	TA-1 (%)	SL (%)	GCM (%)	ABH (%)	TA-2 (%)
H.T	29	31	52	62	88
H.S	31	29	44	51	—
N.M	27	27	42	56	—
T.K	32	*	49	64	85
M.N	43	42	50	*	73
M.S	32	35	49	55	83
A.H	38	38	47	*	—
Y.K	36	*	57	52	80
K.O	39	39	*	49	—
W.R	44	32	47	48	—
Mean	35.1	34.1	48.6	54.6	81.8
SD	± 5.8	± 5.3	± 4.4	± 5.9	± 5.7

TA-1 = tibialis anterior in backward leaning, SL = soleus, GCM = medial gastrocnemius, ABH = abductor hallucis, TA-2 = tibialis anterior in forward leaning.

standing postures (e.g., Murray et al. 1975). Recently, a bulk of data was presented by Hirasawa (1977) on the CFP (% foot length) of Japanese standing in natural balance. The foot positions assumed in these previous reports, however, are by no means uniform. Further, the time of CFP measurements for each subject as well as identification of the foot structure accommodating the CFP appears to be not necessarily sufficient in these reports. Taking into account the intra- and inter-individual variability of the CFP in natural balance with the closed foot position, we disclosed in this paper that the CFP in this posture varied between the anterior half of the navicular and the central portion of the medial cuneiform bone. This implies that the CFP is accommodated in the mid-portion of the medial longitudinal arch of the foot.

Among recent papers correlating the lower limb muscle activity with the CFP shift, few seem to have analyzed the correlation in terms of the foot structure. Yamaji and Miyake (1980) reported that integrated EMGs from the plantar- and dorsi-flexors of the ankle joint exhibited an excellent correlation with the CFP position in % foot length. The results presented here have elucidated the attitudes of muscles around the ankle joint as related with the foot element accommodating the behavior of the CFP. In Figure 4 is illustrated the activity of individual muscles in one of the subjects. The critical CFP positions in the figure are fairly close to the average position given in Table 2.

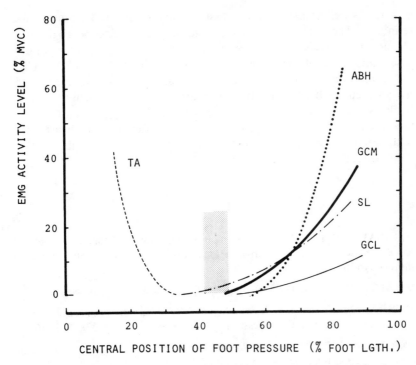

Figure 4—Relative muscle activity for one subject (% maximum contraction) of each muscle as correlated with the CFP position (% foot length). TA = tibialis anterior, GCL = lateral gastrocnemius, GCM = medial gastrocnemius, SL = soleus, ABH = abductor hallucis.

It is obvious from the figure that an area exists between some 30% and 40% foot length from the pternion where muscle activity almost vanishes. Corresponding to the tarsal portion extending from the transverse-tarsal joint to the navicular bone, as mentioned above, this area constitutes the highest portion of the longitudinal foot arch. Human stance in natural balance with the closed foot position, however, is not maintained with the CFP accommodated in this area. Instead, it proved to be maintained with the CFP located somewhat more anterior, thus giving rise to a moderate activity in the soleus and, according to circumstances, also in the medial gastrocnemius.

It may be mentioned here that the activities of the gastro-soleus muscles disappeared when the CFP shifted backward from the natural balance position but still remained in the vicinity of the transverse-tarsal joint, which is considerably anterior to the talo-crural joint. Since this situation should yield a rotational moment to the forward direction in the ankle joint, a cancelling force must be provided for maintenance of balance. Possible elements to provide this moment may be an active pull by those muscles we have not examined, e.g., the tibialis posterior and peronei, as well as a fixation by passive structures like the ligaments and retinuculi. The effect of pronated feet accompanying Romberg's position should not be overlooked for the latter elements in particular. In any case, elucidation of the pertinent element may invite further research.

An additional point of interest is that an alternation of muscle activities occurred between the tibialis anterior and gastro-soleus in the particular situation mentioned above. This appears to suggest that the functional axis of the human foot in dorso-plantar flexion while bearing the body weight lies adjacent to the transverse-tarsal joint rather than within the talo-crural joint.

References

HIRASAWA, Y. 1977. An observation on standing ability of Japanese males and females. Shisei (Proc. 2nd. Symp. on Human Posture), pp. 41-46. Shisei Kenkyusho, Tokyo. (in Japanese)

MORTON, D.J. 1935. The Human Foot. Columbia Univ. Press, New York.

MURRAY, M.P., Seireg, A.A., and Sepic, S.B. 1975. Normal postural stability and steadiness: quantitative assessment. J. Bone and Jt. Surg. 57A:510-516.

OKADA, M. 1972. An electromyographic estimation of the relative muscular load in different human postures. J. Hum. Ergol. 1:75-93.

STEINDLER, A. 1935. Mechanics of Normal and Pathological Locomotion in Man. C.C. Thomas, Springfield, Ill.

YAMAJI, K., and Miyake, A. 1980. Fluctuation of center of gravity and EMG in some postures in man. Rinsho Noha (Clinical Electroencephalography) 22:199-204. (in Japanese)

Sex Differences in Histochemical Properties of the Soleus Muscles of Mice during Growth

Hideki Matoba
Yamaguchi University, Yamaguchi-shi, Japan

Naotoshi Murakami
Yamaguchi University School of Medicine, Ube-shi, Japan

It is generally accepted that males have larger muscle mass and greater muscular strength than females. Contradictory data concerning sex differences in the fiber type composition have been reported, however (Komi and Karlsson, 1978; Jansson, 1980). It seems appropriate to study sex differences in histochemical properties of skeletal muscles using animals ranging widely in age from young to adult, because changes in the fiber type composition and fiber cross-sectional area have been shown to occur during postnatal growth (Kugelberg, 1976). Therefore, the purpose of the present study was to examine sex differences in histochemical properties of the soleus muscles of mice aged five to 40 weeks.

Materials and Methods

ICR strain mice of both sexes were used in this study. They were fed on commercial chow and tap water ad libitum. Animals aged 5, 10, 22, 30, and 40 weeks were sacrificed by an inhalation of ether and weighed. After weighing, the soleus muscles were removed and thrust perpendicularly in the livers. The muscles were then frozen in isopentane precooled with liquid nitrogen. Serial transverse sections of 10 μm thickness were cut through the belly of the muscle with a cryostat and the muscle sections were stained for myosin adenosine triphosphatase (ATPase) and succinate dehydrogenase (SDH) activities. Myosin ATPase activity was demonstrated by the method developed by Padykula and Herman (1955) with slight modifications (Matoba and Murakami, 1981). SDH activity was estimated according to the method of Nachlas et al. (1957). The stained sections were photomicrographed and muscle fibers were classified as fast-twitch oxidative glycolytic (FOG) or slow-twitch ox-idative (SO) type on the basis of myosin ATPase and SDH stains. The fiber type composition was determined by counting the number of fibers of each fiber type in a whole area of the cross-section and expressed as the percentage of SO fibers. The fiber cross-sectional area of each fiber type was calculated by measuring the lesser fiber diameter (Brooke, 1973). Subsequently, the ratio of mean fiber cross-sectional area of SO fibers to that of FOG fibers [the ratio of mean fiber cross-sectional area, SO/FOG] was computed. The percentage of the area occupied by SO fibers to the total area (the percentage area of SO fibers) was calculated by the following equation: the percentage area of SO fibers = [(the percentage of SO fibers × mean fiber cross-sectional area of SO fibers) ÷ (the percentage of SO fibers × mean fiber cross-sectional area of SO fibers + the percentage of FOG fibers × mean fiber cross-sectional area of FOG fibers)] × 100.

Results

The results are presented in Table 1. Males had significantly heavier body weights than females in all age groups except for the group aged 10 weeks, although body weights increased with advancing age in both males and females.

There was little difference in the percentage of SO fibers between sexes 5, 10, and 22 weeks after birth. Females, however, had a significantly higher percentage of SO fibers at 30 and 40 weeks of age. The significant difference in the percentage of SO fibers between sexes observed at later periods after birth resulted from the larger increase in the percentage of SO fibers in females than in males during postnatal growth. As shown in Figure 1, an increase in the percentage of SO fibers was observed throughout the examination period in females, whereas an increase in the percentage of SO fibers in males continued until 22 weeks of age but thereafter it was fairly constant.

A tendency of a higher fiber cross-sectional area of SO fibers than that of FOG fibers was observed in both males and females throughout the period. In addition, the difference in the fiber cross-sectional area of the two fiber types tended to be larger in females than in males. Conse-quently, the ratio of mean fiber cross-sectional area (SO/FOG) was significantly higher in females than in males in all age groups.

Postnatal changes in the percentage of SO fibers and the ratio of mean fiber cross-sectional area (SO/FOG) in both sexes resulted in a signifi-cantly higher percentage area of SO fibers in females than in males in all age groups, as shown in Figure 2. Furthermore, sex difference in the

Table 1

Histochemical Properties of the Soleus Muscles of Male and Female Mice Aged 5 to 40 Weeks

Age (weeks)	Sex	Body weight (g)	% of SO fibers	Fiber cross-sectional area			% area of SO fibers	Total number of fibers
				FOG (μm²)	SO (μm²)	SO/FOG		
5	male	25.3 ± 1.2(n = 7)**	37.4 ± 1.6(n = 7)	564 ± 45(n = 6)	721 ± 64(n = 6)	1.28 ± 0.05(n = 6)**	42.1 ± 1.5(n = 6)***	810 ± 52(n = 7)
	female	20.8 ± 0.7(n = 9)	40.5 ± 1.2(n = 9)	552 ± 20(n = 7)	818 ± 39(n = 7)	1.48 ± 0.04(n = 7)	50.9 ± 1.2(n = 7)	804 ± 27(n = 9)
10	male	31.5 ± 1.6(n = 6)	46.6 ± 2.2(n = 6)	871 ± 34(n = 5)	1002 ± 35(n = 5)	1.16 ± 0.05(n = 5)**	48.7 ± 1.6(n = 5)*	774 ± 27(n = 6)
	female	29.2 ± 1.0(n = 6)	49.9 ± 3.3(n = 6)	747 ± 69(n = 6)	1214 ± 130(n = 6)	1.64 ± 0.12(n = 6)	61.3 ± 4.1(n = 6)	722 ± 34(n = 6)
22	male	42.4 ± 1.3(n = 6)***	49.8 ± 1.8(n = 6)	1087 ± 91(n = 6)**	1229 ± 118(n = 6)	1.13 ± 0.03(n = 6)**	52.8 ± 2.1(n = 6)*	656 ± 39(n = 6)
	female	32.4 ± 0.8(n = 6)	56.4 ± 2.5(n = 7)	809 ± 22(n = 7)	1152 ± 62(n = 7)	1.42 ± 0.06(n = 7)	63.9 ± 3.3(n = 6)	719 ± 46(n = 7)
30	male	43.2 ± 0.8(n = 10)***	48.9 ± 1.1(n = 10)**	1024 ± 51(n = 10)*	1103 ± 74(n = 10)	1.08 ± 0.05(n = 10)**	50.5 ± 1.9(n = 10)***	821 ± 38(n = 10)
	female	33.2 ± 0.5(n = 8)	59.2 ± 2.6(n = 8)	826 ± 50(n = 8)	1082 ± 62(n = 8)	1.32 ± 0.03(n = 8)	65.4 ± 2.6(n = 8)	744 ± 42(n = 8)
40	male	45.9 ± 1.0(n = 10)***	50.0 ± 2.3(n = 10)***	1250 ± 107(n = 9)***	1378 ± 120(n = 9)	1.11 ± 0.05(n = 9)*	50.9 ± 2.5(n = 9)***	766 ± 30(n = 10)
	female	34.1 ± 1.1(n = 10)	65.1 ± 2.2(n = 10)	808 ± 54(n = 10)	1026 ± 67(n = 10)	1.27 ± 0.04(n = 10)	70.1 ± 2.2(n = 10)	701 ± 25(n = 10)

Values are means ± S.E.M.

The number of animals or muscles in each group is given in parentheses.

For calculating fiber cross-sectional area, the lesser fiber diameter was measured on at least 50 muscle fibers in each fiber type per muscle.

*, **, and *** denote $P < 0.05$, $P < 0.01$, and $P < 0.001$, respectively, when compared with females in the same age group.

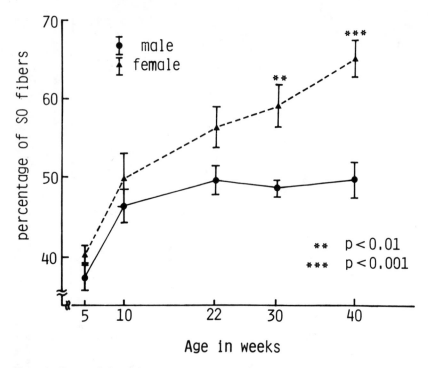

Figure 1 – Postnatal change in the percentage of SO fibers in the soleus muscles of male and female mice. Lines through each point indicate ± 1 S.E.M., ** and *** denote P < 0.01 and P < 0.001, respectively, when means are compared between males and females in the same age group.

percentage area of SO fibers became larger with advancing age.

No significant sex difference in the total number of muscle fibers was observed in each age group.

Discussion

The results indicated that females have a higher percentage of SO fibers than males at later periods after birth, and females also have a higher ratio of mean fiber cross-sectional area (SO/FOG) and a higher percentage area of SO fibers throughout the period examined. These findings imply that the soleus muscle of the mouse is more highly adapted for slow contraction in females than in males.

The mechanism which produces sex differences in histochemical properties is not fully elucidated. Our previous study (Matoba and Niu, 1980), however, showed that males have a significantly lower percentage of SO fibers than females in the soleus muscle of the mouse, and the percentage of SO fibers is increased to the female level by the castration of males

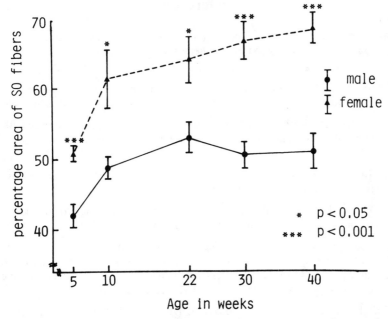

Figure 2 — Postnatal change in the percentage area of SO fibers in the soleus muscles of male and female mice. Lines through each point indicate ± 1 S.E.M., * and *** denote P < 0.05 and P < 0.001, respectively, when means are compared between males and females in the same age group.

and recovered to the control level by testosterone administration to castrated males. Therefore, it can be said that the significantly lower percentage of SO fibers in males than in females at later periods after birth is attributable, at least in part, to the higher testosterone level in males. The finding that the ratio of mean fiber cross-sectional area (SO/FOG) was significantly higher in females than in males in all age groups may reflect genetic differences between males and females. It is also possible, however, that the difference between males and females with respect to the ratio is due to sex difference in the testosterone level or due to sex difference in recruitment patterns of FOG and SO fibers under normal breeding conditions.

In conclusion, sex differences in the fiber type composition were observed at later periods after birth, while sex differences in the ratio of mean fiber cross-sectional area (SO/FOG) and the percentage area occupied by each fiber type were observed throughout the examined period. Testosterone was suggested to be a factor producing sex differences in the fiber type composition at later periods after birth.

References

BROOKE, M.H. 1973. The pathologic interpretation of muscle histochemistry. In: C.M. Pearson and F.K. Mostofi (eds.), The Striated Muscle, pp. 86-122. The Williams and Wilkins Company, Baltimore.

JANSSON, E. 1980. Diet and muscle metabolism in man with reference to fat and carbohydrate utilization and its regulation. Acta Physiol. Scand. Suppl. 487.

KOMI, P.V., and Karlsson, J. 1978. Skeletal muscle fibre types, enzyme activities and physical performance in young males and females. Acta Physiol. Scand. 103:210-218.

KUGELBERG, E. 1976. Adaptive transformation of rat soleus motor units during growth. J. Neurol. Sci. 27:269-289.

MATOBA, H., and Murakami, N. 1981. Histochemical changes of rat skeletal muscles induced by cold-acclimation. Jpn. J. Physiol. 31:273-278.

MATOBA, H., and Niu, N. 1980. The effects of castration and testosterone administration on the histochemical fiber type distribution in the skeletal muscles of the mouse. In: A. Morecki, K. Fidelus, K. Kedzior, and A. Wit (eds.), Biomechanics VII-B, pp. 606-611, University Park Press, Baltimore.

NACHLAS, M.M., Tsou, K., Souza, E. DE, Chen, C., and Seligman, A.M. 1957. Cytochemical demonstration of succinic dehydrogenase by the use of a new p-nitrophenyl substituted ditetrazole. J. Histochem. Cytochem. 5:420-436.

PADYKULA, H.A., and Herman, E. 1955. The specificity of the histochemical method for adenosine triphosphatase. J. Histochem. Cytochem. 3:170-195.

Analysis of Dynamic Force during a
Concentric Contraction in Human Elbow Extensors

Hisashi Aoki, Katsumi Mita,
Kentaro Mimatsu, and Kyonosuke Yabe
Institute for Developmental Research, Aichi, Japan

The characteristics of the dynamic force during a concentric contraction has been investigated on the basis of the force-velocity relation (Hill, 1938). This force-velocity relation was represented by the mean and the maximal values obtained in many trials. Therefore, it could not explain the nature of the dynamic force during a single concentric contraction.

It has been found that the bimodal pattern appeared in the dynamic force curve of a maximal concentric contraction. This bimodal force of a maximal curve was shown by Megaw (1974), but he did not explain it in detail. Therefore the origin and functional significance of the phenomenon are still obscure. The present study was designed to examine the condition for the appearance of the bimodal pattern and to investigate the mechanism of the bimodal dynamic force curve.

Methods

The subjects were five healthy males, ranging in age from 24 to 42 years. Maximal voluntary elbow extension in response to a visual stimulus was adopted as a concentric contraction. The apparatus consisted of an aluminum alloy lever (length of 35 cm) and a steel gear (radius of 11 cm) corrected directly to the level at its axis (see Figure 1A). The weight was suspended by a chain which was mounted around the gear. As the subject pushed the lever, the gear rotated, pulling the weight up. In this system, the force applied to the wrist should be constant throughout every elbow joint angle so that the position of the elbow joint was adjusted that of the gear axis. All the parts of this system were combined so as to minimize the possible compliance.

223

A

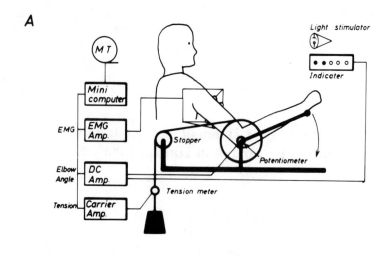

B

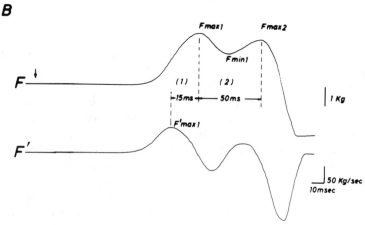

Figure 1 — A = Experimental arrangement. B = Dynamic force curve and its differential curve during maximum elbow extension. F = Dynamic force curve. F' = Differential force curve. $F'max_1$ = the first peak of differential force curve. $Fmax_1$ = the first peak of dynamic force curve. $Fmin_1$ = the first minimum value of the dynamic force. $Fmax_2$ = the second peak of dynamic force curve. (1) = mean time difference between $F'max_1$ and $Fmax_1$. (2) = time betwen $Fmax_1$ and $Fmax_2$. ↓ = light stimulus.

As seen from Figure 1A, the subject was seated so that the elbow joint was flexed to about 60° and the shoulder joint was flexed to about 45°. The subject was asked to maintain this position for 1-2 sec before the visual stimulus (xenon lamp) was given. Thereafter, the subject responded to the visual stimulus by extending the right elbow joint as quickly as possible. The elbow extension ended when the forearm reached the table surface. The dynamic force during elbow extension was calculated from the tension on the chain measured by a strain gauge transducer. The surface EMGs of the triceps brachii and the biceps brachii were recorded simultaneously by means of small surface electrodes (10 mm in diameter). The EMGs were applied for judgment of correct elbow extension.

The dynamic force and EMGs were sampled at the rate of 2000 samples/sec and converted to 12-bit digital signals, which were stored on digital magnetic tape for later analysis. The stored signals were low-pass filtered and differentiated digitally, and further displayed on the CRT (see Figure 1B). Thereafter, the operator discriminated the pattern of the dynamic force curve and measured the maximal values and the minimal values visually. The entire data-taking, display, storage and analysis process were controlled by a programming system called RECMT residing on a NEAC-3200 digital computer (32k core memory).

Results

Appearance of the Bimodal Pattern in the Dynamic Force Curve

Figure 2A illustrates three typical patterns of the dynamic force curve. The upper traces (F) are dynamic force curves and the bottom traces (F') are differential force curves. The pattern of Type I shows a bimodal wave of which the first peak ($Fmax_1$) is greater than the second peak ($Fmax_2$). Type II is a bimodal wave whose $Fmax_1$ is smaller than $Fmax_2$. Type III has a unimodal shape with only one peak. The pattern of the bimodal force curve is defined by the number of the zero-points on the differential force curve and by the value of the two peaks ($Fmax_1$, $Fmax_2$). It was found that there were three zero-points on the differential force curve in Type I and Type II. On the other hand, one zero-point was seen in Type III.

The frequency of the appearance in each pattern with load is shown in Figure 2B. The load was expressed as a percentage of the maximal isometric force. In this way, there was no further influence of the maximal isometric force of the subject.

The mean value of the appearance of the pattern of Type I was 60% with the light load, and it had a decreasing tendency with increasing load. On the contrary, the appearance of Type III increased from 0 to 55% when the load was increased. The percentage appearance of Type II was lower than those of the other patterns.

The Relation Between the Rate of Force Drop and the Maximal Rate ($F'max_1$) in the Bimodal Force Curve

The rate of force drop considered in this study was defined as ($Fmax_1$ − $Fmin_1$)/$Fmax_1$, where $Fmax_1$ was the first peak of the bimodal force

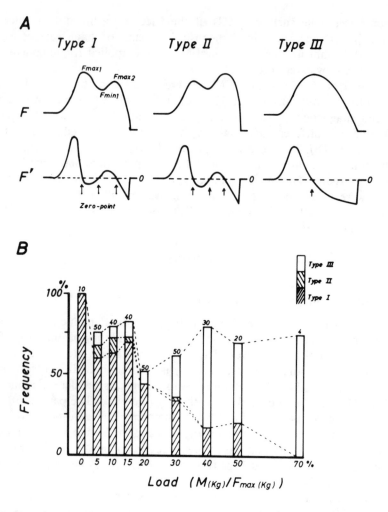

Figure 2—A = Typical patterns of dynamic force curve. ↑ = zero-point of differential force curve. B = Averaged frequency of three patterns appearance with load. Abscissa = Load was expressed as percent of maximum isometric force. Ordinate = Percentage appearance of three patterns.

curve and $Fmin_1$ was the first minimal value of the bimodal force curve. $F'max_1$ represents the slope of the dynamic force curve at which the differential force was maximum. Figure 3 showed an example of the correlation obtained in 100 trials of a subject. The load employed was 15% of the maximal isometric force, because it was the suitable condition for the bimodal force curve occurrence.

As shown in Figure 3A, there was a positive correlation between the

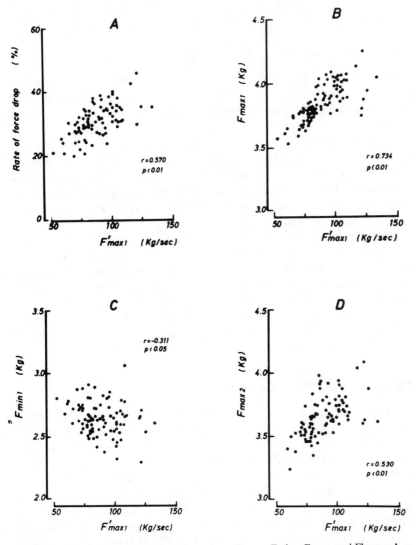

Figure 3—Interrelation of the rate of force drop, F_{max_1}, F_{min_1}, F_{max_2} and F'_{max_1}. A = Correlation between the rate of force drop and F'_{max_1}. B = Correlation between F_{max_1} and F'_{max_1}. C = Correlation between F_{min_1} and F'_{max_1}. D = Correlation between F_{max_2} and F'_{max_1}. Rate of force drop = $(F_{max_1} - F_{min_1})/F_{max_1}$.

rate of force drop and F'_{max_1} ($r = 0.570$, $p < 0.01$). And there were positive correlations between $F_{max_1} - F'_{max_1}$ (see Figure 3B) and between $F_{max_2} - F'_{max_1}$ (see Figure 3D) with a significant correlation coefficient. On the other hand, a negative correlation coefficient was found between $F_{min_1} - F'_{max_1}$ (see Figure 3C: $r = -0.311$, $p < 0.05$).

Discussion

Function of the Synergy of Elbow Extensors

A voluntary maximum elbow extension to visual stimulus was adopted as a concentric contraction. The extension of the elbow joint was caused by the action of the two muscles; the anconeus muscle and the heads of the triceps brachii muscle.

According to Hill's characteristic equation, when a muscle reached its maximum shortening velocity, it did not develop any force so that it can then be considered as "eliminated" (Pertuzon, 1972). Since the maximal shortening velocity was approximately proportional to its length (Wilkie, 1950), the anconeus muscle, whose length was very short relative to that of the triceps, can well be eliminated.

For the maximal elbow extension, the time intervals between the onsets of activities of the three heads of triceps were less than 10 msec (Le Bozec, 1980). However, the time interval between the first peak and the second peak of the bimodal force curve was about 50 msec (see Figure 1B). This time interval was greater than between those of the three heads in triceps brachii. So it was suggested that the delay time between the onsets of the three heads did not take part in the occurrence of the bimodal force curve.

The Visco-elastic Properties of Muscle

A damped oscillation with period of 70 msec (14 Hz) was shown on the twitch of a cat muscle with inertial mass of 300 g and elastic load of 66 g/mm (Bawa et al., 1976). It was difficult to compare the damped oscillation in cat isolated muscle and the bimodal force curve in human intact muscle. It cannot be denied, however, that the visco-elastic properties of muscle might influence an occurrence of the bimodal force curve. If the bimodal force curve could be produced by the damped oscillation which results from the interrelation of the muscle elasticity and the external mass, the bimodal pattern should appear in all trials having the same maximal dynamic force with the same load. It was found that the average frequency of the appearance of the bimodal force curve was 60% with a light load although the peak value of the bimodal force curve was similar to that of the unimodal force curve. From these results, it was suggested that the bimodal force curve was not caused by the damped oscillation resulting from the visco-elastic properties of muscle.

The Inhibitory Phenomenon Caused by Negative Feedback

There was a significant correlation between the rate of force drop and the maximal rate of the bimodal force curve (see Figure 3A). As a result of this significant correlation, it was considered that the maximal rate of the bimodal force curve should influence the occurrence of the bimodal force curve. It was supposed that the force drop was caused by negative feedback loop. Furthermore, the time difference between the latency of $F'max_1$ and $Fmax_1$ was about 15 msec (see Figure 1B). This time difference, which was considered as the latency of the feedback loop, was too short in comparison with the latency of the feedback loop of the upper neural levels (Jones et al., 1978). Therefore, the feedback loop related to the occurrence of the bimodal force curve seemed to consist of lower level elements, e.g., tendon organ, Renshaw cell and reciprocal innervation. Additionally, the bimodal force curve was easily obtained in trials with greater acceleration. It was suggested from these results that the bimodal force curve was caused by negative feedback loop which was sensitive to the acceleration of elbow extension.

Summary

The present study examined the appearance of the bimodal force curve during maximal elbow extension. The results obtained were as follows:

1. The appearance of the bimodal force curve decreased as the load increased. On the contrary, that of the unimodal force curve increased as the load increased.

2. A significant correlation coefficient was found between the rate of force drop and the maximal rate of the force curve.

3. These results implied that negative feedback loop related to spinal reflex was sensitive to the acceleration of the concentric contraction.

References

BAWA, P.A., Mannard, A., and Stein, R.B. 1976. Predictions and experimental tests of a visco-elastic muscle model using elastic and inertial load. Biol. Cybernetics 22:139-145.

HILL, A.V. 1938. The heat of shortening and dynamic constants of muscle. Proc. Roy. Soc. B126:136-195.

JONES, S.J., and Small, D.G. 1978. Spinal and sub-cortical evoked potential following stimulation of the posterior tibial nerve in man. Electroenceph. Clin. Neurophysiol. 44:299-306.

LE BOZEC, L., Maton, B., and Cnockaert, J.C. 1980. The synergy of elbow extensor muscles during dynamic work in man. I. Elbow extension. Eur. J. Appl. Physiol. 44:255-269.

MEGAW, E.D. 1974. Possible modification to a rapid on-going programmed manual response. Brain Research 74:425-441.

PERTUZON, E. 1972. La contraction musculaire dans le mouvement volontaire maximal. Thèse. Doctorat. d'Etat, Université Lille, vol I, 1, p. 208 (from Le Bozec et al., 1980).

WILKIE, D.R. 1950. The relation between force and velocity in human muscle. J. Physiol. 110:249-280.

Correction of Error Reaction
Caused by Feinting Stimulus in
Repeated Wrist Extension and Flexion Movement

Tatsuyuki Ohtsuki and Shoko Kawabe*
Nara Women's University, Nara City, Japan

In many sports such as basketball, soccer, badminton, tennis, and fencing, a quick change of movement is one of the most important factors for good performance. Especially, when a player is "feinted" by his opponent, he must correct his error movement as quickly as possible in order to minimize the effect given by the "feint." The present study was designed to determine the minimum time necessary for changing an ongoing movement into a new movement.

Methods

Subjects

Six normal healthy female university students served as subjects (22 to 24 years of age). They were all right-handed in writing.

Apparatus

Two small lamps (9 mm in diameter), 80 cm apart horizontally, were mounted on a board located about 1 m in front of the subject's face. The subject's right hand, with the palm turned to the midsagittal plane and the fingers and thumb extended in parallel, was held immobile by a hand holder. The hand holder could rotate horizontally around the subject's wrist joint so that the palmar flexion of the wrist produced the leftward rotation of the hand holder. The angle of joint rotation was recorded by

*Now at College of Liberal Arts, Kobe University, Tsurukabuto 1 chome, Nada-Ku, Kobe City, Japan.

the electrogoniometer attached to the axis of rotation of the hand holder. An aluminum disc of 90 mm in diameter and 10 mm in thickness was mounted horizontally under the hand holder. The center of the disc coincided with the axis of rotation of the hand holder. A steel wire was wound around the disc and fastened to the border of the disc at its midpoint and the two free ends of the wire were connected to force transducers made of steel bars on which paper strain gauges were bonded. Each force transducer was connected to the fixed structure by means of a spring to measure the strength of extension and flexion separately. The strength of the spring was such that 0.7 kp was necessary to stretch the spring by 1 cm. EMGs were recorded from the flexor and extensor muscles of the right forearm by bipolar surface electrodes.

Procedures

Each subject was instructed to extend (or flex) her wrist as fast as possible when the right (or left) lamp turned on, and in case of an error response, correct it as quickly as possible. The required amplitude of response was defined as the maximum angle of the hand holder rotation (extension 70°, flexion 80°).

One bout of experiments was performed in the following way: two stimulus lamps were alternately turned on at a given time interval for some time as right-left-right-left. These stimuli were called regularly alternating stimuli (RAS). The right lamp indicated extension and the left lamp flexion. And then, in an unforeseen moment between the 5th and 10th trial inclusive, one lamp was lighted twice consecutively followed by the original right-left alternation for at least two trial periods, e.g., as R-L-R-*R*-L-R. The second of the two consecutive stimuli was termed the "feinting stimulus" (FS) in the present experiment. Two or three FS were given in each series of stimuli. The second or third FS occurred between the third and tenth trial after the preceding FS. Four stimulus intervals, 0.5, 1, 1.5, and 2 sec, were employed. Eight series were given for each subject so that 10 FS were given for each extension and flexion at each stimulus interval. In addition to this regularly alternating stimulus condition, simple and choice reaction times to the single stimulus were measured. In the simple reaction task, one lamp (right or left) was turned on at random stimulus intervals between 3 and 15 sec. Twenty-two trials were performed for each extension and flexion. In choice reaction, either the right or the left lamp was turned on at random stimulus intervals between 3 and 15 sec. Twelve extensions and twelve flexions were performed in a random order.

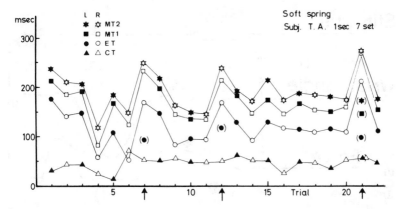

Figure 1—Change of reaction times (ET, MT1, and MT2) and CT in one stimulus series (stimulus interval, 1 sec). See text.

Results

Reaction Time for the Regularly Alternating Stimulus

Figure 1 shows the typical trend of reaction times measured in three ways. ET is a latency from stimulus onset to the appearance of EMG. MT1 indicates the latency of change in force applied to the force transducer. MT2 means the latency of rotation of the hand holder (wrist joint). CT is the time lag from EMG onset to the force application.

After a few trials, reaction times shortened, e.g., in terms of ET, from 180 msec in the first trial to 50 msec in the sixth trial. But in response to the FS (↑) the reaction times increased. For extension, the mean ± SD values of ET for the RAS obtained from the trials preceding the FS trial were, for example, 146.5 ± 44.0, 118.0 ± 43.9, 108.2 ± 43.0, and 146.4 ± 98.7 msec for 2-, 1.5-, 1-, and 0.5-sec stimulus intervals, respectively. For flexion the corresponding values were 163.0 ± 34.1, 149.3 ± 42.8, 133.6 ± 34.6, and 155.8 ± 166.0 msec. ETs of simple and choice reaction were 198.5 ± 38.8 and 243.3 ± 40.8 msec, respectively for extension, and 197.3 ± 30.6 and 237.9 ± 40.7 msec for flexion. Thus, ETs for the RAS were much shorter than the simple reaction. MT1 and MT2 showed the same trend. On the contrary, CT was about 40 msec for all conditions.

Reaction Times for the Feinting Stimulus

The mean ± SD values of ET for the FS in extension were 208.7 ± 63.2, 206.5 ± 70.4, 176.6 ± 65.3, and 282.3 ± 113.6 msec for 2-, 1.5-, 1-, and 0.5-sec stimulus intervals, respectively, and 233.1 ± 44.3, 231.2 ± 62.3,

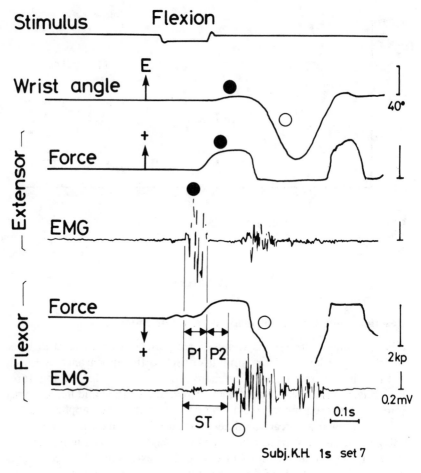

Figure 2—Typical example of correction of error reaction caused by the feinting stimulus. Calibrations are the same for the extensors and flexors. See text.

221.5 ± 68.0, and 312.2 ± 82.7 msec in the same order for flexion. They were about 70 to 150 msec larger than those for RAS.

Switching Time

As shown in Figure 1, for the feinting stimulus not only an increase in reaction time but also movements or indications of movement (force change or EMG) in the direction opposite to that of the lamp stimulus were often observed. This occurred in 35.6% of the total number of FS. Error reaction times are plotted in Figure 1 in parentheses. It is obvious that the error reaction times are the same as the reaction times for the RAS. Figure 2 shows a typical example of the reaction for the FS. For the stimulus indicating flexion, error reaction (extension, ● in Figure 2) occurred and then correction was immediately made to produce correct reaction (○). The time difference from error EMG to correct EMG is termed the "switching time" (ST in Figure 2) after Ohtsuki and Ishiwari (1978). The intensities of error reaction were classified into four types as follows:

I. Only EMG appeared in the antagonist muscle.

II. EMG appeared, and force also changed.

III. EMG plus force change appeared, and the wrist joint rotated within 20°.

IV. The wrist joint rotated by 20° or more.

Figure 3 (number of observations is indicated above ∞).

As is indicated also in Figure 2, the time period from the disappearance on, i.e., error reaction was extension and switching was from extension to flexion. Open circles represent switching from flexion to extension. Mean switching time was 100 to 120 msec for intensity I, 140 to 150 msec for II, 160 msec for III, and 280 to 340 msec for IV. For intensity IV the error movement was sometimes so intense and complete that the subject could not perform the correct reaction. This case is represented by ∞ in Figure 3 (number of observation is indicated above ∞).

As is indicated also in Figure 2, the time period from the disappearance of error EMG to the appearance of correct EMG was electrically silent (in 91.8% cases of the total number of error reactions). Therefore, the switching time was divided into two phases: the first phase (P1) is the duration of error EMG, and the second phase (P2) is the silent period. Figure 4 shows the P1 and P2 components of switching time in relation to the intensity of error reaction. P1 increased from 50 to 60 msec (intensity I) to 230 to 250 msec (intensity IV) in proportion to the intensity of error reaction. On the contrary, most of mean P2s ranged between 40 and 60 msec and did not show a clear increase as compared to P1.

Increase in Reaction Times to the Feinting Stimulus When No Error Reaction Occurred

Even when no error symptoms appeared in the antagonist muscles, the correct reaction was considerably delayed by the FS. The difference between the reaction time for the FS in such cases and that for the RAS immediately before the FS was 58.6 ± 63.6 msec (mean $\pm$ SD) for extension, and 61.9 ± 67.3 msec for flexion.

Discussion

The present result that the reaction times for the regularly alternating stimuli were shorter than simple reaction times clearly indicates that the

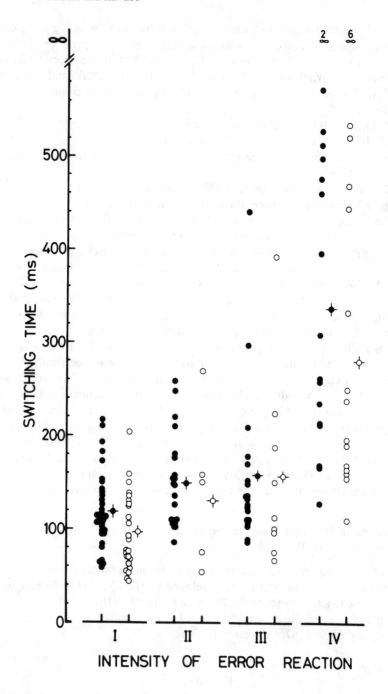

Figure 3—Values of switching time at different intensities of error reaction. ✦and ✧ represent mean values.

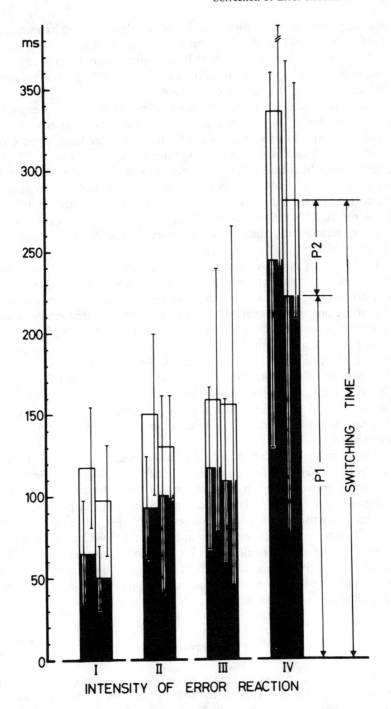

Figure 4—Mean ± SD values of P1 and P2. Left bars represent switching from extension to flexion. Right bars, from flexion to extension.

voluntary reaction to a visual stimulus is greatly shortened by establishing the anticipation for the direction and timing of the stimuli in the subject's mind. This is also confirmed by the appearance of error reaction. This result coincides with Ohtsuki and Ishiwari (1978) who employed alternating right-left stepping of the foot.

Switching time calculated as the time difference between the error ET and correct ET in the present experiment was about 100 msec when the error reaction appeared only as EMG, but the more intense the error reaction, the longer was the switching time up to over 300 msec. This increase in switching time was brought about by the increase in P1, i.e., the duration of error EMG, which increased from about 50 to 250 msec in proportion to the intensity of error reaction. On the other hand, the silent period (P2) with no electrical activity in the agonist or in the antagonist muscles was almost always observed, and its duration did not change in relation to the error intensity.

The switching of movement consists of two processes, i.e., 1) canceling of the ongoing motor programs, and 2) issuing new motor commands. The present finding that new movement cannot begin until the cancellation of the ongoing movement has been perfectly inhibited indicates that these two processes of switching do not occur independently of each other but with some temporal relations.

Even when no error reaction appeared, the 60 msec delay in reaction (ET) for the FS was observed. This delay is close to the P2 values in Figure 4. Thus, this is probably a special case of error reaction in which the duration of error EMG was reduced to zero. Therefore, this delay time may be regarded as the minimum value of the switching time in the human brain.

It is interesting from the biomechanical viewpoint that there exists the time period (P2) in which no voluntary effort (biological process) can control the movement even in an urgent change of direction of maximal dynamic movement with large momentum (mechanical phenomenon). The more intense the ongoing movement, the larger is the inertia of the body. Therefore, the time necessary to change the direction of movement would be more and more elongated for the intense movement even if the muscles are activated as early as in the less intense movement. In this sense, the P2 of the switching time may be given the name of "true" dead time during the processing of voluntary movement.

References

OHTSUKI, T., and Ishiwari, M. 1978. "Dead time" in reaction to unexpected changes of situation with reference to "feinting skill." In: E. Asmussen and K. Jørgensen (eds.), Biomechanics VI-A, pp. 165-171. University Park Press, Baltimore.

Biomechanical Analysis
of Muscle Output at Various Speeds

Masayasu Suzuki, Akira Kijima,
Kazunori Sagawa, and Kihachi Ishii
Nippon College of Health and Physical Education,
Tokyo, Japan

In walking or running, velocity is composed of two components, step length and step frequency. On the other hand, jumping is composed of only one component, frequency. Walking, running, and jumping have been analyzed by many researchers but few studies of simpler movements have been conducted.

In this study, the movement of the full squat was modified. This modification involved full flexion of the knee but with the heels remaining in contact with the floor, with hands on the hips. The freedom of the movement of this full squat was relatively simple compared with either walking or jumping, because the exercise of the full squat was performed with the feet in a stationary position. Therefore, the full squat was studied as a one-dimensional movement, along the vertical axis.

In this study, the human body was dealt with as a rigid segment. The purpose was to calculate the mechanical power of the vertical component of the center of gravity, to determine the relationship between the force and the velocity of the center of gravity, and to try to assess the characteristics of the output of this fundamental movement of the human body.

Procedures

Four frequencies (30, 40, 50 and 60 times per minute) of the full squat which were controlled by a metronome were performed by the subjects. The subjects were nine healthy male college students with mean height of

169.0 ± 3.69 cm, and weight of 63.9 ± 5.44 kg. It was assumed that these subjects had the average physical characteristics of contemporary young Japanese men. Each subject performed the frequencies of the full squat on the force platform (Kistler) but only the vertical component of the ground reaction force was recorded by an electronic visiograph (San-ei 5MIIB) and at the same time, the force was integrated by an integrator (San-ei Type 1310) to obtain a velocity curve. For the calibration and correction of the integrated velocity curve, a 16 mm high speed camera was used to film the full squat performance of each subject (50 f/sec). Each subject had an electrogoniometer attached at the knee joint which allowed the matching of each phase of the ground reaction force with the velocity during the full squat. Output signals from the goniometer were amplified and placed along the trace of the vertical components of the force and the velocity. Analyses of the 16 mm films were done on a Nac film motion analyzer (GP 2000). The coordinates of the center of gravity obtained by the motion analyzer were finite-differentiated by 50 msec, and smoothed.

Results

Because there were no significant differences among the maximum angles of knee flexion during each frequency, it was assumed that almost the same movement was performed in each squat.

Mean curves of force, velocity and power obtained from the nine subjects are shown in Figure 1. The upper curves show mean force curves for each frequency. The force curve was divided into levels of body weight for each subject, namely, the upward phase and the downward phase. Comparing the peak forces, the greater the frequency of the squat, the greater the force. The velocity-time relationship is shown on the middle curves in Figure 1. The velocity in the downward phase was negative, in the upward phase positive. With the increase in the frequencies of each full squat, the peak velocity was expressed as a higher value in each force curve. The lowest curves were the mechanical power-time relationships which showed four peaks in one cycle of the full squat: the first peak was the positive power in the downward phase (A phase), which was the work done due to gravity; the second was the negative power in the downward phase (B phase), which meant that the muscles resisted movement caused by gravity; the third peak was the positive power in the upward phase (C phase), which was the force of resisting the work done by the muscles; the last was the negative power in the upward phase (D phase), during which the center of gravity of the human body had a negative acceleration.

Force (F), velocity (V), and power (P) are shown in Table 1 for the four peaks of each phase, A, B, C, and D. With the increase in the frequencies of the full squat, mechanical power increased.

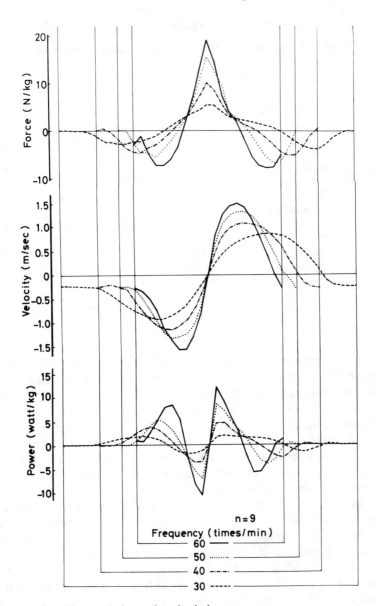

Figure 1 — Mean force, velocity, and mechanical power curves.

Discussion

From the results obtained, it was concluded that the power exerted depended upon the frequency of each squat, and instantaneous power was influenced by the force and velocity which was composed of the product of the two factors. As mentioned already, power in each cycle

Table 1

Absolute Value of Force, Velocity, and Power to the Four Frequencies

Phase		Frequency (times/min)			
		30	40	50	60
A.	Force (N/kg)	−2.10	−4.04	−5.48	−6.03
	Velocity (m/sec)	−0.83	−0.93	−1.00	−1.38
	Power (watt/kg)	1.95	3.74	5.35	8.35
B.	Force (N/kg)	2.87	4.75	10.43	12.78
	Velocity (m/sec)	−0.64	−0.75	−0.67	−0.81
	Power (watt/kg)	−1.79	−3.66	−7.16	−10.66
C.	Force (N/kg)	4.06	5.77	12.58	14.53
	Velocity (m/sec)	0.46	0.80	0.66	0.81
	Power (watt/kg)	2.01	4.69	8.62	12.02
D.	Force (N/kg)	−3.48	−4.15	−5.31	−5.03
	Velocity (m/sec)	0.48	0.63	0.77	1.18
	Power (watt/kg)	−1.51	−2.73	−3.91	−5.88

n = 9

during each frequency was divided into four phases (A, B, C, D), the increment ratio of the value of F, V, P, and $\sqrt{P}$ calculated for each frequency was compared with the value for the 30-times/min exercise which was expressed as 100%, where $\sqrt{P}$ revealed the contribution of F and V to P (see Figure 2).

The A phase was the part of the downward phase which included the time from the start of the decrease in the ground reaction force, to the recovery of the force to the level of body weight. The B phase was the latter part of the downward phase which was the interval from the level of body weight to the peak value. F and P increased largely with an increase in the frequency. On the other hand, the change of V was not very large. As a result, in this phase the increment of F had a more important influence on increase in P. The C phase was the first part of the upward phase, from the full squat position to a recovery of the ground reaction force to the level of body weight while the force was decreasing. In this phase, the work done by the muscles was positive, resisting the gravitational force; however, in the B phase the work done by the muscles was negative in order to sustain the body weight. P in this phase increased linearly with the increase in the frequencies, and the percentages of the increases in P were much like those in the C phase. The D phase was the finish of the upward phase, in which as the frequency of the full squat increased, V showed gradually higher values and at the frequency of 60 times/min, V reached its highest value.

With such a fundamental movement of the human body, even though

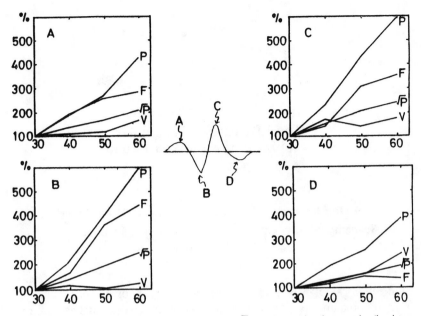

Figure 2—The ratio (vertical axis) of F, V, P, and $\sqrt{P}$ with increasing frequencies (horizontal axis), where $\sqrt{P}$ revealed a similar contribution of force and velocity to mechanical power.

it was simple, according to the results of the analysis, the P, F, and V of the center of gravity in the human body showed complex changes. The fact that P always increased with an increment in the frequency indicates the role of F and V in determining P. The value of F and/or V were shown to vary in the A, B, C, and D phases, respectively (see Figure 2). From the results obtained, it is assumed that V played an important role in the downward phase; on the other hand, F played a larger role during the upward phase in the development of power.

References

CAVAGNA, G.A., and Margaria, R. 1966. Mechanics of walking. J. Appl. Physiol. 21:271-273.

CAVAGNA, G.A. 1975. Force platforms as ergometers. J. Appl. Physiol. 39:174-179.

FUKUNAGA, T., and Matsui, A. 1980. Effect of running velocity on external mechanical power output. Ergonomics 23: 123-136.

MCLAUGHLIN, T.M., Lardner, T.J., and Dillman, C.J. 1978. Kinetics of the parallel squat. Res. Quart. 49:175-189.

THYS, H., Faraggiana, T., and Margaria, R. 1972. Utilization of muscle elasticity in exercise. J. Appl. Physiol. 32:491-494.

The Efficiency of Muscular Stabilization in the Wrist Joint

Stefan Kornecki, Jerzy Zawadzki and Tadeusz Bober
Academy of Physical Education, Biomechanics Laboratory,
Wroclaw, Poland

Biomechanical investigations of muscles mainly involve strength, magnitude of force and its transient performance rather than mechanical factors of the exertion of force on external objects. The latter is closely related with the movement technique (Luhtanen and Komi, 1978; Hay et al., 1980). The important feature of the force production process is the dual role of muscles which simultaneously exert directional force F_D on a given object and control the needless degrees of freedom. Previous studies showed that fixation of one movable link in the system by the stabilizing force results in a 28% drop in directional force ($\triangle F_D$) compared with the systems requiring no stabilization (Kornecki et al., 1981; Bober et al., 1981). The studies dealt with an upper extremity pushing task, and since the wrist joint was found to be the main point requiring stabilization it was assumed that the drop in force was due to this joint. Generally, if balance of such a system is to be maintained the body activates the neuromuscular feedback loop. The present study examined the functions of this loop and sought to increase the efficiency of force production in an unstable system. To achieve this a special practice was applied. It was assumed that the practice should improve the proprioceptors as a key link of the neuromuscular system in such a task.

Material

Two experiments were carried out. Experiment I involved a twelve-day practice period with nine 22-year-old students whose mean height was 178 cm and whose weight was 76 kg. Thirteen male subjects performed in Experiment II. Their average age was 28 years, height 176 cm, and weight 78 kg.

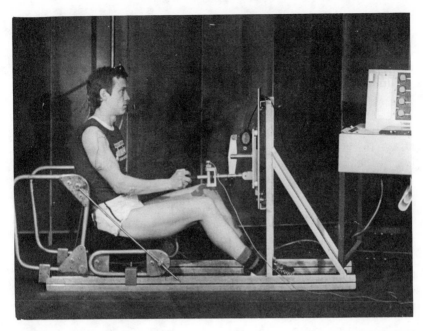

Figure 1—The test apparatus (external system) with subject in a testing position. The subject keeps the handle of external system stable with changeable movability measured by a strain gauge and elgon. EMG is 8 channel RFT 8EEG-III; Filters, Low 0.1 Hz and High 2000 Hz. An amplifier and recorder complete the instrumentation.

Method

Throughout the studies the special device, the so-called 'external system' shows in Figure 1 with two replaceable links of 0 and 1° movability was used (Kornecki et al., 1981). The instrumentation enabled synchronous measurements of the reaction force to the push F(t), the deflection of the moving link from equilibrium $\varphi(t)$, and EMG data to be carried out. The electrical activity of muscles was determined as a percentage amplitude of the maximum static contraction. The anterior fibers of the deltoid muscle were selected and surface electrodes placed after palpation of the greater tubercle of the humerus. The electrodes for the triceps were placed over the fibers of the lateral head. The flexor and extensor muscle group electrodes were located over the largest muscle masses 10 to 15 cm distal to their respective epicondyles of common origin. For testing and practice, the subject was seated in the frame in an upright position with the arm adducted and flexed approximately 20°. An angle of 110° was maintained at the elbow joint. The subject grasped the handle of the external system with the hand in medial-rotation of 90°.

The conditions of the experimental design were as follows: 0 movability for the stable external system, and 1°H and 1°V, i.e. one degree movability in horizontal and vertical planes, respectively, for the unstable system.

Variation of force produced: F_{max} = a push with maximum force, and F_{80}, F_{60}, F_{40} = pushes at respective percentages of maximum force. Parameters measured:

F_D - the reaction force exerted on the external system [N],
φ - the deflection of the mechanical link [°],
U_i - the instantaneous activity of muscles [μV],
U_{max}- the maximum activity of muscles in static condition [μV].

Experiment I

The actual experiment lasted 12 days, with an intermediate and pre- and post-examining days for testing. Three ten-trial series were performed on the 1°H link at maximum push. Each of the 5-sec trials was followed by a voluntarily regulated break while the intervals between the series were 3 to 5 min.

Experiment II

During this experiment subjects were similarly tested in a pushing task on the external system with stable and unstable (1°H) links. In each condition the task was performed with and without visual control.

Results and Discussion

The practice applied was meant to increase the efficiency of force in the upper extremity pushing task (Experiment I). In fact a 35% increase of force production was obtained (see Table 1). Such a result can be explained by a relatively low force figure at the beginning of tests which was improved upon via skill development. The increase was proportional for both stable and unstable external systems, which means that the efficiency expressed by $\triangle F$ remained virtually unchanged. Deflection of the external link ($\triangle \varphi$) resulted in diminished force ($\triangle F_D$) (see Figure 2). If these effects were only mechanical in nature then the function $\triangle F_D = f(\triangle \varphi)$ would be represented by curve No 6. Empirical curves (Numbers 1 to 5) suggest that there are also some other reasons such as activation of the neuromuscular feedback loop with the discrimination threshold $\triangle \varphi$ min = 2 to 6°.

The lack of evident improvement in efficiency leads to the conclusion that the practice did not influence the process of neuromuscular regula-

Table 1

Directional Force of the Maximum Push (F_{Dmax}) for Stable (0) and Unstable (1° of Movability) External System and the Relative Differences ($\triangle F_D$)
(Groups 3 to 6 are from Kornecki et al., 1981)

No	Group	n	0 [N]	1°V and H [%]	$\triangle F_D$
1	Experiment I	9			
	before practice		363	266	27
	after 6 days		464	339	27
	after 12 days		496	356	28
2	Experiment II	13			
	with visual control		525	350	33
	without visual control		524	358	32
3	Judo athletes	27	456	341	25
4	Acrobats	9	456	341	25
5	Boxers	31	370	287	22
6	Students	20	342	259	24

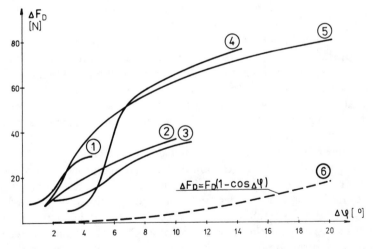

Figure 2 — Instantaneous drop of directional force ($\triangle F$) as a function of deviations of external system ($\triangle\varphi$) in condition 1°V and H. Empirical examples subject (1 to 5). Theoretical drop due to mechanical deviations (6).

tion. We can suppose that in order to keep given segments of the body in a state of equilibrium, such as posture, the regulatory system becomes autonomous, and to some extent involuntary. This kind of regulation is attributed to the lower level of the central nervous system, i.e. The cere-

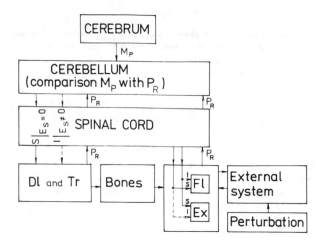

Figure 3—Hypothetical neuromuscular control ring of the stabilizing process in the wrist. M_p = motoric program, S = stimulation, I = inhibition, E_s = error signal, P_R = proprioception.

bellum and the grey substance of the spinal cord. On this level, motions are accomplished beyond telereceptors as feedback loop links, a conclusion which has been supported by the results of Experiment II—There is no difference between the tests with or without visual control (see Table 1).

The frequency of deflections at 5 to 7 Hz also suggests an involuntary nature of the stabilizing function of muscles. In cases of conscious corrections the frequency of stabilizing actions would be 1 to 3 Hz, depending on the mass involved (Morecki, 1976). The above leads to the conclusion that the stabilizing process for the wrist joint under a given motor task bears some resemblance to involuntary action, yet being autonomous it is not simple at all.

The likely links of the feedback loop (see Figure 3) directing these processes are proprioreceptors (tendon and muscle receptors). The deflection of the external link from equilibrium stretches a certain group of muscles in the forearm. Lidder and Sherrington (after Best and Taylor, 1959) stated that a stretch of as little as 0.8% of a muscle's actual length is sufficient to activate the muscle's receptor. The signal from this receptor evokes contraction of the stretched muscle and inhibits activation of the antagonists. The tendon's receptor is activated while the muscle shortens. The next links of this neuromuscular loop are afferent pathways, the spinal cord, and the nuclei of the central nervous system. The afferentation probably takes place through the tractus spino-cerebellaris anterior and posterior to the cerebellum, and the motor response comes from the red nucleus via the tractus rubro-spinalis. It is likely that the response is not limited just to activation of one group of the forearm muscles and the inhibition of another. We must remember

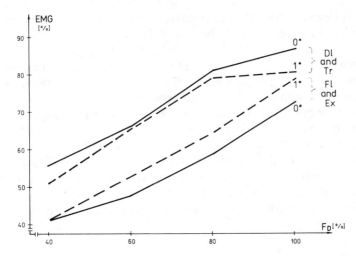

Figure 4 – EMG activity for summed deltoid and triceps and summed flexor and extensor group of forearm for 0 and 1° of movability for all three testing days in respect to force produced.

that the forearm muscles tend to balance the wrist joint under great tension produced by the triceps brachii and the synergy of the deltoid. In the studied case it led to the required drop in contraction of the triceps and deltoid, and then to the diminishing of the force acting on the external system via the wrist joint. This, in turn, enabled the stabilizing muscles of the forearm (Fl and Ex) to balance the system; however, the process was momentary and recurrent. At any rate, the total level of the acting force (F_D) was always lower in an unstable system than in a stable one.

The previously discussed results also correspond with the effect of plaster of Paris applied to the wrist joint. Such immobilization enables the subject to reach an even higher force figure when acting on an unstable external system than on a stable one but without a band (Kornecki et al., 1981).

If we assume that the electrical activity of a muscle corresponds with the muscle's acting force, then some additional evidence on tension controlled directional muscles is found in EMG data (see Figure 4). There is an almost linear dependence between EMG and the force produced over the full range. In addition, a lower activity in directional muscles (Dl and Tr), and a higher activity in the stabilizing muscles (Fl and Ex) were observed while changing the task from stable to unstable systems.

Conclusion

The nervous system, even in its simplest structures, contains certain logical elements that allow for control of the biomechanical parameters in order to efficiently fulfill the aim of the motor task. The criterion of economy can be expressed by the efficiency index ($h\eta$) as a ratio of the utilized force (F_u), the force exerted under standard conditions, and the potential force (F_p) (Bober, 1981). This ratio tends to unity. However, we cannot say that the difference between F_u and F_p denotes wasted force. Probably, due to neuromuscular control of motion, muscles do not exert the maximum force under such conditions that do not allow its full utilization. Because of nervous processes underlying the control of motor tasks this index can be termed the biomechanical efficiency index. From our tests it follows that for a system requiring stabilization of 1° of freedom the index assumes values between 0.6 and 0.8. We do not presume that the optimum value of the biomechanical efficiency may reach unity.

References

BEST, C.H., and Taylor, N.B. 1959. Fizjologiczne podstawy postepowania lekarskiego (Physiological basis of medical practice) PZWL Warszawa. pp. 1123-1153.

BOBER, T. 1981. Biomechanical Aspects of sports technique. In: A. Morecki and K. Fidelus (eds.), Biomechanics VII-B, pp. 497-506. PWN Warszawa.

BOBER, T., Kornecki, S., Lehr, R.P., Jr., and Zawadzki, J. 1981. Biomechanical analysis of upper extremity stabilization during force production. (unpublished paper)

HAY, J.G., Andrews, J.G., and Vaughan, C.L. 1980. Some biomechanical aspects of strength training. A paper presented at the International Symposium on Biomechanics of Sport, Cologne, West Germany.

KORNECKI, S., Zawadzki, J., Bober, T. 1981. Effect of forced stabilization on magnitude of revealed muscle force. In: A Morecki and K. Fidelus (eds.), Biomechanics VII-B, pp. 105-111. PWN Warszawa.

LUHTANEN, P., and Komi, P.V. 1978. Segmental contribution to force in vertical jump. Europ. J. Applied Physiol. 38:181-188.

MORECKI, A. 1976. Manipulatory bioniczne (Bionical manipulators) PWN Warszawa. pp. 140-152.

Function of Limb Speed on Torque Patterns of Antagonist Muscles

L.R. Osternig, J.A. Sawhill, B.T. Bates, and J. Hamill
University of Oregon, Eugene, Oregon, U.S.A.

Athletic injuries to certain muscles in the body occur with greater frequency than to others. A number of hypotheses have been put forth to account for such phenomena. One suggestion is that an imbalance between antagonist muscles causes the weaker muscle group to be more vulnerable to stress. While strength relationships between agonist and antagonist muscles have been determined isometrically, little is known about the effect of movement speed on these relationships. The purpose of this study was to determine the dynamic torque relationships between contralateral antagonist muscle groups at various speeds of motion.

Method

The maximum dynamic torque values of contralateral hamstring/quadriceps muscle groups of 20 male and female college athletes were tested on an isokinetic dynamometer. The subjects were tested at eight velocities ranging from 50 to 400°/sec throughout a range of 120 to 0° of knee flexion.

A position potentiometer was adapted to an Orthotron isokinetic dynamometer to supplement the existing force transducer and provide simultaneous recordings of muscular torque and joint range of motion. The transducer and potentiometer which instrument the isokinetic system were powered by direct electrical current. The input signal was modulated by these variable resistance devices and the output was calibrated to represent values of torque and angular position. Changes in muscular force and/or position transmitted a variable analog electrical signal to a Tektronix 4051 graphics computer via a TransEra analog-to-digital converter. Data were sampled at 100 Hz.

Flexion/extension torque was measured with the Orthotron positioned so that the axis of rotation coincided as closely as possible with the knee's frontal axis according to the procedure of Lindahl et al. (1969). It should be noted that since the knee is not a true ball-and-socket joint, the thigh-leg angle may not have precisely reflected the true tibio-femoral angle. Kettlekamp and associates (1970) studied human knee motions utilizing electrogoniometers and reported potential differences existed between actual knee motion and the motion of the goniometer arms. Errors of up to 2° were found during extension and flexion and were attributed to difficulty in the placement of the goniometer with respect to the true knee axes.

The subject was seated on an exercise chair with both legs hanging freely over the edge. The chair back formed an angle of 85° to the seat. Two straps were used to stabilize the buttocks and thighs of the subject. The lever arm against which the leg was extended and flexed was positioned at a point 12 inches below the axis of knee rotation. Voluntary maximum isokinetic forces were recorded for knee flexion/extension at eight velocities ranging from 50 to 400°/sec. The measured angles of knee flexion/extension were based on a range of knee extension with full extension equalling zero degrees.

Results

The results revealed contralateral limb symmetry for both men and women in flexor/extensor peak torques, antagonist torque ratios and peak torque joint position at all speeds tested (see Figures 1 and 2). No significant differences were found between left and right limbs for any of the parameters tested. Mean absolute differences between limbs were less than 5% for a given individual for all parameters throughout the range of motion.

As the speed of motion increased, hamstring/quadriceps torque ratios increased from 0.58 at 50°/sec to an average of 0.77 at 400°/sec. These increases were more dramatic in the women than in the men (see Figure 3).

The angle at which peak hamstring torque occurred also shifted from 32° to 61° from 50 to 400°/sec, respectively. Similarly, peak torque joint position for the quadriceps also changed from 87 to 63° from 50 to 400°/sec. As the speed increased, the optimal joint positions for flexor/extensor torque generation tended to converge near the 60° position (see Figure 4).

As the speed setting of the dynamometer was increased, the time for the limb to reach dynamometer speed increased linearly. The time to engage the speed of the dynamometer and the extent of constant velocity of performance were examined by computing the derivations of the posi-

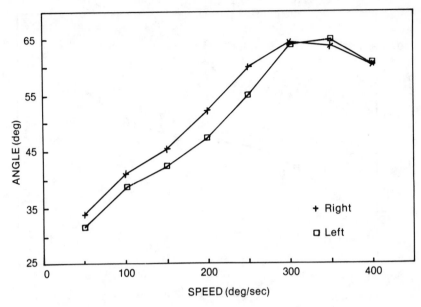

Figure 1—Contralateral knee angle position at which peak flexor torque was generated at different speeds.

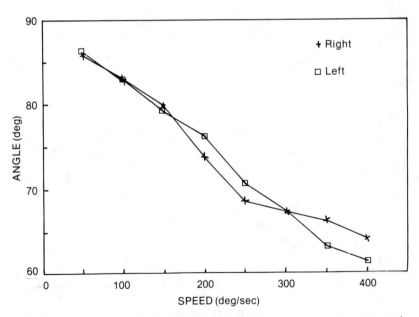

Figure 2—Contralateral knee angle position at which peak extensor torque was generated at different speeds.

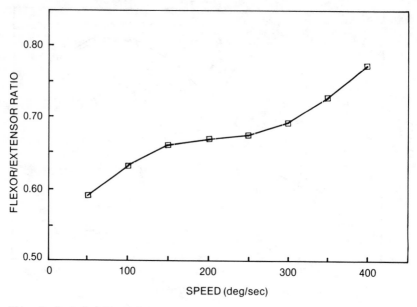

Figure 3—Peak flexor/extensor torque ratios in relation to changes in exercise speed.

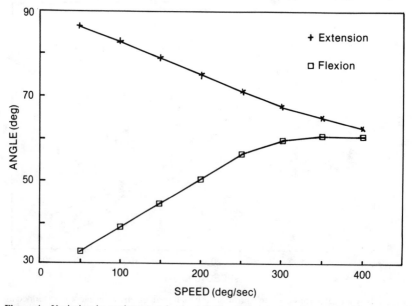

Figure 4—Variation in peak torque knee angle in relation to exercise speed.

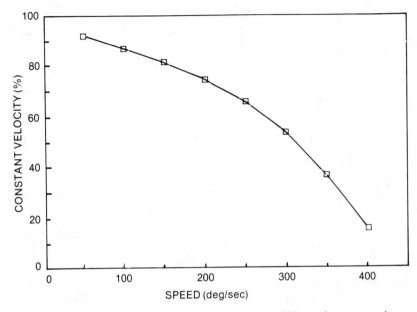

Figure 5—Duration of constant velocity (isokinetic) cycle within total movement time at different speeds.

tion/time curves over all rotational speeds. Data, fitted by cubic spline technique, revealed a decreasing isokinetic function with respect to increased speed settings. Acceleration and deceleration phases at the beginning and ending of the isokinetic movement were observed and indicated that from 92 to 16% of the entire movement at speeds from 50 to 400°/sec were in a constant velocity mode. Hence, a smaller portion of the entire movement was isokinetic as the dynamometer speed was increased (see Figure 5).

Discussion

The results of this study corroborate previous research indicating symmetry between contralateral limbs for maximum isokinetic torque (Osternig et al., 1980). The relatively small absolute differences and average variations between contralateral limbs for peak torque joint position at all speeds tested suggests that right and left limbs respond to speed changes in similar fashions (see Figures 1 and 2). This finding supports further the commonly practiced clinical technique of comparing the strength of injured limbs with healthy contralateral counterparts to estimate rehabilitative progress.

Maximal isometric hamstring/quadriceps strength ratios are generally considered to be about 0.60 (Scudder, 1980). The finding of increased

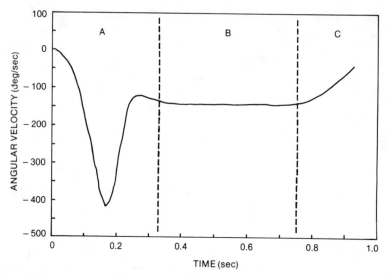

Figure 6—Angular velocity during leg extension at 150°/second. (A) Oscillatory phase. (B) Isokinetic phase. (C) Deceleration phase.

flexor/extensor torque ratios as a function of increased speed suggests that antagonist muscles tend to be more equally balanced at speeds approximating actual performance. Previous reports have revealed somewhat conflicting results on this parameter. Wilkerson and associates (1980) found flexor/extensor ratios changed from 0.71 to 1.50 at isokinetic speeds ranging from 30 to 300°/sec. Scudder (1980) found no significant ratio changes at speeds ranging from 3 to 15°/sec. However, the speed variations in Scudder's study were possibly too small to evoke changes in antagonist torque ratios.

The suggestion that antagonist strength imbalance may result in vulnerability to muscular strains must be evaluated in light of the speed at which muscles were tested. Hence, apparent antagonist imbalances tested isometrically may be dramatically altered when muscles are tested at high speeds.

The changes in peak torque joint position accompanying speed variations may be due to a longer acceleration period at the faster speeds since it is necessary for the limb to "catch up" to the predetermined speed of the dynamometer. Indeed, the time of the acceleration phase was linearly related to the speed setting. Isokinetic torque can only be recorded after the exercising limb reaches the set speed of the dynamometer. The subjects accelerated through an arc of movement and exceeded the preset dynamometer speed before being rapidly decelerated by the isokinetic device (see Figure 6A). High torque values were found to occur immediately following the acceleration phase of movement. This was possibly due to high resistance imposed on the exercising muscles as the limb rapidly decelerated to accommodate the dynamometer speed. This oscillation to dynamometer accommodation and the deceleration phase at the end of a movement resulted in a limited period in which the movement was truly isokinetic, i.e., in a constant speed mode (see Figure 6).

In conclusion, the findings in this study suggest that changes in movement speed can significantly affect functional antagonist muscle balance and can also change the peak torque joint position. Judgments regarding the relative safety of isokinetic exercise must be made in light of the observation that substantial portions of a specific exercise on isokinetic dynamometers may place acceleration and deceleration demands on muscles.

References

KETTLEKAMP, D., Johnson, R., Schmidt, G., Chao, E., and Walker, M. 1970. An electrogoniometric study of knee motion in normal gait. J. Bone Joint Surg. 50A:775-790.

LINDAHL, O., Movin, A., and Ringqvist, I. 1969. Knee extension: Measurement of isometric force in different knee-joints. Acta Orthop. Scand. 40:79-85.

OSTERNIG, L., Bates, B., and James, S. 1980. Patterns of tibial rotary torque on knees of healthy subjects. Med. Sci. Sports. 12:195-199.

SCUDDER, G. 1980. Torque curves produced at the knee during isometric and isokinetic exercise. Arch. Phys. Med. Rehabil. 61:68-73.

WILKERSON, J., Martin, D., and Sparks, K. 1980. Leg muscle strength, power and endurance in elite marathon runners. Abstract. Med. Sci. Sports Exer. 12(2):142.

Utilization of Stored Elastic Energy
in Leg Extensors

Senshi Fukashiro, Hitoshi Ohmichi,
Hiroaki Kanehisa, and Mitsumasa Miyashita
University of Tokyo, Tokyo, Japan

Recently, the elastic properties of muscles have been studied extensively by many researchers (Cavagna, 1977). Their approaches could be divided mainly into two as follows: 1) increment of the maximal positive work done by a previously stretched muscle. 2) increment of a mechanical efficiency in submaximal exercise of alternating positive and negative work.

Asmussen and Bond-Peterson (1974) and Komi and Bosco (1978) compared the utilization of stored elastic energy of two vertical jumps, with and without countermovement. However, they only analyzed the kinetic energy before and after the supporting-phase. In order to determine the mechanical behavior of the prestretched muscle during jumping, the time must be considered more precisely in the supporting-phase from touch-down to take-off.

The purpose of this study was to investigate the possibility for reusing the mechanical energy, stored in the muscles as elastic energy during a phase of negative exercise, in a subsequent phase of positive exercise during a vertical jump.

Methods

The subjects were three male adults, aged 21 to 25 yrs, whose heights ranged from 167 to 181 cm and whose weights ranged from 54 to 70 kg, and three female adults, aged 22 to 23 yrs, whose heights ranged from 154 to 163 cm and whose weights ranged from 44 to 52 kg.

The study consisted of two experiments. In experiment 1, each subject was asked to jump vertically with his or her best effort in the following two ways: 1) from a squatting position without preparatory counter-movement (SJ), and 2) from an erect standing position from different

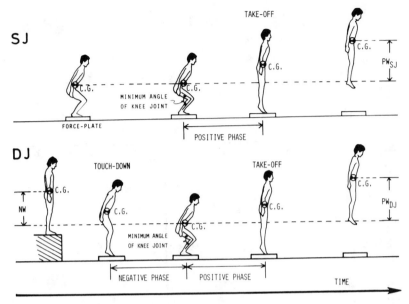

Figure 1—Vertical jumps with and without counter-movement.

elevations (from 0 to 90 cm height) and then dropping directly onto the force-plate (Kistler Inc., Switzerland) with a subsequent jump upwards with counter-movement (DJ) in almost the same manner as was described by Asmussen and Bond-Peterson (1974). In all test conditions, the subjects kept their hands at their sides (see Figure 1).

The maximum height of the body center of gravity during the flight was calculated from the flight time (Asmussen and Bond-Peterson, 1974). Electrogoniometers were fixed on the knee and ankle joints. From the electrogoniogram, the displacement of the body center of gravity was assessed during the landing. To observe the possible errors in the calculations from the goniogram, nine different jumps were filmed with a Photo-Sonics 16-mm Model 16-1P high-speed cine-camera (Photo-Sonics, Inc., California) set to operate at 100 frames/sec. The location of the body center of gravity was computed from the film on a motion analyzer (NAC Inc., Tokyo) using Miura's et al (1974) model for the body segment parameters. Comparison of the displacement data determined from films yielded an error of ± 4% for the computation from the method of using a goniometer and force-plate.

In experiment 2, the mechanical efficiencies during repetitive vertical jumps were determined. Each subject was asked to repeat a vertical jump of a certain height as constantly as possible for one min. Five trials of different heights were required for each subject. When the flexion of the knee joints on the platform was small, the vertical displacement of the body center of gravity during the jump was small, and vice versa.

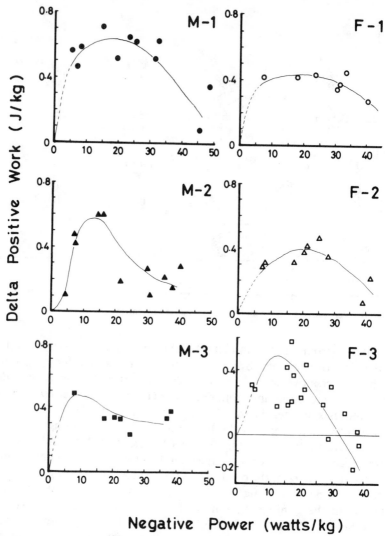

Figure 2—△ positive work related to negative power.

Results and Discussion

Since the negative linear relationships were observed in SJ between the total amount of work done and the minimum angle of the knee joint (when the legs were fully extended, the angles of knee joints were 180°), the linear regression equation was determined for each subject: correlation coefficients ranged from 0.96 to 0.99 (P < 0.001) for all subjects. Therefore, the positive work (PW) done in the corresponding SJ could be estimated from these equations at a respective minimum angle of knee

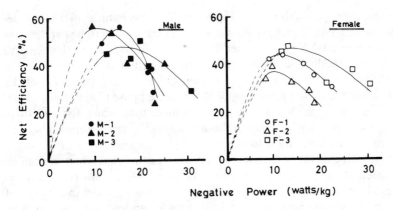

Figure 3—Relationship between net efficiency and negative power in repetitive vertical jumps.

joints in DJ. Then the delta positive work ($\triangle$PW) was calculated by the following equation:

$$\triangle PW = PW_{DJ} - PW_{SJ} \tag{1}$$

where PW_{DJ} or PW_{SJ} was the positive work done in DJ or in SJ (see Figure 1).

In order to estimate the elastic energy, PW done by a previously stretched muscle can be hypothesized to be the sum of PW done by a contractile component (CC) and a series-elastic one (SEC). Based on this hypothesis, $\triangle$PW in the present study could be considered as the utilized elastic energy which had been stored during a negative phase of exercise. Cavagna et al. (1974, 1975), however, described that not only SEC but CC in previously stretched muscle made PW increase. According to their statements, $\triangle$PW in this study might be overestimated in comparison with the 'true' PW done only by the SEC.

The negative power (NP) was calculated from the displacement of the body center of gravity from the erect standing position on the elevated platform to the maximum squatting position (the minimum angle of the knee joint) and the time of negative phase during supporting-phase (see Figure 1). $\triangle$PW was plotted against NP in Figure 2. Figure 2 shows the convex relationship between $\triangle$PW and NP, although there were considerable interindividual differences. Namely, increasing NP was not followed by a further increase in PW, but rather by a decrease after a certain point. Asmussen and Bond-Peterson (1974) discussed the reason for this, in that the forces developed during the breaking of the fast downward movement were so great that they might endanger the jumpers, who consequently did not exert themselves maximally.

The relationship between the net efficiency and NP of a vertical jump in experiment 2 is shown in Figure 3. The mechanical efficiency decreased with NP from 10 to 30 watts/kg. There seemed to be two reasons for these results. First, NP linearly increased in relation to the negative work (r = 0.80, P < 0.001). When the amplitude of the movement is great (as in the exercise of deep flexion of the knees), the recoil of the SEC affects only the beginning of the positive phase (Thys et al., 1972). In other words, Thys et al. (1975) stated that when the repetitive vertical jump was performed without bending the knees appreciably, the positive work done by elastic energy might amount to ½ to ⅔ of the total positive work. Secondly, Hill (1964) reported that when shortening occurred during a maintained isotonic contraction, the efficiency remained very constant throughout, even over a considerable range of length. However, when the load (or muscle tension) was too small or great, the efficiency decreased. In the case of the great NP, the muscle tension was so great that the efficiency of the CC decreased.

Each subject also showed a peak value in the relationship between the efficiency and NP in Figure 3. The peak values in Figures 2 and 3 appear at NP of about 10 to 15 watts/kg in spite of the difference between maximal (experiment 1) and submaximal exercise (experiment 2). These results indicate that there is an optimum NP for a person to utilize for the next jump.

The present experiments indicated the importance of storage of elastic energy in the contracted muscle during eccentric conditions as was already reported by several researchers. Furthermore, it can be said from the present results that the usable amount of stored energy depends on NP previously exerted.

References

ASMUSSEN, E., and Bond-Peterson, F. 1974. Storage of elastic energy in skeletal muscles in man. Acta Physiol. Scand. 91:385-392.

CAVAGNA, G.A. 1977. Storage and utilization of elastic energy in skeletal muscle. In: R.S. Hutton (ed.), Exercise and Sport Sciences Reviews 5:89-129. Journal Publishing Affiliates.

CAVAGNA, G.A., and Citterio, G. 1974. Effect of stretching on the elastic characteristics and the contractile component of frog striated muscle. J. Physiol. 239:1-14.

CAVAGNA, G.A., Citterio, G., and Jacini, P. 1975. The additional mechanical energy delivered by the contractile component of the previously stretched muscle. J. Physiol. 251:65P-66P.

HILL, A.V. 1964. The efficiency of mechanical power development during muscular shortening and its relation to load. Proc. Roy. Soc. 159(B):319-325.

KOMI, P.V., and Bosco, C. 1978. Utilization of stored elastic energy in leg extensor muscles by men and women. Med. Sci. Sports Exercise 10:261-264.

MIURA, M., Ikegami, Y., and Matsui, H. 1974. Calculation of the center of gravity by segmental method. J. Health Phys. Edu. Rec. 24:517-524. (in Japanese)

THYS, H., Faraggiana, T., and Margaria, R. 1972. Utilization of muscle elasticity in exercise. J. Appl. Physiol. 32:491-494.

THYS, H., Cavagna, G.A., and Margaria, R. 1975. The role played by elasticity in an exercise involving movements of small amplitude. Pfluger Arch. 354: 281-286.

Influence of Pre-Stretch on Armflexion

M. Van Leemputte, A.J. Spaepen,
E.J. Willems, and V.V. Stijnen
Katholieke Universiteit te Leuven, Heverlee, Belgium

Among others, Asmussen and Sorensen (1971) and Cavagna (1977) have drawn attention to the fact that in human movements a contraction of a muscle produces more energy if it is preceded by an active stretch of that muscle ('wind-up' movement).

Different methods have been used to search for determining factors that influence the effect of pre-stretch. A first method, which is employed both in vivo and in vitro, consists of changing only one factor. In this way, for instance, the influence of the length of a muscle is examined at a constant speed, or the influence of speed is examined for constant lengths. The question arises whether the results of this kind of analysis may be transferred to freely executed, global movements. In a second method, however, those global movements are the main subject of study. In activities, such as walking, running and jumping, the counter movement is caused and limited by gravity. The effect of pre-stretch is not easily related to the performance of the muscle, because several muscles or groups of muscles may be active at a certain joint simultaneously. Nevertheless, several investigations led to an insight into the influence of mechanical factors such as length of the muscle (Cavagna et al., 1968), speed of stretch (Cavagna et al., 1968; Asmussen and Sorensen, 1971), maximal force (Cavagna, 1977) and mechanical resistance (Bober and Erdmann, 1976).

The design of this experiment was intended to evaluate those factors in a freely organized movement with one active muscle group only. The arm flexors were subjected to a series of tests with and without a counter-movement independent of gravity.

Procedure

An apparatus (see Figure 1) was used which consisted of a flywheel (F)

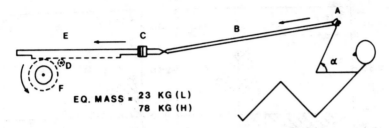

EQ. MASS = 23 KG (L)
78 KG (H)

Figure 1—Experimental lay-out. For explanation see text. $\alpha = 70°$ to $175°$.

with known inertia and connected to the hand of the subject by a toothed bar (E). The upper arm was supported in a horizontal position. Flexion or extension of the elbow within a range from $70°$ to $175°$ induced a rotation of the flywheel. The force on the handle (A) was registered by a strain gauge (C). A potentiometer (D) measured the displacement of the toothed bar. Both signals were sampled at a rate of 500 Hz and stored in a computer for later processing. The error, due to the inclination of the bar (B), was limited to 1%.

Using this apparatus, the authors tested 45 subjects (mean age = 20 years) who performed two movements. The first (wind-up movement) started with an elbow angle of $70°$; the arm extensors initiated a rotation of the flywheel that was decelerated by the eccentric work of the arm flexors (E neg) and was immediately followed by a concentric effort (E pos) of the same muscle group. During the movement the maximal extension in the elbow was recorded. The second movement (concentric movement) was executed starting from the recorded angle in the elbow. The task in both cases was to attain a maximal velocity of the flywheel at the end of the movement (E pos). Thus, the subjects were totally free in the organization of those movements. The latter were executed with a high (eq. mass 78 kg) and with a low (eq. mass 23 kg) inertial resistance.

After test and retest the total group (N = 45) was divided into two equivalent subgroups, based on the results of the preceding test (E pos). Both subgroup A (N = 23) and subgroup B (N = 22) attended a period of practice exercises, consisting of nine sessions, three per week. In every session 30 wind-up movements were performed with high inertial load. Subgroup A was totally free in the organization of the movements. Subgroup B, however, was limited to a maximal angle at the elbow of $165°$. After this training period all subjects were tested again for both movements and both resistances.

Results

Figures 2A and B show a typical example of a registered wind-up move-

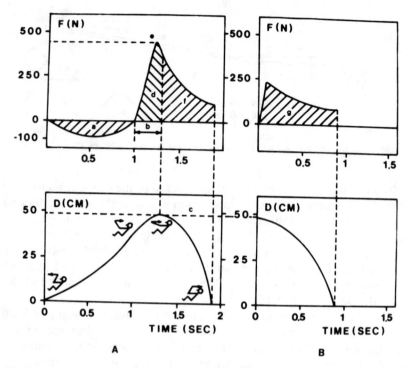

Figure 2—Typical example of a registered wind-up movement (A) and a concentric movement (B). Force (F) and displacement (D) as a function of time.

ment and a concentric movement. Evaluated parameters are marked on the curves. Parameter a denotes the area under the force-time curve during the pushing phase of the wind-up movement. This area is proportional to the velocity, and for the same inertial load, also related to the energy produced prior to the beginning of the eccentric phase. This energy or velocity is decelerated (E neg) during the time given by parameter b. Velocity there is zero and the angle in the elbow is maximal (parameter c). Area d is equal to area a. Parameter e refers to the maximal force. This force is reached almost simultaneously with the maximal extension in the elbow. During the concentric phase the force decreases with increased speed of movement. Area f is related to the velocity at the end of the movement and the resulting energy (E pos). Finally, the resulting energy of the second movement is denoted by parameter g.

Resulting Energy of the Wind-up Movement and the Concentric Movement

Table 1 shows E pos (in Joules) of both movements in the six experimental situations and the relative gain in energy between those two move-

Table 1

Mean Values of Resulting Energy (J) for Six Experimental Situations

| | Before Exercise | | After Exercise | | | |
	H TOT	L TOT	H A	B	L A	B
E pos in wind-up movement	79.15	49.70	95.09	88.69	54.12	52.03
E pos in concentric movement	70.12	44.89	68.57	69.53	45.40	44.58
Gain in E pos in %	12.9	10.7	38.6	27.5	19.2	16.7

H = high inertial load (eq. mass 78 kg), L = low inertial load (eq. mass 23 kg), A = without extension limit, B = with limit

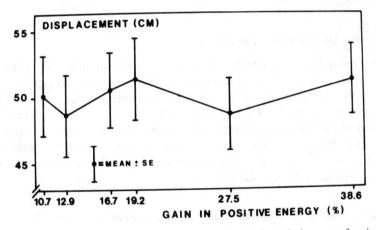

Figure 3—Maximal displacement as a function of the relative gain in energy for six experimental situations.

ments. The following conclusions could be drawn:

1. In the six situations, E pos of the wind-up movement is higher than E pos of the concentric movement. The gain in E pos is increased after exercise.

2. In the three groups and for both movements, E pos is higher for the high inertial load.

3. The relative gain in E pos is also higher with high resistance.

4. In the wind-up movement, E pos after exercise is significantly higher with regard to E pos before exercise.

In the following paragraphs the relation between the gain in energy and the evaluated parameters is examined.

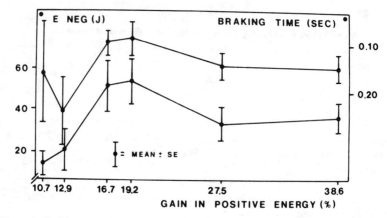

Figure 4—Braking energy (E neg) and time as a function of the relative gain in energy for six experimental situations.

Maximal Displacement (Parameter c)

The correlation between the maximal displacement and the gain in energy was calculated for the total group before exercise. No significant correlation could be found. However, after exercise the gain in energy was less in group B, suggesting that a limitation of the displacement results in a smaller gain in energy. In Figure 3 the mean value and standard deviation of the elbow angle at maximal extension is plotted against the relative gain in energy for the six experimental situations. These parameters seem to be unrelated.

Braking Energy (E neg) and Braking Time (Parameters a and b)

The correlation between both parameters was found to be $-.72$. A high kinetic energy is decelerated in a shorter time. Both braking energy ($r = .76$) and time ($r = -.76$) correlate with the maximal force but not with the gain in energy. Indeed, Figure 4 shows that the curves for braking energy and time are similar but not linearly related to the gain in energy.

Maximal Force (Parameter e)

The maximal force, as seen in the previous section, correlates with braking energy and time. Furthermore, higher maximal forces correspond to a higher gain in energy ($r = .65$). This relation is shown in Figure 5.

Discussion

The ratio between the positive energy in the wind-up movement and the

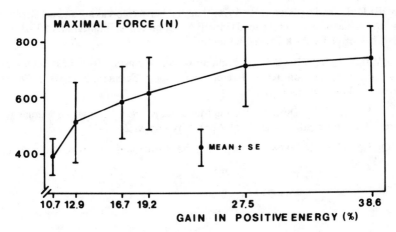

Figure 5 — Maximal force as a function of the relative gain in energy for six experimental situations.

positive energy in the concentric movement is highest for group A after exercise and equal to 1.4. The mean speed of stretch was 73 cm/sec. These numbers are closely related to Cavagna et al.'s findings (1968) where for similar movements a ratio of 1.3 was found at a constant speed of stretch of 64 cm/sec. According to the same author this ratio increased with increasing speed and attained 1.8 at 200 cm/sec. Results mentioned by Asmussen and Sorensen (1971) seem to support these data.

Our results confirm the relation between high speed of stretch and maximal force, the latter correlating with the gain in energy. A similar view applies to braking time. Bober and Erdmann (1976) and Hochmuth (1968) pointed out that a short braking time results in a high energy, while we found that a short braking time refers to a high force, and that this force results in a high energy.

As it appears from the correlations, a high speed of stretch together with a short braking time, resulting in a high maximal force, are necessary to obtain a high gain in energy.

These findings probably depend on the experimental situation, which utilized loads that were independent of gravity. As a consequence of this, the range of braking time and negative energy were expanded. The subject could freely choose between both parameters to obtain maximal output. In this way the study of the combined effect of the parameters mentioned was possible.

References

ASMUSSEN, E., and Sorensen, N. 1971. The "wind-up" movement in athletics. Le Travail Humain. 34:147-156.

BOBER, T., and Erdmann, W. 1976. Relation between breaking and accelerating a movement with different loads. In: P.V. Komi (ed.), Biomechanics V-B, pp. 53-57, University Park Press, Baltimore, MD.

CAVAGNA, G.A. 1977. Storage and utilization of elastic energy in skeletal muscle. In: R.S. Hutton (ed.), Exercise and Sport Sciences Reviews 5:89-129, Franklin Press, Philadelphia.

CAVAGNA, G.A., Dusman, B., and Margaria, R. 1968. Positive work done by a previously stretched muscle. J. of Appl. Physiology. 24:21-32.

HOCHMUTH, S. 1968. Biomechanische Prinzipien. Medicine and Sport 2:155-260.

Function of the Knee Extensor Muscles During Fatigue

Jukka T. Viitasalo
University of Jyväskylä, Jyväskylä, Finland

Some characteristics of myoelectrical activity (EMG) of human skeletal muscle are sensitive to fatigue. The power of the EMG frequency spectrum has been shown as moving towards the lower frequencies during the course of fatigue, while the changes in integrated EMG (IEMG) probably depend on the level of contraction used. At submaximal levels, IEMG has been shown to increase as a function of fatigue time (Edwards and Lippold, 1956; Lippold et al., 1960; De Vries, 1968) whereas, at the maximal level of contraction IEMG has been reported to decrease or remain at the original level (Stephens and Taylor, 1970; 1972; Komi and Rusko, 1974; Komi and Viitasalo, 1977).

In fatigue studies, where the myoelectrical activity has been monitored in agonist muscles, it has been suggested that the migration of activity from one muscle to another takes place during the course of fatigue (Lippold, 1955; Lippold et al., 1960; Komi and Viitasalo, 1977).

In order to explain this phenomenon in more detail the present study was designed to investigate the existence and amount of migration of activity between the three superficial knee extensor muscles during fatigue.

Methods

Male students (n = 31) in physical education volunteered for the study. All the subjects participated in the first fatigue test, where the subjects were instructed to maintain a 70% isometric contraction level as long as possible. The testing was terminated after an abrupt decrease in force. After one week, 10 subjects performed the second fatigue test consisting of 100 maximal isometric contractions (2 sec) as reactions to simultaneous auditory and light signals, which were given with random interpauses between 1.4 and 4.0 sec (for details see Viitasalo et al., 1980).

In both the test situations, the right quadriceps femoris musculature was fatigued with isometric knee extensions in a sitting position with a hip and knee angle of 90° and 120°, respectively. The dynamometer has been described in greater detail elsewhere (Viitasalo, 1980).

During each contraction, EMG was recorded bipolarly from the motor point area of m. vastus lateralis, vastus medialis and rectus femoris with Beckman miniature skin electrodes, amplified with Brookdeal 9432 preamplifiers (60 dB, 1 Hz to 1 kHz) and stored on magnetic tape (Racal Store 7) simultaneously with the force signal. EMGs were integrated for each individual isometric contraction (test 2). In the case of the continuous isometric extension (test 1), EMGs were limited to 11 samples of 2 sec at constant intervals covering the entire fatigue time. In each case IEMGs were expressed per sec.

Results

In the first test, the subjects were able to maintain the 70% isometric force level 47.9 ± 10.9 sec (26 sec to 73 sec). Integrated EMG of the three superficial knee extensor muscles showed a continuous increase (p < .05) during the fatigue time (see Figure 1A), the increase for m. vastus medialis being the greatest. The 100 isometric contractions brought about a 32% (p < .01) decrease in maximal force with no significant changes in the absolute IEMG values (see Figure 1B).

In order to evaluate the possible migration of activity between the muscles during fatigue, the IEMG activity of each muscle at each phase of fatigue was expressed as a percentage calculated from the sum of IEMG values of the three muscles. Figure 2 shows the effects of the two fatigue loadings on the relative IEMG values.

In the continuous isometric fatigue situations (see Figure 2A) the relative activity in m. vastus medialis increased and in m. vastus lateralis decreased significantly (p < .001) during the course of the fatigue time, while in m. rectus femoris the initial increase (p < .01) in the relative activity was followed by a decrease (p < .05) during the second half of the fatigue time. In the intermittent fatigue test (see Figure 2B) the relative activity in m. vastus lateralis increased (p < .05), while that in m. vastus medialis showed a tendency to decrease during the 100 contractions.

Discussion

The increase of IEMG during fatigue at submaximal contraction levels such as those in the present study has been assumed to be due to failure in contractility of individual active muscle fibers, which is compensated for by the recruiting of new motor units and/or the increasing of the firing frequency of the active motor units (Edwards and Lippold, 1956; Lip-

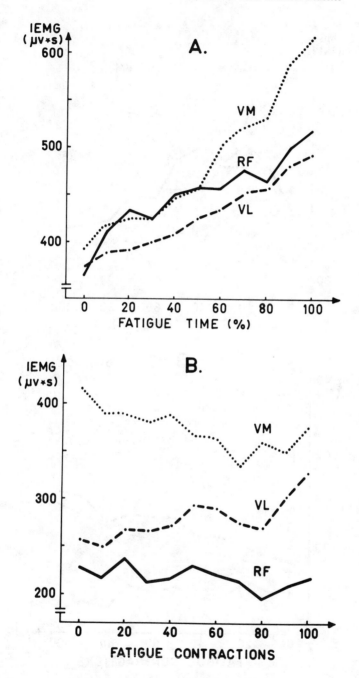

Figure 1—Absolute IEMG values for m. vastus lateralis (VL), vastus medialis (VM) and rectus femoris (RF) during the fatigue loading of constant 70% isometric tension (A) and during the 100 maximal isometric contractions (B).

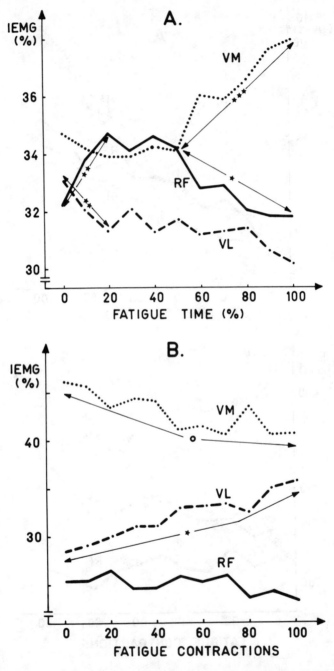

Figure 2—Relative IEMG values for the selected muscles during constant 70% (A) and intermittent (B) fatigue loadings. The percentage IEMG values for each muscle have been calculated from the sum of IEMG values of the three muscles at each phase of fatigue time. *** = p < .001, ** = p < .01, * = p < .05, o = p < .1.

pold et al., 1960; Person and Kudina, 1972). In this respect an increase of synchronization (Person and Kudina, 1968) and a possible change of type of active MUs (fast-slow, e.g., Gydikov and Kosarov, 1973; Gydikov et al., 1976) are supposedly also of importance. However, the present results showed that the increase of myoelectrical activity during fatigue did not happen similarly in the three superficial knee extensor muscles. When the activities were compared on a relative scale (see Figure 2A) it was found that, during the first quarter of fatigue time, the activity migrated from m. vastus lateralis to m. rectus femoris, while during the second half of fatigue time the relative decrease in the activity of m. rectus femoris was compensated by an increase in the activity of m. vastus medialis. These migrations can be explained by changes in measurement positions or by changes in the function of the neuro-muscular system. The hip and knee of the subjects were fixed during the fatigue tests and the subjects were instructed to maintain a constant position throughout the course of testing. No pronounced and systematic change in the measurement position was observed. Thus, the reason for the migration of activity was most probably in the function of the neuromuscular system, showing that the working intensity of the muscles was different at the beginning and during the course of the submaximal fatigue test, and/or that the fatigability of the three knee extensor muscles was different.

It has been suggested that two-joint muscles, such as m. rectus femoris in the present study, are less resistant to fatigue than muscles which act over just one joint (Kondo, 1960; Kondo and Sato, 1960), although no significant differences have been reported (Sato, 1965). Greater fatigability of m. rectus femoris, when compared to m. vastus lateralis and m. vastus medialis, was also suggested in a previous study (Viitasalo et al., 1982), where effects of prolonged cross country skiing were studied. Thus, the differences in fatigability, which are supposedly connected to the fiber distributions, may explain the intermuscular differences (migration) in IEMG activity found in the present study during the second half of fatigue time in the continuous isometric extension test.

The migration of activity between the muscles was also seen when maximal intermittent isometric contractions (test 2) were used. Contrary to what was found in test 1, however, the decrease of relative and also absolute electromyographic activity of m. vastus medialis (Figures 1B and 2B) was compensated by a simultaneous increase of IEMG in m. vastus lateralis. This increase of activity in m. vastus lateralis was similar to the trend found in an eccentric type of fatigue test by Komi and Viitasalo (1977). The literature is not in full agreement concerning the effects of fatigue on IEMG at maximal force levels; increases (Thorstensson and Karlsson, 1976; Nilsson et al., 1977), no changes (Merton, 1954), as well as decreases (Stephens and Taylor, 1972; Komi and Rusko, 1974; Ochs et al., 1977; Viitasalo et al., 1981) have been reported. In the present study a trend both to increase (m. vastus lateralis) and to decrease (m. vastus medialis) was found, showing that there exist intermuscular differences among agonist muscles in the sensitivity to fatigue.

Thus, the current results suggest that the migration of activity between the superficial knee extensor muscles takes place during the course of isometric fatigue, and that the migration is affected by the type of fatigue test (continuous-intermittent) and/or by the level of muscular contraction (submaximal-maximal).

References

DE VRIES, H. 1968. Method for evaluation of muscle fatigue and endurance from electromyographic fatigue curves. Amer. J. of Phys. Med. 47, 3:125-135.

EDWARDS, R.G., and Lippold, O.C.J. 1956. The relation between the force and the integrated electrical activity in fatigue muscles. J. Physiol. 132:667-681.

GRIMBY, L., and Hannerz, J. 1976. Disturbances in voluntary recruitment order of low and high frequency motor units on blockades of proprioceptive afferent activity. Acta Physiol. Scand. 96:207-216.

GYDIKOV, A., Dimitrov, G., Kosarov, D., and Dimitrova, N. 1976. Functional differentiation of motor units in human opponens pollicis muscle. Experimental Neurology 50:36-47.

GYDIKOV, A., and Kosarov, D. 1973. Physiological characteristics of the tonic and phasic motor units in human muscles. In: A.A. Gydikov, N.T. Tankov, and D.S. Kosarov (eds.), Motor Control, pp. 75-94. Plenum Press, New York.

KOMI, P.V., and Rusko, H. 1974. Quantitative evaluation of mechanical and electrical changes during fatigue loading of eccentric and concentric work. Scand. J. Rehab. Med. Suppl. 3:121-126.

KOMI, P.V., and Viitasalo, J.T. 1977. Changes in motor unit activity and metabolism in human skeletal muscle during and after repeated eccentric and concentric contractions. Acta Physiol. Scand. 100:246-254.

KONDO, S. 1960. Anthropological study on human posture and locomotion mainly from the view point of electromyography. J. Fac. Sci., Univ. Tokyo, sec. V 2:189-260.

KONDO, S., and Sato, M. 1960. Functional differentiation between one- and two-joint muscles. Proc. 13th Annual Meet. Jap. EMG Soc.

LIPPOLD, O.C.J. 1955. Fatigue in finger muscles. J. Physiol. 128:33P.

LIPPOLD, O.C.J., Redfearn, J.W.T., and Vuco, J. 1960. The electromyography of fatigue. Ergonomics 3:121-131.

MERTON, P.A. 1954. Voluntary strength and fatigue. J. Physiol. 123:553-564.

NILSSON, J., Tesch, P., and Thorstensson, A. 1977. Fatigue and EMG of repeated fast voluntary contractions in man. Acta Physiol. Scand. 101:194-198.

OCHS, R.M., Smith, J.L., and Edgerton, V.R. 1977. Fatigue characteristics of human gastrocnemius and soleus muscles. Electromyogr. Clin. Neurophysiol. 17:297-306.

PERSON, R.S., and Kudina, L.P. 1968. Cross-correlation of electromyograms showing interference pattern. Electroenceph. Clin. Neurophysiol. 25:58-68.

PERSON, R.S., and Kudina, L.P. 1972. Discharge frequency and discharge pattern of human motor units during voluntary contraction of muscle. Electroenceph. Clin. Neurophysiol. 32:471-483.

SATO, M., Hyami, A., and Sato, H. 1965. Differential fatigability between the one- and two-joint muscles. J. Anthrop. Soc. Nippon 73:82-90.

STEPHENS, J.A., and Taylor, A. 1970. Changes in electrical activity during fatiguing voluntary isometric contraction of human muscle. J. Physiol. 207:5-6P.

STEPHENS, J.A., and Taylor, A. 1972. Fatigue of maintained voluntary muscle contraction in man. J. Physiol. 220:1-8.

THORSTENSSON, A., and Karlsson, J. 1976. Fatigability and fibre composition of human skeletal muscle. Acta Physiol. Scand. 98:318-322.

VIITASALO, J. 1980. Neuromuscular performance in voluntary and reflex contraction with special reference to muscle structure and fatigue. Studies in sport, physical education and health, 12, University of Jyväskylä, Jyväskylä.

VIITASALO, J.T., Komi, P.V., Jacobs, J., and Karlsson, J. 1982. Effects of prolonged cross-country skiing on neuromuscular performance. In: Paavo V. Komi (ed.), Exercise and Sport Biology, pp. 191-198, Human Kinetics Publishers, Champaign, IL.

VIITASALO, J.T., Saukkonen, S., and Komi, P.V. 1980. Reproducibility of measurements of selected neuromuscular performance variables in man. Electromyogr. Clin. Neurophysiol. 20:487-501.

B.
Neural
Activation

Segmental Stretch Reflex Activity During Hopping Movements in Man

Y. Yamazaki

Nagoya Institute of Technology, Nagoya, Japan

G. Mitarai

Research Institute of Environmental Medicine,
Nagoya University, Nagoya, Japan

T. Mano

Hamamatsu University School of Medicine,
Hamamatsu, Japan

Many papers reporting experiments with animals and humans have emphasized that the segmental stretch reflex (SSR) is unable to regulate a limb position against unexpectedly imposed load changes, and experimenters have stressed the role of the long-loop reflex for load compensation. Thus, Melvill Jones and Watt (1971) could not record the SSR from the gastrocnemius muscle during rhythmic hopping movements on one foot, while they observed that the electromyograms showed the long-loop reflex and the anticipatory activities that begin before the foot-ground contact.

However, some recent observations have prompted other researchers to reconsider the role of the SSR in motor control. Dietz et al. (1981, 1979) showed that both the SSR and the anticipatory muscle activities contribute to the reaction to high velocity stretches after impacts, in landing from a forward fall and in running.

Since the role of the SSR in different motor tasks is still open to arguments, the present experiments were intended to study the SSR and the anticipatory activities during hopping movements with both knees extended.

Method

The experiments were carried out on sixteen 18- to 32-year-old healthy male volunteers, who were asked to make rhythmic hopping movements with both knees extended. They were instructed, first, to hop in "preferred frequencies" corresponding to three different amplitudes (small, moderate and large), and then in different frequencies (1, 3, and 4 Hz) synchronized to auditory stimuli delivered by headphones. The auditory stimuli were 1 kHz sine wave tone bursts of a duration of 25 msec and an intensity of 80 dB.

Electromyograms (EMGs) of the soleus (SOL) and the gastrocnemius (GC) muscles of the right leg were recorded with bipolar surface electrodes. The EMGs were high-pass filtered (at 30 Hz) and full-wave rectified, digitized with a sampling rate of 1 kHz, and averaged by means of a data processing computer. The raw EMGs triggered by the foot-ground contact were also averaged. The ankle joint angle was recorded on a potentiometer. The ground reaction force was measured by strain gauges and the ground contact signal by an electrical touch switch. These data were stored in a 9-channel FM tape recorder for further off-line analysis.

The H (Hoffman)-reflexes were elicited in an upright position by percutaneous electrical stimulations of a duration of 1 msec, applied to the tibial nerve at the popliteal fossa. The T (tendon)-reflexes were evoked by percussions of the right Achilles tendon.

The plate (52 × 40 × 16 cm), on which a subject was hopping with a preferred frequency, was suddenly removed from under the unaware subjects, in an experiment designed to distinguish between the anticipatory EMG activities and the stretch-evoked activities.

Six volunteers were subjected to the ischemic blocking of the tibial nerve for 20-30 minutes by applying to both thighs tourniquets inflated to a pressure of 300 mmHg. The blocking effects were monitored by the H- and T-reflexes. During the ischemic block, each subject's hopping movements were synchronized with 3 Hz auditory stimuli.

Results

The subjects' constant hopping movements with preferred frequencies measured 3.05 ± 0.034 (SD) Hz. The rectified and averaged SOL EMGs started 16 ± 26.3 (SD) msec before and ended 102 ± 20.3 (SD) msec after the foot touched ground. The EMGs of the GC started 70 ± 24.3 (SD) msec before and ended 107 ± 18.0 (SD) msec after ground contact. Figure 1 is an example of recordings obtained during rhythmic hoppings with preferred frequencies. After the foot-ground contact, which is indicated by an arrow, the average ankle joint angle shows high velocity stretchings of both the SOL and GC. The SOL and GC shows a high

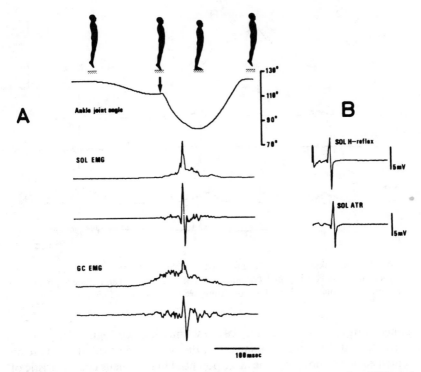

Figure 1 — The ankle joint angle and the EMG activities recorded from the soleus (SOL) and the gastrocnemius (GC) muscles during preferred frequency hopping. A — From top to bottom: ankle joint angle, rectified and averaged EMG, averaged raw EMG of the soleus (SOL EMG), rectified and averaged EMG, and averaged raw EMG of the gastrocnemius (GC EMG). The arrow indicates the foot-ground contact. B — Hoffman reflex (SOL H-reflex) and Achilles tendon reflex of the soleus (SOL ATR).

amplitude, synchronized potential (large potential) with short and constant latencies succeeding the stretch onset (Figure 1-A). In 15 out of 16 subjects, the SOL presented the large potential, while in 9 subjects the GC did not present the large potential. The large potential was more frequently observed in athletes than in non-athletes. The mean latency of the large potential was 33.4 ± 2.54 (SD) msec in the SOL and 31.2 ± 3.15 (SD) msec in the GC, when measured from the stretch onset. The mean latencies of the H- and T-reflexes were 29.4 ± 0.96 (SD) msec and 32.3 ± 3.15 (SD) msec, respectively. The averaged raw EMGs of the SOL and GC triggered by the ground contact of the foot showed constant wave forms, which were similar to those of the H- and T-reflexes under the same recording condition (see Figure 1).

During hoppings with three different amplitudes (small, moderate and large) in preferred frequencies the EMG activities showed an enhancement of the large potential in the soleus and gastrocnemius, in accor-

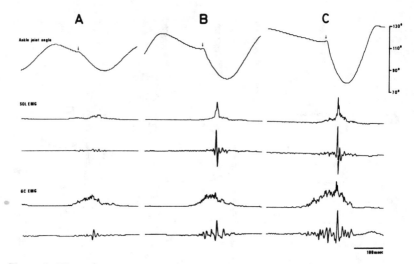

Figure 2—The ankle joint angle and EMG activities of the soleus (SOL) and the gastrocnemius (GC) during small (A), moderate (B), and large (C) amplitude hoppings with the preferred frequency. Recording conditions are the same as in Figure 1.

dance with the stretch velocity of these muscles (see Figure 2).

In three of the subjects, unaware of the sudden removal of the plate on which each was hopping, the large potential in the soleus characteristic of each foot-ground contact (first two arrows in Figure 3) did not reappear after the anticipated but unrealized foot-ground contact (the dotted line in Figure 3), nor did the large potential of the SOL EMG at each foot-ground contact, after the removal of the plate that is indicated by the third arrow in Figure 3.

Figure 4 is an example of recordings made during hoppings synchronized with 3 Hz auditory stimuli before (A) and 20 minutes after the ischemic blocking of the tibial nerve (B). After the ischemic block, the amplitude of the H- and T-reflexes of the SOL were reduced to about one third of the values before blocking. The amplitudes of the large potential in the same muscles also became remarkably reduced after the ischemic blocking.

Figure 5 is an example of recordings made while the subjects were asked to hop in synchronization with auditory stimuli of a frequency of 1 (A), 3 (B) and 4 (C) Hz. The ratio between the EMG discharges which started before the foot-ground contact (anticipatory activities) and the large potential in the triceps surae muscle varied according to the hopping frequencies. The EMG patterns of the 3 Hz hoppings were similar to those of the preferred frequencies, but during the 4 Hz hoppings the anticipatory activities were more and the large potentials less apparent than during the 3 Hz hoppings. The EMG patterns of the 1 Hz hoppings were considerably different from both the 3 and 4 Hz hoppings, as both the

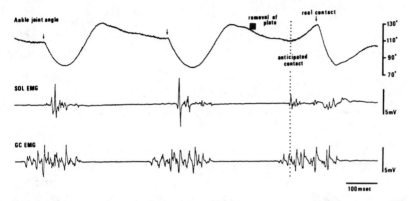

Figure 3 — The effect of unexpected removal of the plate on which the subject was hopping with preferred frequencies. The arrow represents the foot-ground contact. SOL EMG-raw EMG of the soleus, GC EMG-raw EMG of the gastrocnemius. The dotted line indicates the anticipated time of the foot-ground contact if the plate had not been removed.

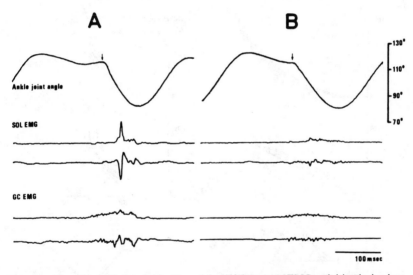

Figure 4 — The effect of ischemic blocking of the tibial nerve on EMG activities during hoppings synchronized with 3 Hz auditory stimuli. Recording conditions were the same as in Figure 1.

anticipatory and the large potentials became minimal.

Figure 6 is an example of X-Y recordings of the ankle joint angle and the platform force during hoppings. Five consecutive traces were superimposed in this figure. It should be noticed that during hopping with both knees extended, the triceps surae muscle of both legs is the prime mover, and that the length of the muscle correlates to the ankle joint angle. Figure 6 represents, therefore, the approximate length-tension

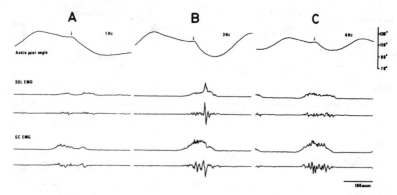

Figure 5—The ankle joint angle and EMG activity during hoppings synchronized with auditory stimuli. A-1 Hz, B-3 Hz, C-4 Hz. Recording conditions were the same as Figure 1.

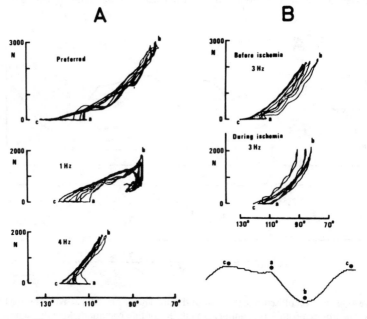

Figure 6—The relationship between the ankle joint angle and the platform force during hopping movements. A-hoppings with preferred, 1 Hz, and 4 Hz frequencies. B-hoppings with frequency of 3 Hz before and 20 minutes after ischemic block of the tibial nerve. a-touch down of the foot, b-maximum of the stretch phase, c-peak of the plantar flexion. Ordinate-platform force in Newtons (N). Abscissa-joint angle in degrees.

relationship of the triceps surae. During the 4 Hz hoppings, the muscular stiffness (the ratio of force change to length change) was high and relatively constant during both the stretching and shortening phases, while in the short range the stiffness was limited, compared with that during the preferred frequency hopping (see Figure 6-A). In the 4 Hz and the preferred frequency hoppings the trajectories between touchdown and take-off of the foot were almost the same. The relationship between the length and the tension was found to be analogous to that of a spring. In contrast, the 1 Hz hoppings did not show this spring-like property, because the force rapidly decreased after the maximum of the stretch phase without the length variation of the muscle. During the 3 Hz hoppings the spring-like property was visible, while after the ischemic blocking of the tibial nerve the 3 Hz hoppings showed the deviation from the spring-like property (see Figure 6-B).

Discussion

During rhythmic hopping with preferred frequencies, the SOL and GC EMGs showed high amplitude, synchronized potentials (large potentials) after the foot-ground contact. This type of EMG activity is attributable to the segmental stretch reflex (SSR) because of the short and constant latencies, the dependency on the stretch velocity and the wave form similar to the H- and T-reflexes. These large potentials do not seem to be produced by the subject's anticipation, as the sudden removal of the plate on which the unaware subject was hopping cancelled these potentials after the anticipated but unrealized ground-foot contact movement. As the input of the segmental stretch reflex has been known to be mediated by the group Ia (GIa) afferent fibers, the ischemic blocking procedure was adopted to reduce the GIa input. The amplitude of the large potentials was reduced after the ischemia, as happened with the H- and T-reflexes. This result supports the assumption that the large potentials are mediated by GIa afferent fibers.

The large potential depends on the stretch velocity of the triceps surae as Figure 2 shows. In the hopping movements of our experiments, the stretch velocity must be considered in relation to the anticipatory muscle activities which begin before the foot-ground contact. In the 4 Hz hoppings, the stretch velocity was low, due to the high muscular activation before the contact, and the amplitude of the large potential became relatively low. Thus, the output impedance of the stretch source seems to cause the appearance of the SSR. In addition to the amount of afferent input depending on the stretch velocity, the SSR magnitude may depend on the reflex gain controlling mechanisms (Gottlieb and Agarwal, 1979), as no SSR (large potential) could be recorded while the stretch velocity was moderate, i.e., during the 1 Hz hoppings and at foot-ground contact after the unexpected removal of the hopping plate.

Melvill Jones and Watt (1971) could not record the SSR from the GC muscle during single foot hopping. The hopping is, however, a multi-joint movement involving three joints—the ankle, knee and hip joints.

The knee joint flexion reduces the length of the GC muscle even if the ankle joint becomes small.

Thys et al. (1975) showed that the efficiency of the rhythmic jumping was higher than that of jumping with a short pause, a fact that was interpreted by the difference of potential energy release stored in the muscular system during the stretch phase before the upward acceleration. It seems that when the potential energy is stored, the muscle is activated and shortened prior to the stretch, and the interval between the stretch and the successive shortening is short. Since all our hopping movements, except the 1 Hz hoppings, fulfilled these conditions and the dynamic length-tension relationship showed a spring-like property of the muscle during the hoppings, a utilization of potential energy may exist. In the 4 Hz hoppings, the high anticipatory activation of the muscle may be related to the utilization of a potential energy, which, in short length range, makes the muscle act like a spring, because of the length-tension property of the preactivated muscle. Though during the preferred and the 3 Hz hoppings the muscle length showed longer variations, the length-tension curves were approximately linear and showed the spring-like property both in the shortening and in the lengthening phases. In cooperation with the anticipatory muscle activities, the SSR may facilitate rendering the muscle property linear and elastic (Crago et al., 1976).

Acknowledgment

A part of this work was supported by Grant no. 579070 from the Ministry of Education, Science and Culture of Japan (Y. Yamazaki).

References

CRAGO, P.E., Houke, J.C., and Hasan, Z. 1976. Regulatory actions of human stretch reflex. J. Neurophysiol., 39(5):925-935.

DIETZ, V., Noth, J., and Schmidtbleicher, D. 1981. Interaction between preactivity and stretch reflex in human triceps brachii during landing from forward falls. J. Physiol. (London), 311:113-125.

DIETZ, V., Schmidtbleicher, D., and Noth, J. 1979. Neuronal mechanisms of human locomotion. J. Neurophysiol. 42(5):1212-1222.

GOTTLIEB, G.L., and Agarwal, G.C. 1979. Response to sudden torques about the ankle in man: Myotatic reflex. J. Neurophysiol., 42(1):91-106.

MELVILL JONES, G., and Watt, D.G.D. 1971. Observations of the control of stepping and hopping movements in man. J. Physiol. (London), 219:709-727.

THYS, H., Cavagna, G.A., and Margaria, R. 1975. The role played by elasticity in an exercise involving movements of small amplitude. Pflügers Arch., 354:281-286.

Switching Mechanism of Neuromuscular Activity
in Top World Athletes

Kiyonori Kawahats
University of Kyoto, Kyoto, Japan

It has been indicated that, when a rapid voluntary contraction is performed following slightly sustained contraction, an EMG silent period appears before the synchronized discharges of the voluntary movement in the agonist muscle. This premotion silent period might be chiefly caused by supraspinal control because its early appearance is beyond the latency due to a peripheral origin, and also by the non-existence of mechanical and electrical changes in the agonist muscle (Ikai et al., 1973; Yabe, 1976; Kawahats et al., 1977, 1979). Nothing very definite is known about the mechanism, however. Supposedly, it is plausible to postulate that this premotion silent period is electromyographic transitional response due not only to the neural switch directing excitation into the pure alpha activity (Granit, 1970) from sustained contraction, but also to the switching of spinal interneurons differentially influencing γ and α motoneurons (Matthews, 1972) or to the switching of the central nervous system from using the spindle as an element in a postural servo to using it to provide a degree of servo-assistance in a large rapid movement (Matthews, 1972). The neural and electromyographic activity of specific muscle groups of trained and untrained subjects and how these relate to performance characteristics have not been studied.

The main purpose of this study was to determine whether or not the frequency of appearance and duration time of the EMG premotion silent period were of the same order in top world class athletes as compared with ordinary sport students.

Methods

Subjects participating in this study were divided in two groups. One group consisted of three top female athletes (a sprinter [100 m/11.2 sec]

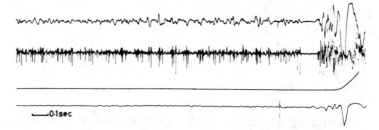

Figure 1 – An EMG premotion silent period recorded from m. vastus lateralis femoris by using surface and needle electrodes. Top trace = action potentials recorded by surface electrodes. Second trace = action potentials recorded by a needle electrode. Third trace = goniogram of the knee extension. Bottom trace = electric stimulation as a signal to the skin of the arm.

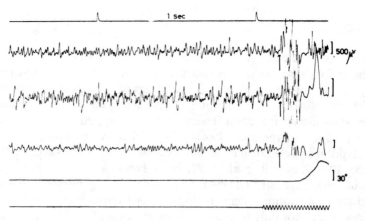

Figure 2 – Surface-recorded electromyographic premotion silent period (↑) with short duration from three knee extensors of a top sprinter. Top trace = time mark. Second trace = action potentials from m. rectus femoris. Third trace = action potentials from m. vastus medialis femoris. Fourth trace = action potentials from m. vastus lateralis femoris. Fifth trace = goniogram, showing knee extension. Bottom trace = visual signal asking rapid extension of lower extremities.

and two high jumpers, both 1.95 m) and the other group consisted of three female physical education students. The subjects were asked to respond as quickly as possible to an acoustic signal with an extension of the lower extremities, from a slightly squatted stationary posture. They were instructed, in advance, to avoid any counteractive movement prior to the previously described extension. The displacement curve was recorded with the use of an electrogoniometer attached to the lateral side of the right knee joint. EMG activity was registered from the right knee extensors, m. quadriceps femoris, during the squatted stationary posture and during the reaction movement. Bipolar surface electrodes (10 mm diam silver discs) were placed on the skin around the probable motor point of

Table 1

The EMG Premotion Silent Period in Time

Subject	rec. f.	v. med. f.	v. lat. f.
Sport student			
A	66.6 ± 20.7 (n = 36)	61.7 ± 16.7 (n = 32)	65.9 ± 20.5 (n = 37)
B	75.3 ± 18.1 (n = 20)	71.8 ± 12.4 (n = 28)	84.1 ± 24.6 (n = 28)
C	76.0 ± 19.9 (n = 67)	62.7 ± 18.4 (n = 37)	64.5 ± 18.6 (n = 62)
X̄	72.6	65.4	71.5
Top athletes			
D	29.5 ± 12.8 (n = 114)	32.8 ± 12.3 (n = 152)	32.8 ± 13.6 (n = 152)
E	39.5 ± 17.0 (n = 70)	37.0 ± 16.6 (n = 134)	39.0 ± 16.6 (n = 125)
F	40.2 ± 24.7 (n = 24)	41.6 ± 17.4 (n = 41)	39.1 ± 13.7 (n = 28)
X̄	36.4	37.1	37.0

mean ± SD (msec)

the muscle at the longitudinal distance of 3 to 5 cm. In the case of implanted needle electrodes, a coaxial (40 μm diam) or unipolar (Siemens Co.) copper needle was applied in order to record the activity from the limited motor units. The EMG, the mechanical curves, and the signal were amplified and displayed on an inkwriting recorder system (Mingograf 81, Elema-Schønander Co., Stockholm). All experiments were carried out during the same year when the top athletes made their new records already mentioned.

Results

Figure 1 shows a typical example of complete cessations of each action potential obtained by using surface and needle electrodes techniques. In other words, the silent periods were observed simultaneously from limited motor units and integrated units on a sport student. Figure 2 shows a striking example of considerably short duration time in the silent period on a sprinter. Most of the frequency distributions in the silent periods were in the ranges of 11 to 30 msec for m. rectus femoris, 21 to 40 msec for m. v. medialis f., and 21 to 30 msec for m. v. lateralis f. (left in Figure 3). It was generally longer than 40 msec in a sport student, however, as in the figure (right). Table 1 shows the comparison of the duration of the silent period in each muscle between athletes and sport students. These times were remarkably shorter in athletes as compared to students.

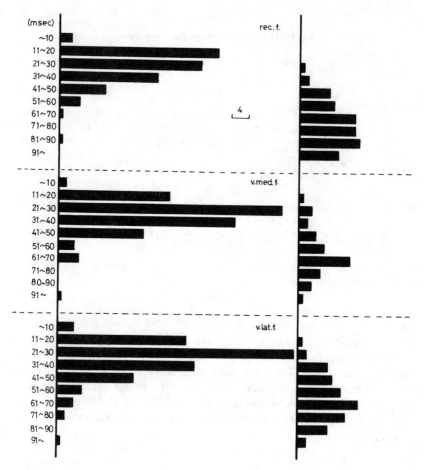

Figure 3 – Distribution of the EMG premotion silent duration in a sprinter (left) and a sport student (right).

Discussion

Considering that the movement in this study was a reaction with a large rapid switch from tonic to phasic voluntary contraction of the muscle, it seemed appropriate that the premotion silent period could be interpreted as an occurrence of the electromyographic transitional phase, which is attributable to a switching mechanism originating in the central nervous activity. This work revealed that the duration time of the premotion silent period was comparably shorter in the top athletes than that in ordinary sport students and in adult subjects previously reported (Ikai, 1955; Ikai and Shibayama, 1965; Gatev, 1972; Ikai et al., 1973; Yabe et al., 1975, 1976; Kawahats et al., 1977). These results raise the question of how this regulation of the short silent period occurs. It may be that the switch from tonic to phasic excitation in the nervous center occurs more quickly in the world's top athletes than in ordinary sport students and adult subjects.

The electromyographic premotion silent period reveals only a nervous ability related to a rapid voluntary muscle contraction, i.e., independent of the mechanical capacities and characteristics in muscle itself. The short silent period appearing in the top athletes could be considered as a means of estimating the motor control from a neural aspect. The short silent period must have an effect upon agility and coordinated voluntary movement.

References

GATEV, V. 1972. Role of inhibition in the development of motor coordination in early childhood. Develop. Med. Child. Neurol. 14:336-341.

GRANIT, R. 1970. The Basis of Motor Control, pp. 163-186. Academic Press. Lond.

IKAI, M. 1955. Inhibition as an accompaniment of rapid voluntary act. J. Physiol. Soc. Japan 17:292-298.

IKAI, M., and Shibayama, H. 1965. Dosa no binshosei — Sono seiriteki haikei. Science of Physical Education 15:149-156. (in Japanese)

IKAI, M., Yabe, K., Yamamoto, T., Kawahats, K., Watanabe, K., and Tezuka, M. 1973. The mechanism of "Silent period" preceding the act in rapid voluntary movements. Res. J. Physical Educ. Tokyo 18:127-133.

KAWAHATS, K., Kurek, B., and Hollmann, W. 1977. Über eine elektrische Behinderung vor einer schnellen, willkürlichen Bewegung beim Menschen. Sportarzt und Sportmed. 28:204-212.

KAWAHATS, K., Kurek, B., and Hollmann, W. 1979. Die Bedeutung einer aktionsfreien Phase im Elektromyogramm vor schneller Willkürbewegung beim Menschen. Deutsche Z. f. Sportmed. 30:271-277.

MATTHEWS, P.B.C. 1972. Mammalian muscle receptors and their central actions. In: H. Davson, et al. (eds.), Monographs of the Physiological Society. No. 23, pp. 546-606. Edward Arnold Ltd., Lond.

YABE, K., and Murachi, S. 1975. Role of the silent period preceding the rapid voluntary movement. J. Physiol. Soc. Japan 37:91-98.

YABE, K. 1976. Premotion silent period in rapid voluntary movement. J. Appl. Physiol. 41:470-473.

The Bimodal Muscular Control
for Very Fast Arm Movements

Ken Mishima, Takao Kurokawa, and Hiroshi Tamura
Osaka University, Osaka, Japan

Voluntary movements are carried out under centrally preprogrammed open loop control and/or peripheral feedback control. While the motor center requires a certain amount of time to receive information regarding the working muscles, very fast movements must depend solely on the open loop control. Such movements that are generated only by the central program have been called ballistic. When the program is executed, the central motor command is transmitted to the corresponding muscles along the motor nerve and the motoneuronal axons and induces action potentials which can be recorded as an electromyogram (EMG). EMG, therefore, directly reflects the central command during ballistic movements, for peripheral feedback signals do not mingle with it. In the present study we extracted from EMG the central command which controls a very fast forearm movement and suggest a concept of the bimodal control policy of the motor center for generating very fast ballistic movements.

Method

Five right-handed male adults participated as subjects in the experiment on the fast forearm extensions. Each subject was seated in a chair and grasped a lever handle with his right hand, the forearm being supported by the lever. The lever was pivoted just at the elbow joint and could be moved in the horizontal plane. At the lever tip a red LED was lighted to indicate the arm position. Along the outside of the radius of the rotating lever, a curved display panel was placed in order to indicate the target angle Θ_T by lighting one of the 100 red LEDs attached to it. The subject performed a step movement as fast as possible toward the visual target after he received the instruction to start. A precision potentiometer mounted at the axis of the rotating lever was used to monitor the forearm position Θ. Angular velocity $\dot{\Theta}$ and acceleration $\ddot{\Theta}$ were obtained by differentiating the Θ signal electronically. The EMGs were simultaneously recorded from both the triceps brachii (the agonist in the present study) and the biceps brachii (the antagonist) by means of bipolar hooked-wire electrodes. The target angles presented were 5°, 11°, 24°, 45°, and 60°. In addition, three of the subjects were requested to try 1°, 3°, 15°, and 30° target angles. These targets were lighted in a random order and for each target more than five trials were performed with about 15 sec rest time between them.

The early course of the movement was highly stereotyped. Such a stereotyped and fast course has been regarded as a ballistic portion of the movement (Stark, 1968; Megaw, 1974). This portion was divided into two phases: Phase 1 (positive acceleration phase) and Phase 2 (negative acceleration phase). Kurauchi et al. (1980) have described these phases as the positive peak acceleration time $t(\ddot{\Theta}_1)$ and the first zero cross time $t(\dot{\Theta}_1)$ as indicated in Figure 1a. The fact that nervous information is represented by variation in firing (pulse) density allowed us to extract the motor command from the EMG (Mishima et al., 1981). The extracting process is shown in Figure 1 in the case of the agonist EMG. The EMG was AD-converted every 0.2 msec and noise was reduced by applying an appropriate threshold (b). Then each single action potential in the EMG was transformed into a pulse by a computer algorithm which detects different waveforms of action potential. The resultant pulse train is illustrated in Figure 1c. The pulse density function was computed from the pulse train and similar functions obtained from several trials under the same conditions were averaged (d). The antagonist motor command was also extracted in the same way.

We can assume that the average pulse density function obtained reflects the central command with reasonable accuracy as long as Phases 1 and 2 exist because of the stereotyped waveform of the movement and the open loop property during these phases. In order to quantify the central command the initial portion of the density function was optimally approximated with a trapezoid in a sense of least-squares (e). Of the parameters characterizing the approximated motor command forms, the present study is concerned with the amplitudes (A_T, A_B) and the time difference t_{TB} between the command onsets.

Results

Dependence of the peak acceleration $\ddot{\Theta}_1$ and the corresponding time $t(\ddot{\Theta}_1)$ on the target angle Θ_T is shown in Figure 2. Three types of dependence are recognizable according to ranges of Θ_T. In Range A of the smallest

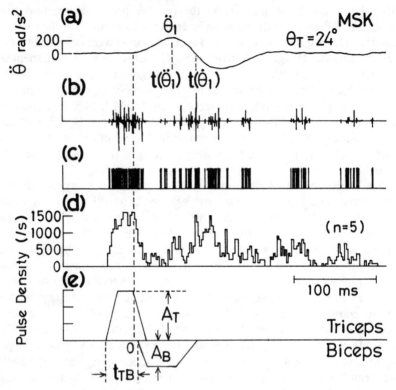

Figure 1—Typical acceleration waveform (a) and extraction of motor command from EMG (b-e). See text for details.

movements, $\ddot{\Theta}_1$ steeply increases with Θ_T while $t(\ddot{\Theta}_1)$ remains almost constant. In the second Range B corresponding to movements of medium amplitude, both $\ddot{\Theta}_1$ and $t(\ddot{\Theta}_1)$ increase in proportion to Θ_T. When target angles become large (Range C) the two parameters stay constant. As $\ddot{\Theta}_1$ is a monotonic function of Θ_T as in Figure 2, $\ddot{\Theta}_1$ can be discussed in terms of ranges A, B and C. These three ranges were found in all subjects. Acceleration waveforms (inset of Figure 2) show that their time courses were specific to each of the three ranges: separate courses in Range A, common courses before the positive peak time in Range B, and common courses up to the peak time in Range C.

Figure 3 depicts how one of the motor parameters, $\ddot{\Theta}_1$, is related to the command amplitude A_T and to the onset time difference t_{TB} in all subjects. In low $\ddot{\Theta}_1$ values $\ddot{\Theta}_1$ is linearly related with A_T (Range A). A_T, however, saturates when $\ddot{\Theta}_1$ becomes larger (Ranges B and C). On the other hand, the relation between $\ddot{\Theta}_1$ and t_{TB} exhibits a striking contrast to the relation between $\ddot{\Theta}_1$ and A_T. In lower range of t_{TB} $\ddot{\Theta}_1$ is independent of t_{TB} (Range A). $\ddot{\Theta}_1$ is positively correlated with t_{TB} when t_{TB} is relatively large (Range B), although $\ddot{\Theta}_1$ becomes saturated in Range C. In Range C,

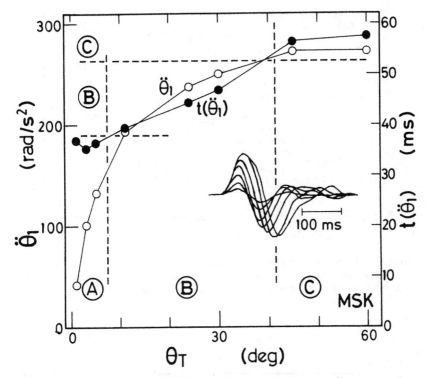

Figure 2—Positive peak acceleration $\ddot{\Theta}_1$ and corresponding peak time $t(\ddot{\Theta}_1)$ as functions of target angle Θ_T. Inset shows average acceleration waveforms for seven targets.

t_{TB} is well correlated with $t(\ddot{\Theta}_1)$. Some other parameters, such as rising time and duration of command forms, were also examined. They were either independent of or less correlated to motor parameters. These results not only strongly support that the amplitude and the onset time difference are important parameters to specify the motor command but suggest that they are main factors to control very fast extensions of the forearm. This suggestion is stressed by Figure 4a which shows the command property in terms of the amplitude and the onset time difference. All curves have similar features; at relatively small target angles (Range A) they are parallel with the vertical axis and at larger ones (Ranges B and C) they become parallel with the horizontal axis. This means that one of the command parameters is always set to its extreme value. Small movements might be controlled by changing the command amplitude while large ones might be controlled by changing the time difference between the agonist and the antagonist command onsets. We can designate these basically different control ways as the amplitude mode and time difference mode, respectively. The same property is seen in the antagonist command controlling Phase 2 (Figure 4b). It is clear that one mode is switched to the other at the same target angle in both commands.

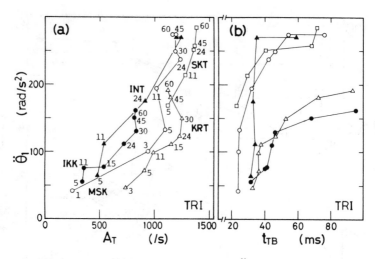

Figure 3—Relations between positive peak acceleration $\ddot{\Theta}_1$ and two parameters A_T (a) and t_{TB} (b) of central command in five subjects. Each subject is expressed with the same symbol in the following figures. Numerals indicate target angles in degrees.

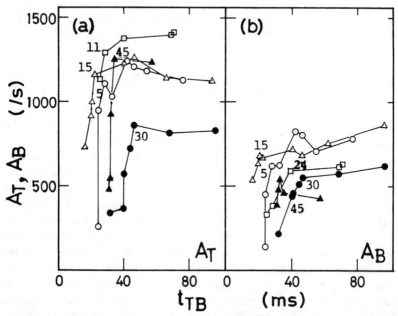

Figure 4—Motor command property in terms of command amplitudes A_T (a) and A_B (b) and onset time difference t_{TB}. This suggests two-mode control policy for forearm ballistic movement. Numerals indicate target angles where control mode is switched.

Discussion

If the EMG, corresponding to a ballistic portion of movements, is regarded as a direct reflection of the central command, we can extract and examine the central command. The results strongly suggest that very fast elbow extension is executed under one of two control modes according to the amplitude to be attained. This bimodal muscular control is reasonable for performing very fast movements because of the wide range of target position. As the subjects were asked to cover the target angle with maximum effort, it is purposive to maintain the extreme amplitude of motor command for relatively large target angles. This necessarily requires varying the other parameter, the onset time t_{TB} of the antagonist command. But for small angles of the target this control policy is unsuitable, because the onset of the antagonist command falls within the rising period of the agonist command and so the motor center is called for sophisticated control of the command onset time. The requirements for the motor center can, however, be reduced by employing the amplitude mode.

As shown in Figure 4, the amplitude mode and the time difference mode are mutually exclusive and the switching from one mode to the other occurs simultaneously in the agonist and antagonist commands. We are certain that these also enlighten the role of the motor center.

Many authors have indicated that ballistic movement can be understood using a simple on-off control model or a pulse height model. Freund and Büdingen (1978) and Ghez and Vicario (1978) have suggested that the amplitude of the central command to muscles is the main factor of movement control. Lestienne and Bouisset (1971) have stated that the onset time difference of central commands is an important parameter in controlling forearm movements. Their dissimilar results might be attributed to the difference of muscles used and narrow ranges of movements studied. It is suggested that their results respectively describe separate aspects of the general property of ballistic movements. We believe that the present results could lead to better understanding of the nature of the agonist-antagonist synergy in fast ballistic movements.

References

FREUND, H.-J., and H.J. Büdingen. 1978. The relationship between speed and amplitude of the fastest voluntary contractions of human arm muscles. Exp. Brain Res. 31:1-12.

GHEZ, C., and Vicario, D. 1978. The control of rapid limb movement in the cat. II. Scaling of isometric force adjustments. Exp. Brain Res. 33:191-202.

KURAUCHI, S., Mishima, K., and Kurokawa, T. 1980. Characteristics of rapid positional movements of forearm. Jap. J. Ergonomics 15:263-270.

LESTIENNE, F., and Bouisset, S. 1971. Quantification of the biceps-triceps synergy in simple voluntary movements. Visual Information Processing and Control of Motor Activity: Bulgarian Academy of Science: 445-448.

MEGAW, E.D. 1974. Possible modification to a rapid on-going programmed manual response. Brain Res. 71:425-441.

MISHIMA, K., Kurokawa, T., and Tamura, H. 1981. A method of estimating motor commands from electromyogram and its application to fast forearm movements. Trans. Soc. Instr. Control Engnrs. 17:574-581.

STARK, L. 1968. Neurological Control Systems. Plenum, New York.

Activation History of Muscle and Changes
in Human Reflex Excitability

R.S. Hutton, R.M. Enoka, and S. Suzuki
University of Washington, Seattle, Washington, U.S.A.
Kyorin University, Tokyo, Japan

In both humans and quadripeds, previous studies have shown that stretch reflex pathways are transiently facilitated in relaxed muscle in the aftermath of an isometric contraction while, paradoxically, human Hoffmann (H-) reflexes show depression during this same period (Hutton, Smith, and Eldred, 1975; Enoka, Hutton, and Eldred, 1980). Assuming that EMG responses to tendon (T-) taps are reflecting a central plus peripheral contribution (from muscle spindles), then the actual T-wave response will be reduced by the amounts reflected in the excitability curve for H-responses, a measure of a central component (alpha motor neuron excitability). Correcting for this by subtracting out the H-wave excitability curve, it has been proposed that a marked enhancement in T-wave excitability is attributed to peripheral input which is present at the initial stages of the post-contraction period followed by a decay of facilitation thereafter (Enoka et al., 1980). This observation is consistent with the post-contraction discharge and increase in stretch sensitivity of spindle receptors seen after contraction of a muscle in experimental animals (Eldred, Hutton, and Smith, 1976). To test whether H-reflex depression may have resulted from a decrease in the size of the subliminal fringe of alpha motor neurons, these investigations were replicated under conditions of low-level contractions, voluntarily sustained at a constant force during reflex elicitation. The effects of contractile magnitude and duration were also tested.

Procedure

The experimental conditions have been described previously by Enoka et al. (1980). Hoffmann reflexes (EMG) from medial gastrocnemius (MG) and soleus (Sol) muscles were tested in 16 subjects before (10 control responses) and immediately after (10 experimental responses) a 2.5 or 5.0 sec duration voluntary contraction at 25% or 100% of maximum (MVC). The ISI between reflexes was 5.0 sec. After each reflex sequence, subjects were asked to dorsiflex the foot to stretch the triceps surae. This would be expected to reset spindle discharge as influenced by the previous reflex contraction (cf. Hutton et al., 1973). During reflex testing, subjects superimposed two oscilloscope traces initially separated at a distance equivalent to 5% of their 100% MVC tension by sustained plantar flexing against a force transducer. Nine of these subjects were available on a separate day for re-testing using tendon tap (T-) reflexes.

Each percent and duration of MVC sequence of H- and T- reflexes were performed twice, but in reverse order for a given subject. Peak-to-peak electromyographic responses per trial were measured from 35 mm film records. A separate analysis of variance was performed on standardized EMG responses grouped according to reflex type, magnitude of contraction, or contraction duration. Depending on the groupings, tests were performed for the main effect of contraction, percent MVC, contraction duration, trials, treatment order and for possible interaction effects ($p < 0.05$).

Results and Discussion

As found in previous observations, H-reflexes were significantly depressed ($p < 0.05$) following contraction. Since no main effects were observed for treatment order or contraction duration, the data were collapsed for these variables and the percent of mean pre-contraction responses were plotted against time (trials) for 25% and 100% MVC. These results for Sol EMG are shown in Figures 1 and 2, respectively. H-reflex depression was evidenced following either %MVC ($p < 0.05$), except 25% MVC in soleus muscle, but the amount of depression was greater for a 100% MVC than for the 25% MVC ($p < 0.05$) except in one condition involving MG H-reflexes. Two-way interaction effects were evidenced in both muscles (except MG at 5 sec duration) for magnitude of contraction and pre-contraction — post-contraction comparisons. This interaction was primarily attributed to the greater depression caused by 100% MVC over 25% MVC rather than variations in the pre-contraction values.

In contrast, except for a significant effect of trials in Sol EMG and a pre-contraction — post-contraction trials interaction for MG EMG, no main effects were found for any of these variables for T-reflexes. This observation may, in part, be attributed to the smaller sample size.

Since a constant contractile force was maintained during testing, these

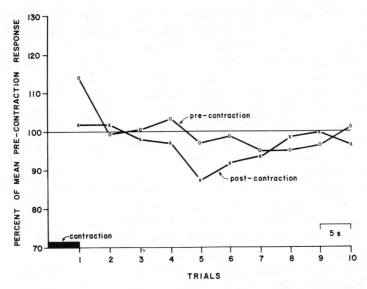

Figure 1–Percent of mean pre-contraction H-wave soleus responses plotted against trials (time) for control (pre-contraction) and post-contraction (25% MVC) reflexes.

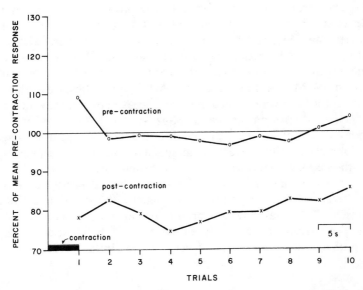

Figure 2–Percent of mean pre-contraction H-wave soleus responses plotted against trials (time) for control (pre-contraction) and post-contraction (100% MVC) reflexes.

observations argue against the possibility that the lingering H-reflex depression also found in relaxed muscle was associated with depressed excitatory input to alpha motoneuronal pools following contraction. That T-reflexes showed no similar trend suggests that altered gamma-loop or stretch receptor activity may have compensated for this central effect, but this conclusion is tempered by differences in subject sample size and the nature of the triggering stimulus. For example, Gassel and Diamantopoulos (1966) have shown that group I volleys elicited by an H- as opposed to a T-tap stimulus differ in temporal dispersion along the conducting pathways, thus suggesting that equivalency of afferent input between the two reflexes may be influenced by differences in spatial and temporal summation occurring at the input:output coupling stage. With the techniques employed, there is no way to refute this possibility.

Since the contractions were relatively brief, such changes in reflex excitability would appear dissociated from mechanisms related to either muscle temperature or muscle fatigue. However, in the case of the H-reflex, the depth of depression was influenced by the magnitude of the previous contraction.

That T-reflexes were not similarly depressed may be attributed to compensation provided through additional excitation by aftereffects of gamma loop activation (Brown, Goodwin, and Matthews, 1969; Hutton and Suzuki, 1979). Since gamma motoneurons are known to be recruited early and saturate in frequency of firing at low levels of excitability relative to extrafusal thresholds, the lack of post-contraction facilitation of the T-reflex under conditions of a sustained 5% MVC is not necessarily disparate from previous observations (Hutton, Smith, and Eldred, 1973; Post, Rymer, and Hasan, 1978; Enoka et al., 1980; Burke, Hagbarth, and Skuse, 1978). In an additional experiment with these same subjects, it is interesting to note that when they were asked to estimate without visual feedback the amount of force associated with a sustained 5% MVC following a 100% MVC, a significant overshoot in estimated force occurred which gradually decayed to control values over a 30 sec period (Hutton, Enoka, and Suzuki, 1981). This overshoot and temporal decay (cf. Enoka et al., 1980) would be predicted if an increase in peripheral excitation were transient and additive to a supraspinal input set at the necessary level to produce a constant force.

In conclusion, these findings further demonstrate the existence of transient alterations in reflex pathways associated with short-term activation of muscle. They may have important clinical implications for muscle facilitation treatments presently employed by rehabilitation medicine therapists and athletic trainers (cf. Moore and Hutton, 1981).

References

BROWN, M.C., Goodwin, G.M., and Matthew, P.B.C. 1969. After-effects of fusimotor stimulation on the response of muscle spindle primary afferent endings. J. Physiol. 205:677-694.

BURKE, D., Hagbarth, K.-E., and Skuse, N.V. 1978. Recruitment order of human spindle endings in isometric voluntary contractions. J. Physiol. 285:101-112.

ELDRED, E., Hutton, R.S. and Smith, J.L. 1976. Nature of the persisting changes in afferent discharge from muscle following its contraction. In: S. Homma (ed.), Understanding the Stretch Reflex. Progress in Brain Research. 44:157-170. Elsevier, Amsterdam.

ENOKA, R.M., Hutton, R.S. and Eldred, E. 1980. Changes in excitability of tendon tap and Hoffmann reflexes following voluntary contractions. Electroenceph. Clin. Neurophysiol. 48:664-672.

GASSEL, M.M., and Diamantopoulos, E. 1966. Mechanically and electrically elicited monosynaptic reflexes in man. J. Appl. Physiol. 21:1053-1058.

HUTTON, R.S., Enoka, R.M., and Suzuki, S. 1981. Constant force errors in force production following a maximum contraction. Med. Sci. Sports and Exercise 18(2):77.

HUTTON, R.S., Smith, J.L., and Eldred, E. 1973. Postcontraction sensory discharge from muscle and its source. J. Neurophysiol. 36:1090-1103.

HUTTON, R.S., Smith, J.L., Eldred, E. 1975. Persisting changes in sensory and motor activity of a muscle following its reflex activation. Pflügers Arch. 353:327-336.

HUTTON, R.S., and Suzuki, S. 1979. Postcontraction discharge of motor neurons in spinal animals. Exp. Neurol. 64:567-578.

MOORE, M.A., and Hutton, R.S. 1980. Electromyographic investigation of muscle stretching techniques. Med. Sci. Sports and Exercise 12:322-329.

POST, E., Rymer, W.Z., and Hasan, A. 1978. Relation between extrafusal and intrafusal activity in the decerebrate cat model: A role for beta fibers. Soc. for Neurosci. Abstracts 4:303.

Neuromuscular Adaptation to Training

G.A. Wood, F.S. Pyke, P.F. Le Rossignol, and A.R. Munro
University of Western Australia, Nedlands, Western Australia

Physiological changes in response to exercise are well documented, and it is clear that much of the histochemical, biochemical, and ultra-structural alteration that human skeletal muscle undergoes is specific to the training stimulus (Edgerton, 1976). Thus, endurance training favors adaptation of the oxidative machinery of muscle, while strength training enhances the contractile apparatus. And these adaptive responses are not confined to muscle alone, as changes to the motor neuron and neuromuscular junction have also been reported (Gerchman et al., 1975; Crockett et al., 1976).

It is less clear, though, how these changes affect neuromuscular functioning, and what neural mechanisms undergo change in addition to the more obvious morphological changes.

Motor unit firing patterns displaying greater synchrony have been observed with strength training (Milner-Brown et al., 1975; Stepanov and Burlakov, 1961), an effect which has been attributed to an enhancement of long-loop reflex activity (Milner-Brown et al., 1975). Spinal reflex times have also been shown to improve with strength training (Francis and Tipton, 1969), while the force-time characteristics of muscle appear to show little change (Sukop and Nelson, 1974).

Whether or not parallel changes occur with endurance training is unclear, although a commonly held fear among trainers is that long distance running will erode the muscle power capability of their athletes, and perhaps even predispose them to injury (Smart et al., 1980). Recently, Rochcongar et al. (1979) have presented electromyographic evidence to suggest that aerobic training increases the role of motor units receiving Ia afferent input. The ratio of a maximum Hoffmann (H) reflex response to maximum direct muscle (M) response used by these investigators has been found in our laboratory to be correlated with muscle fiber type composition (Wood et al., 1979), and the reported changes may herald an important alteration in neuromuscular functioning. The present study

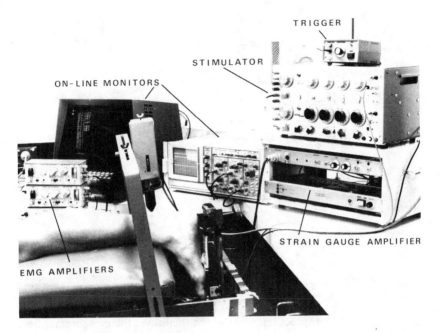

TRIGGER

STIMULATOR

ON-LINE MONITORS

STRAIN GAUGE AMPLIFIER

EMG AMPLIFIERS

Figure 1—Subject testing environment.

was undertaken to further explore neuromuscular changes which might accompany endurance running training.

Methods

Seven college-age students underwent an eight-week endurance running program designed to progressively elevate their aerobic capacity. Prior to and again following the training period selected reflexively evoked and voluntary muscle responses were studied in the right foot plantar flexors. A control group, consisting of a similar number of subjects, was tested at the same time to ensure that any observed changes could reasonably be attributed to the training program per se.

Both Achilles tendon and stretch reflex responses to controlled mechanical disturbances were recorded, together with responses elicited by percutaneous stimulation (square wave pulses of 1 msec duration) of the posterior tibial nerve. Subjects were also required to produce brisk plantar flexion contractions of their own volition. Electromyographic (EMG) records were obtained from bipolar surface electrodes overlying m. soleus and gastrocnemius, and the mechanical event associated with plantar flexion was recorded by means of a strain gauged foot plate (see Figure 1). All signals were processed on-line by means of a laboratory computer (PDP Lab 8e), while subsequent data reduction was performed

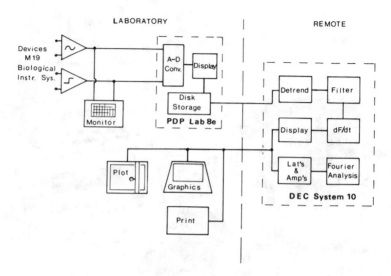

Figure 2 — Data acquisition and processing system.

interactively on a graphics terminal which facilitated the measurement of latencies, amplitudes, and slopes (dF/dt) of all records. The total data acquisition process is depicted in Figure 2.

Results and Discussion

The occurrence of a significant training response was indicated by a mean increase in maximum oxygen uptake ($\dot{V}O_2$) of 3 ml • Kg^{-1} • min^{-1} (P < .05) for the trained group, together with an elevated anaerobic threshold and reduced heart rate.

No consistent changes were observed in reflexively evoked muscular responses. However, voluntary plantar flexion efforts of the trained subjects, while showing no appreciable change in peak force levels (a factor not controlled, other than to require a strong, brisk contraction), displayed a diminished rate of tension development (dF/dt) and longer latency to peak dF/dt (see Figure 3).

These findings were suggestive of an altered motor unit involvement and to further explore that possibility a frequency-domain analysis of the electromyographic activity was undertaken. While there was no apparent shift in EMG spectra as a result of training (see Figure 4), a significant reduction in the gastrocnemius-to-soleus mean power spectral density ratio was evident ($\bar{X}$:3.37 − 0.845). While comparisons between the myoelectric activity of different muscles is rather tenuous, this finding did suggest a greater contribution to plantar flexion force provided by m. soleus.

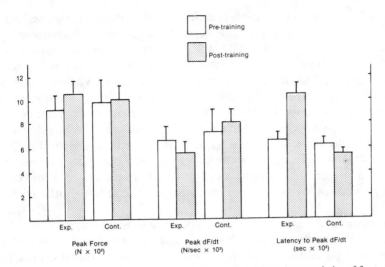

Figure 3—Effects of endurance running training on force-time characteristics of foot plantar flexors ($\overline{X}$ + $SE_{\overline{X}}$).

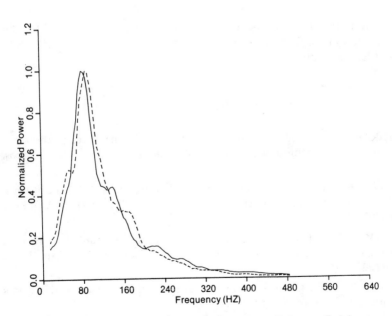

EMG Power Spectral Density Before (—) and After (---) Endurance Training

Figure 4—Power spectra of voluntary response EMG before (—) and after (---) endurance training.

Insofar as the human soleus muscle typically contains a high percentage of type I fibers while m. gastrocnemius is reasonably heterogeneous in composition (Edgerton, 1975), and given the 'slow-twitch' characteristic of a type I fiber, it would appear that this training program brought about an enhanced involvement of slow-twitch fibers. Whether this was due to muscle hypertrophy, hyperplasia, altered neural activity, or a combination of these factors is unknown, but effectively a more dominant role played by slow-twitch units would result in a slower mechanical response and decreased rate of tension development as observed in this study. Interestingly, the opposite effect has been observed with disuse, where a preferential atrophy of slow-twitch fibers occurs (Edström, 1970), resulting in an increase in speed of contraction (Burke et al., 1975).

In summary, it is suggested that endurance-type training can alter the contractile properties of human skeletal muscle through a decreased rate of tension development. In the absence of a demonstrated change in spinal reflex activity, the precise neuromuscular mechanisms remain obscure, but it would seem that slow-twitch units may acquire a proportionately greater effect on total force output.

Acknowledgment

The authors gratefully acknowledge the data processing assistance of Miss Maryanne Dawson and the technical help of Messrs. A.R. Pearce and E. Harrison.

References

BURKE, R.E., Kanda, K., and Mayer, R.F. 1975. The effect of chronic immobilization on defined motor units in cat medial gastrocnemius. Soc. Neurosci. 1:763.

CLARKE, D.H. 1964. The correlation between strength and the rate of tension development of a static muscular contraction. Physiol. Einschl. Arbeitsphysiol. 20:202-206.

CROCKETT, J.L., Edgerton, V.R., Max, S.R., and Barnard, R.J. 1977. The neuromuscular junction in response to endurance training. Exp. Neurol. 51(1):207-215.

EDGERTON, V.R. 1976. Neuromuscular adaptation to power and endurance work. Can. J. Appl. Sport Sci. 1:49-58.

EDGERTON, V.R., Smith, J.L., and Simpson, D.R. 1975. Muscle fiber type populations of human leg muscles. Histochem. J. 7:259-266.

EDSTRÖM, L. 1970. Selective atrophy of red muscle fibers in the quadriceps in long-standing knee-joint dysfunction. Injuries to the anterior cruciate ligament. J. Neurol. Sci. 11:551-558.

FRANCIS, P.R., and Tipton, C.M. 1969. Influence of a weight training program on quadriceps reflex time. Med. Sci. Sport. 1(11):91-94.

GERCHMAN, L., Edgerton, V.R., and Carrow, R. 1975. Effects of physical training on the histochemistry and morphology of the ventral motor neurons. Exp. Neurol. 49:790-801.

MILNER-BROWN, H.S., Stein, R.B., and Lee, R.G. 1975. Synchronization of human motor units: Possible roles of exercise and supraspinal reflexes. Electroenceph. Clin. Neurophysiol. 38:245-254.

ROCHCONGAR, P., Dassonville, J., and Le Bars, R. 1979. Modifications du reflexe de Hoffmann en fonktion de l'entrainement chez le sportif. (translation) Eur. J. Appl. Physiol. 40:165-170.

SMART, G.W., Taunton, J.E., and Clement, D.B. 1980. Achilles tendon disorders in runners—a review. Med. Sci. Sports and Exercise. 12(4):231-243.

STEPANOV, A.S., and Burlakov, M.L. 1961. Electrophysiological investigation of fatigue in muscular activity. Sechenov Physiol. J. USSR. 47:43-47.

SUKOP, J., and Nelson, R. 1974. Effects of isometrical training on the force-time characteristics of muscle contractions. In: R.C. Nelson and C.A. Morehouse (eds.), Biomechanics IV, pp. 440-447. University Park Press, Baltimore.

WOOD, G.A., Richardson, N.R., and Roberts, A.D. 1979. Electromechanical correlates of human muscle fiber type distribution. In: ANZAAS Sports Science I., pp. 39-57. Department for Youth, Sport and Recreation, Western Australia.

Critical Power as a Measure of Muscular Fatigue Threshold and Anaerobic Threshold

A. Nagata
Tokyo Metropolitan University, Japan

T. Moritani
Texas A & M University, U.S.A.

M. Muro
Tokyo College of Pharmacy, Japan

Earlier ergographic experiments have shown that there was a linear relationship between the maximal work (W_{lim}) and the maximal time (T_{lim}) over which it was performed until the onset of local muscle fatigue. According to Monod and Scherrer (1965), this relationship could be expressed by the equation $W_{lim} = a + bT_{lim}$, where W_{lim} was thought to be the result of the use of an energy reserve (a) and an energy reconstitution whose maximal rate was b. It was the purpose of this study to extend this concept to total body work (bicycle ergometer) and to test the hypothesis that anaerobic threshold (AT) might be closely associated with the maximal rate of energy reconstitution, b. In addition, the relationship between critical power and neuromuscular fatigability has been investigated.

Material, Methods, and Procedures

Eight male and eight female college students participated in this study. The determination of anaerobic threshold (AT) was performed for each subject during an incremental exercise test on a bicycle ergometer (Colling's Pedal Mode Ergometer) with a linear incremental load (Davis et al., 1979; Wasserman et al., 1973). In order to elucidate the relationship between neuromuscular fatigability and T_{lim}, surface electrodes were ap-

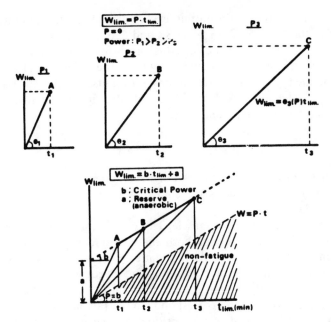

Figure 1—Schematic diagram showing the method for critical power determination.

plied to the vastus lateralis muscle. EMG instrumentation used in this study has been fully described elsewhere (Moritani and deVries, 1978; Moritani and deVries, 1979). EMG power spectral density and mean power frequency (MPF) were obtained according to the methods described by Nagata et al. (1981). Neuromuscular fatigability was estimated as the rate of rise in the IEMG/power as a function of time (Moritani et al., 1981).

Figure 1 illustrates the method for determining the critical power (CP). Here, three dynamic work tests up to the onset of muscular fatigue were performed on the bicycle ergometer which was electrically braked and provided work rates that were independent of the pedalling frequencies between 40 and 80 rpm. For each test, the maximal work (W_{lim}) was performed in the maximal time (T_{lim}) at which the initial power output level could no longer be maintained, i.e., a drop in pedalling frequency below 60 rpm. The quotient of these two variables defined the power of the work (P).

$$P = W_{lim}/T_{lim} \tag{1}$$

so inversely

$$W_{lim} = P \times T_{lim} \tag{2}$$

During each test, the power output level remained constant, but its level (400, 350, 300, & 275 W for male S's and 300, 250, 200, & 175 W for female S's) was chosen sufficiently high to bring about, in a longer or shorter time, the onset of muscular fatigue. In the lower part of Figure 1, W_{lim}, obtained from three different power outputs, are plotted as a function of T_{lim}. The points A, B, and C are situated on a line defined by the relationship between W_{lim} and T_{lim},

$$W_{lim} = a + bT_{lim} \tag{3}$$

Equation 3 indicates that the work done over a period of time before the onset of muscular fatigue is the sum of the two terms, according to Monod (1972):

1. An energy reserve in the muscle. This energy corresponds to factor a in the equation and represents both the energy contained in the high energy phosphagen components and that originating from the use of intramuscular oxygen in aerobic reactions and glycogen.

2. An energy supply established at a constant rate during work via oxidative phosphorylation and corresponds to the factor b in the equation. The factor b or 'critical power' (the slope coefficient) can be considered as a rate of energy supply whose magnitude determines the maximal power output level at which one can perform without fatigue.

When the imposed power output exceeds the critical power of the subject, the muscle must utilize its energy reserves and thus, muscular exhaustion will take place as these energy reserves will be depleted in a time which can be calculated as follows:

$$W_{lim} = P \times T_{lim} \tag{2}$$

and

$$W_{lim} = a + bT_{lim} \tag{3}$$

combining equations (2) and (3), we obtain $P \times T_{lim} = a + bT_{lim}$. Solving for T_{lim},

$$T_{lim} = a/(P - b) \tag{4}$$

Therefore, when the required power output level approaches very closely the critical power (b), then (P − b) will approach zero. Under these conditions, work can be continued—at least theoretically—almost indefinitely. For better understanding of this concept, Figure 2 has been provided in which power output levels and T_{lim} (above) and W_{lim} and T_{lim} (below) relationships are plotted from the same experimental data. The obtained curvilinear relation in the upper part of Figure 2 is defined by

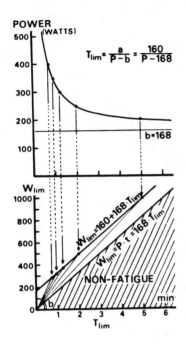

Figure 2—Relationship between the imposed power output and T_{lim} (above) and between W_{lim} and T_{lim} (below) plotted from the same experimental data.

the equation T_{lim} a/(P − b). If we solve this equation for P, we obtain

$$P = a/T_{lim} + b \qquad (5)$$

Theoretically, if we could set the power output level such that one could perform indefinitely, T_{lim} becomes infinite. Therefore, a/T_{lim} will approach zero, thus $P \cong 0 + b$, indicating that the critical power (b) may be considered as the level of power output at which one can perform this type of exercise indefinitely.

Results

Individual Data Analyses

The experimental data plots needed to determine each subject's critical power are shown in Figure 3. The regression analysis for each individual's data revealed that the relationship between W_{lim} and T_{lim} was in fact linear as suggested by Monod and Scherrer (1965) and therefore this relationship could be expressed in the general form, W_{lim} = a + bT_{lim} with R^2 ranging from 0.982 to 0.998 (p < 0.01). However, as is shown in Figure 3 there exist some individual differences in the Y intercept (energy reserve a) and the slope coefficient (critical power b). For example, the male subject, B.R. (long distance runner with $\dot{V}O_{2max}$ of 63.6

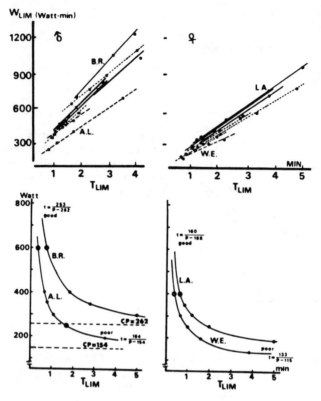

Figure 3—Individual data plots and the relationship between the imposed power output and T_{lim} obtained from trained and untrained individuals.

ml/kg/min) had a critical power of 262 watts. On the other hand, the untrained female subject (W.E.) yielded a critical power of 115 watts. Similar comparisons are made in Figure 3 in which the imposed power outputs were plotted as a function of the corresponding maximal sustaining times (T_{lim}).

Analysis of the $\dot{V}O_2$ equivalent for critical power or $CP\dot{V}O_2$ (by solving for the $\dot{V}O_2$ at the critical power level using the linear regression equation relating $\dot{V}O_2$ and power obtained during the incremental exercise test) showed that the distance runner had the $CP\dot{V}O_2$ of 2.94 l/min which was nearly three times as much as that of the poorly conditioned individual (1.07 l/min, which may be an equivalent O_2 uptake for moderately fast walking at zero grade on a treadmill).

Relationship Between Critical Power and Anaerobic Threshold (AT)

The relationship between critical power and anaerobic threshold as determined by the gas exchange method is shown in Figure 4. It was found that

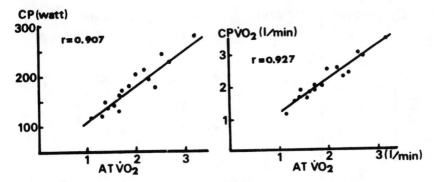

Figure 4— Relationship between anaerobic threshold and critical power.

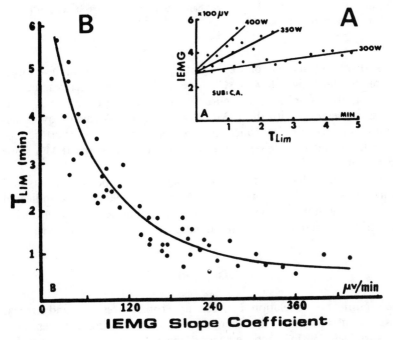

Figure 5—Relationship between fatigability (as estimated by the IEMG slope coefficient) and T_{lim}.

there was a significant correlation between $AT\dot{V}O_2$ and critical power (CP) expressed in watts (r = 0.907, p < 0.01) and between $AT\dot{V}O_2$ and $CP\dot{V}O_2$ (r = 0.927, p < 0.01).

Neuromuscular Fatigability and T_{lim}

Figure 5A shows a typical set of data in which the IEMG's during con-

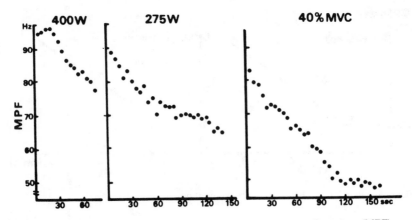

Figure 6—A typical set of data showing the changes in mean power frequency (MPF) as a function of time during cycling and during isometric contraction.

stant power outputs were plotted as a function of time. From these plots, neuromuscular fatigability (IEMG slope coefficient) was determined and then plotted against T_{lim} (see Figure 5B). It is clearly shown that T_{lim} has a close association with the IEMG slope coefficient which may be described as an exponential function. In other words, the shorter the T_{lim} (more fatigable) for a given power output the greater the IEMG slope coefficient.

Analysis of EMG power spectra during constant power outputs and during 40% of MVC (isometric knee extension at 90 degrees) indicated that a decline in mean power frequency was found to be greater during isometric contraction than during dynamic muscular contractions.

Discussion

The significant correlation between anaerobic threshold and critical power seems to indicate that there may be a possible link between these two variables. During exercise beyond anaerobic threshold, the aerobic energy supply must be supplemented by tapping energy reserves available through anaerobic glycolysis with a subsequent increase in lactate formation. Therefore, the subject with very high anaerobic threshold values is most likely to have considerably less obligatory glycogen depletion, since muscle glycogen utilization for ATP regeneration is about 18 times faster via anaerobic glycolysis as compared to oxidative phosphorylation (Davis et al., 1979). In this regard, it can be argued that the physiological significance of anaerobic threshold and critical power may well be similar because critical power would necessarily represent the maximal rate of work ("threshold") beyond which anaerobic reserves (a) will be depleted in time, that is, $T_{lim} = a/(P - b)$. Our data suggest that the rate of energy supply (or power output) as determined by critical power may well represent the maximal level of ATP supply via oxidative phosphorylation beyond which the obligatory lactate accumulation resulting from anaerobic glycolysis will occur.

The observed exponential relationship between the IEMG slope coefficient at various levels of power output and the corresponding maximal sustaining time (T_{lim}) may suggest a possible neurophysiological association between the fatigability of the muscle (vastus lateralis) and its motor unit (MU) activities. For example, increased tissue lactate level with a drop in pH has been postulated to interfere with the excitation-contraction (E-C) coupling with a deficit in the developed force (Fitts, 1977; Fuchs et al., 1970; Nakamura and Schwartz, 1972). If the output had to be maintained under these conditions, the recruitment of some additional MUs with FG and FOG fibers producing large action potentials (Person and Kunida, 1972) or increased firing frequency of already recruited or newly recruited MUs would occur to compensate for the deficit in contractility of some fatigued MUs. Consequently, the exponential increase in the IEMG slope coefficient as a function of T_{lim} may be explained by the different degree of 'phasic MUs' activities followed by synchronization of MU potentials as these MUs may quickly undergo a progressive fatigue (Milner-Brown et al., 1973). The significant decline in mean power frequency observed in this study is in good agreement with the previous findings (Komi and Tesch, 1979; Lindstrom et al., 1970) and can be explained by the effect of lactate upon the action potential conduction velocity (Lindstrom et al., 1970). Our data give strong support for this notion in that despite very exhaustive power outputs a decline in mean power frequency was much less pronounced during the dynamic cycling exercise than that observed during isometric contraction which would produce much greater lactate accumulation in the muscle due to a severely restricted blood flow.

References

DAVIS, J.A., Frank, M.H., Whipp, B.J., and Wasserman, K. 1979. Anaerobic threshold alterations caused by endurance training in middle-aged men. J. Appl. Physiol. 46:1039-1049.

FITTS, R.H. 1977. The effects of exercise-training on the development of fatigue. Ann. N.Y. Acad. Sci. 301:424-430.

FUCHS, F., Reddy, Y., and Briggs, F.M. 1970. The interaction of cations with the calcium-binding site of troponin. Biochi. Biophys. Acta. 221:407-409.

KOMI, P.V., and Tesch, P. 1979. EMG power frequency spectrum, muscle structure, and fatigue during dynamic contractions in man. Eur. J. Appl. Physiol. 42:41-50.

LINDSTROM, L., Magnusson, R., and Petersen, I. 1970. Muscular fatigue and action potential conduction velocity changes studied with frequency analysis of EMG signals. Electromyog. clin. Neurophysiol. 10:341-356.

MILNER-BROWN, H.S., Stein, R.B., and Yemm, R. 1973. The contractile properties of human motor units during voluntary isometric contractions. J. Physiol. 228:285-306.

MONOD, H. 1972. How muscles are used in the body. In: G.H. Bourn (ed.), Structure and Function of Muscle, pp. 23-74. Academic Press, New York.

MONOD, H., and Scherrer, J. 1965. The work capacity of a synergic muscular group. Ergonomics 8:329-338.

MORITANI, T., and deVries, H.A. 1978. Reexamination of the relationship between the surface integrated electromyogram (IEMG) and force of isometric contraction. Am. J. Phys. Med. 57:263-277.

MORITANI, T., and deVries, H.A. 1979. Neural factors versus hypertrophy in the time course of strength gain. Am. J. Phys. Med. 58:115-130.

MORITANI, T., Nagata, A., and Muro, M. 1981. Electromyographic manifestations of neuromuscular fatigue of different muscle groups during exercise and arterial occlusion. Jap. J. Phys. Fitness Sport Med. 30:183-192.

NAGATA, A., Muro, M. Moritani, T., and Yoshida, T. 1981. Anaerobic threshold determination by blood lactate and myoelectric signals. Jap. J. Physiol. 31:585-597.

NAKAMURA, Y., and Schwartz, A. 1972. The influence of hydrogen ion concentration on calcium binding and release by skeletal muscle sarcoplasmic reticulum. J. Gen. Physiol. 59:22-32.

PERSON, R.S., and Kunida, L.P. 1972. Discharge frequency and discharge pattern of human motor units during voluntary contraction of muscle. Electroenceph. clin. Neurophysiol. 32:471-483.

WASSERMAN, K., Whipp, B.J., Koyal, S.N., and Beaver, M.L. 1973. Anaerobic threshold and respiratory gas exchange during exercise. J. Appl. Physiol. 35:236-243.

Scale Effects in Limb Movement

Takeji Kojima
University of Tokyo, Tokyo, Japan

In the sphere of dimensional analysis of animal locomotion, Hill (1950) reported, using assumptions of his own, that 1) the intrinsic muscle shortening velocity (IMV) is inversely proportional to the length of body size (L), 2) the time spent in a single contraction of a muscle is proportional to L, and 3) the maximum speed of running is independent of L. But, the experimental results of Close (1972) suggest that Hill's conclusion on the first point may be erroneous. The purpose of this study was to investigate the scale effect by mathematical model, with stress on IMV.

Model

Figure 1 is the model for elbow flexion (biceps brachii is an equivalent flexor, according to Bouisset, 1973). The scale of body size (SBS) varied from 0.5 to 1.0 by 0.1. The elbow angle starts at 20° and the time occupied from start to 120° (movement time, MT) and the angular velocity at 120° (AV) were calculated under the condition of non-gravity. The movement velocity, $V = L_2 \cdot AV$. The used loads for each SBS were four. The weights were 0, 250, 500 and 1000 g, when SBS was 1.0. The model of muscle contraction was based on Hill's as reported by Jewell and Wilkie (1958). The muscle characteristics were assumed to be as follows: 1) the muscle stress is constant in the range of elbow flexion (Niwa, 1970) and independent of L, 2) the maximum muscle force (P_0) is 1500 N and the maximum muscle shortening velocity (v_0) is 1.5 m/sec when SBS is 1.0, 3) the value of a/P_0 of Hill's equation (Hill, 1938) is 0.25, 4) the compliance of the series elastic component is proportional to L^{-1}; approximating the compliance reported by Cnockaert et al. (1978), leads to the Equation $L = 6 \times 10^{-6} \cdot P$ for the relationship between the length of series elastic component (L in meters) and the muscle force (P

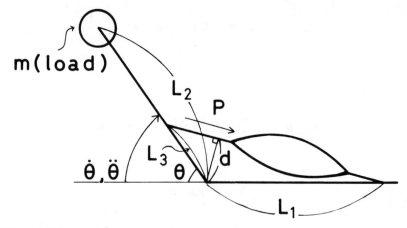

Figure 1—Mathematical model for elbow flexion. $L_1 = 0.3$ m, $L_2 = 0.32$ m, $L_3 = 0.046$ m, inertia of forearm $I = 0.05$ kg $\cdot$ m², $P_0 = 1500$ N, $v_0 = 1.5$ m/s and loads (m) are 0, 250, 500 and 1000 g when the scale of body size is 1.0. θ = angle, $\dot{\theta}$ = angular velocity, $\ddot{\theta}$ = angular acceleration, P = muscle force.

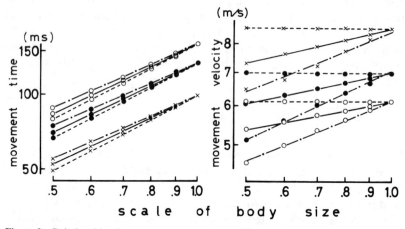

Figure 2—Relationships between movement time and body size (left), movement velocity and body size (right). IMV $\propto$: --- L^{-1}, — $L^{-0.5}$, —·— L^0 load, x 0 g, • 500 g, o 1000 g.

in Newtons) when SBS is 1.0, 5) the excitation of the muscle is maximum throughout a stroke, 6) three kinds of IMV are used, those are proportional to L^{-1}, $L^{-0.5}$, and L^0.

The following equation was solved numerically each 5 msec with a digital computer:

$$P \cdot d = (I + m \cdot L_2^2) \cdot \ddot{\theta}$$

(For an explanation of P, d, I, m, L_2, and $\ddot{\theta}$, see Figure 1)

Table 1

Values of b; Power in Allometric Equation

	Load (g)	b: IMV		
		−1.0	−0.5	0
MT	0	1.009	.920	.840
	250	1.006	.924	.830
	500	1.004	.927	.836
	1000	1.002	.932	.849
V	0	−.011	.233	.401
	250	−.006	.215	.466
	500	−.004	.203	.454
	1000	−.002	.187	.413
W	0	2.979	3.467	3.802
	250	2.988	3.430	3.931
	500	2.992	3.406	3.908
	1000	2.995	3.374	3.827
MP	0	1.970	2.547	2.962
	250	1.982	2.506	3.101
	500	1.988	2.479	3.072
	1000	1.993	2.442	2.978

Results

Figure 2 shows the relationships between MT and L, V and L. MT and V
are plotted against SBS in logarithmic coordinates. Those data were fitted to the allometric equation of $Y = a \cdot L^b$ or $\log Y = \log a + b \cdot \log L$
(Y; MT, V). MT increased with L in all three IMVs. As the power of
IMV equation increased, that of L or the inclination of the regression
line (i.e., b) decreased. V was nearly constant when IMV was inversely
proportional to L. In other cases, V increased with L. The values of b for
work (W) and the mean power (MP) were calculated from the b's of MT
and AV as $W = (I + m \cdot L_2^2) \cdot AV^2/2$ and $MP = W/MT$. The b's of
MT, V, W and MP are shown in Table 1. All b's corresponded with Hill's
model (1950) very well when IMV was proportional to L^{-1}. But b's did
not correspond so well when IMV was proportional to $L^{-0.5}$ or L^0. The
b's of V, W and MP were greater than those of his model.

Discussion

For the scale dependence of animal locomotion, Hill (1950) gave three assumptions: 1) geometric similarity is maintained in similar animals, 2) muscle stress is independent of L, and 3) the maximum work that a muscle can do in a single contraction is proportional to L^3. From these assumptions, Hill concluded that V was proportional to L^0 and that MT was proportional to L. The results of this elbow flexion model corresponded to his conclusion very well when the IMV of the model was proportional to L^{-1}; b's of MT were nearly 1.0 and b's of V were nearly 0. But b's of MT were less than 1.0 and b's of V were greater than 0 when b of IMV was greater than -1.0.

Åstrand and Rodahl (1977) also were led to the result that MT was proportional to L from the assumptions 1) and 2) as follows: Force (P) = acceleration • mass. $P = a • m$, so $a = P/m \propto L^2/L^3 = L^{-1}$, and a $\propto L • t^{-2}$, or $t^2 \propto L • a^{-1}$; so $t^2 \propto L • (L^{-1})^{-1} = L^2$ concludes t $\propto$ L. But these expressions are not always acceptable. In a dynamic condition, the muscle force changes with the muscle shortening velocity, so the muscle force may not be proportional to L^2. In elbow flexion, angular acceleration $\theta = P • d/(I + m • L_2^2) \propto L^2 • L/L^5 = L^{-2}$, when the muscle shortening velocity (Vm) is 0. After a short time ($\triangle t$), angular velocity of movement $AV_m = \theta • \triangle t \propto L^{-2} • L = L^{-1}$, then $V_m \propto AV_m • L_2 \propto L^{-1} • L = L^0$, on the assumptions that IMV $\propto L^{-1}$, i.e., maximum $V_m \propto L^0$ and $\triangle t \propto L$. Then P is proportional to L^2 in a dynamic condition because V_m is independent of L. However, P is not proportional to L^2 and MT is not proportional to L in a dynamic condition when IMV is not proportional to L^{-1}, i.e., maximum V_m is not proportional to L^0.

Hill (1950) reported the running velocities of whippets, greyhounds and horses. According to his report, their velocities were nearly the same but increased slightly with body size. Asmussen and Christensen (1967) reported the relationship between maximum running velocity and the body height of 18-year-old boys. According to their report, it seems that the running velocity increases slightly with the body height. Close (1972) reported that IMV was nearly proportional to $L^{-0.5}$ in geometrically similar animals. In the elbow flexion model, V was proportional to about $L^{0.2}$ when IMV was proportional to $L^{-0.5}$. It is interesting that the law of body size dependence of one joint movement seems also to control the running composed of multi-joint and cyclic movement.

References

ASMUSSEN, E., and Christensen, E.H. 1967. Kompendium i Legemsövelsernes Specielle Teori. Köbenhavns Universitets Fond til Tilvejebringelse af Läremidler, Köbenhavn. :refer to Åstrand et al. (1977)

ÅSTRAND, P.–O. and Rodahl, K. 1977. Textbook of Work Physiology, 2nd ed., pp. 367-388. McGraw-Hill, New York.

BOUISSET, S. 1973. EMG and muscle force in normal motor activities. In: J.E. Desmedt (ed.) New Developments in Electromyography and Clinical Neurophysiology. 1:547-583. Karger, Basel.

CLOSE, R.I. 1972. Dynamic properties of mammalian skeletal muscles. Physiological Reviews 52:129-197.

CNOCKAERT, J.C., Pertuzon, E., Goubel, F., and Lestienne, F. 1978. Series-elastic component in normal human muscle. In: E. Asmussen, and K. Jorgensen (eds.) Biomechanics VI-A, pp. 73-78. University Park Press, Baltimore, MD.

HILL, A.V. 1938. The heat of shortening and the dynamic constants of muscle. Proc. Roy. Soc. B 126:136-195.

HILL, A.V. 1950. The dimension of animals and their muscular dynamics. Proc. Roy. Inst. Great Britain 34:450-471.

JEWELL, B.R., and Wilkie, D.R. 1958. An analysis of the mechanical components of frog's striated muscle. J. Physiol. 143:515-540.

NIWA, N. 1970. Relations between joint angles and muscle strengths — In the case of arm strength. Jap. J. Physical Education 14:201-206. (in Japanese)

III.
REHABILI-
TATION

Keynote Lectures

Biomechanics in the Rehabilitation of Human Movement

David A. Winter

University of Waterloo, Waterloo, Ontario, Canada

As technology has improved, the potential for exploiting gait assessments has drastically increased, and is currently at the state where more advanced analyses may now be performed. The roles that biomechanics can play may be summarized as follows:

1. Development of a baseline of variables for normals.

2. Monitoring of changes as a result of therapy, surgery, or alteration of rehabilitation aids.

3. Assessment of newly developed devices, especially prosthetics and orthotics.

4. More accurate and objective diagnosis of the causes of movement pathologies.

The major problem in all aspects of these assessments is the fact that we are dealing with a multiple input and multiple output system. The resultant explosion in variables or combination of variables has overshadowed all efforts to get at the most sensitive and important ones. In a 12-segment model of a human walking in the sagittal plane there are 108 kinematic variables required to describe the trajectory of those 12 segments in space. Considering the kinetic, energy, power and momentum analyses that can be performed, the number of variables can soon be doubled and if one has an interest in the EMG patterns they are limited only by the number of channels of the data collection system. Thus, the major problem that has not been addressed is the identification of diagnostically sensitive variables and the elimination of redundant variables.

Baseline Variables on Normals

The assessment of normal walking has resulted in more descriptive data

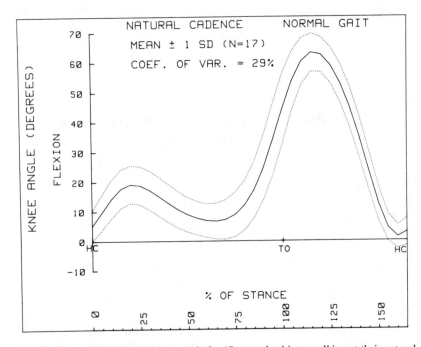

Figure 1—Ensemble average of knee angle for 17 normal subjects walking at their natural cadence. Dotted lines represent ± 1 standard deviation, and the coefficient of variability over the stride period was 29%.

than in any other area of human movement. Yet, even now, there are many gaps in the body of knowledge which prevent one from making objective comparisons with a given patient. The simplest temporal and kinematic variables have been quantified (Grieve and Gear, 1966; Murray et al., 1964; Statham and Murray, 1971; Murray et al., 1970; Winter et al., 1974; Foley et al., 1979; Sutherland et al., 1980) and serve as a basis for comparison with atypical walking. Most of these measures were reported for adults walking at a natural cadence; a few report on specific population groups, such as children (Grieve and Gear, 1966; Statham and Murray, 1971; Foley et al., 1979; Sutherland et al., 1980). Almost no studies have been reported for the older, somewhat overweight and slower walking population from which most patients are drawn. Also, most of the important diagnostic variables such as joint moments of force (Pedotti, 1977; Winter, 1980) and mechanical energy have had few patterns identified, even at comfortable cadences. Electromyographic patterns for most muscles have been measured but are usually quoted for normal walking speeds and are in units that are not transferable between rehabilitation centers (Sutherland et al., 1980; Sutherland, 1966; Battye and Joseph, 1966). In fact, all that is usually reported is the time of onset and end of activity for each muscle as defined by some undefined

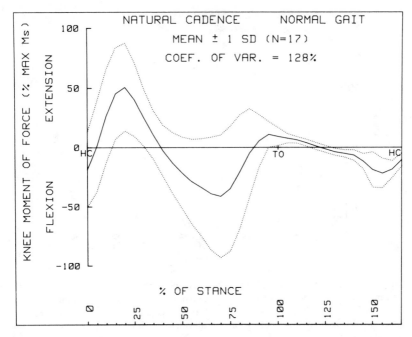

Figure 2—Ensemble average of the knee moments for the same 17 subjects as reported in Figure 1. Moment was normalized against the maximum support moment, Ms, during stance. Dotted lines represent ± 1 standard deviation, and the coefficient of variability over the stride was 128%.

threshold. Dubo et al. (1976) attempted to get around the problem by reporting the level of activity as a percent of maximal voluntary contraction (MVC) level for each muscle. Also, only a few EMG studies have considered the dramatic influence of cadence (Milner et al., 1971; Longhurst, 1980) on the patterns of activity.

Certain variables have low variability in their patterns, as seen in the plot of knee angle (see Figure 1) for 17 normal subjects as a function of stride period; the coefficient of variance over the stride period was only 29%. The patterns of the kinetics associated with this movement are considerably more variable, however. For example, Figure 2 shows the average knee moment of force for these same 17 subjects expressed as a percent of each subject's maximum support moment (Winter, 1980). The coefficient of variability for the knee moments was 128%, considerably higher than that for the knee angle curves. For these same trials the hip moments also showed extremely high variability (153%). Such variability is characteristic of a multiple input system in which many possible combinations of muscle forces at several joints can result in the same or very similar joint angle histories. Also, within-subject variability at different cadences can be quite high, as shown in Figure 3 where the EMG

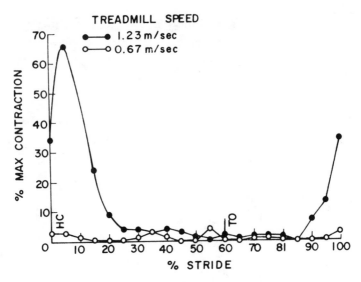

Figure 3—The ensemble average of the linear envelope of the EMG of the vastus medialis during moderate and very slow walking speeds of a normal subject on the treadmill. The activity dropped drastically prior to and shortly after heel contact, indicating that normal cadence EMG patterns may not be a useful reference for comparison with slower pathological gaits.

amplitude (linear envelope, expressed as a percent of maximum voluntary contraction) of the vastus medialis is plotted at natural and slow cadences.

It is evident from these examples that biomechanics has a major role to play in the development of useful profiles of key variables on normal individuals at different walking speeds. These profiles must somehow be normalized so as to partition out anthropometric variables such as body height and weight and kinematic variables such as walking velocity. The distribution of the resultant variable during the movement period is also essential in order that objective comparisons can be made with each patient's pattern. Figure 4 shows a typical EMG pattern normalized in both time (100% of gait stride) and amplitude (% MVC) and showing the variability at each point in the stride. The ensemble average of 10 strides for each of the eight subjects is shown on the left, for the soleus muscle of the right leg. The curve on the right is the ensemble average of the eight subject averages along with a band of one standard deviation.

Monitoring of Rehabilitation Patients

Monitoring is the term used to describe the function whereby the prog-

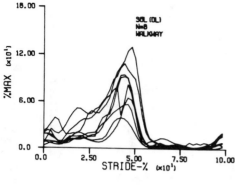

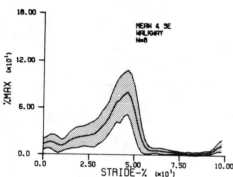

Figure 4 — EMG activity of soleus muscle of eight subjects. Left plot shows the ensemble average of 10 strides from each subject, with the amplitude expressed as a percent of individual subject's maximal voluntary contraction. Right plot shows ensemble average of the eight subject averages with a band indicating one standard error.

ress of the patient is quantified during the rehabilitation process (therapy, surgery, bracing). As such the variables usually employed are those that are readily measured, such as gait asymmetry, stride length, knee angle or force plate signals. The progress of the patient is usually quantified by a descriptive comparison with similar measures in normals. An example of monitoring is shown in Figure 5a and b where this joint replacement patient's pattern of knee vs hip angle can be compared with that of a normal subject. Changes away from or towards the normal pattern can be readily monitored. Caution should be taken not to use these monitoring variables as a diagnostic function, however. Each of the monitored variables is a function of many input variables. The knee angle during stance, for example, is under the influence of at least 15 major muscles, so it would be folly to attempt to infer cause (diagnose) from an abnormal knee angle pattern, unless supported by additional EMG data or moment of force analyses. Similarly, abnormal ground reaction force patterns, as presented in Figure 6, give no clue as to the abnormal muscle activity that caused that pattern in this knee arthroplasty patient. Only by a proper link-segment analysis were we able to diagnose the cause of this abnormal ground reaction pattern; the details of that diagnosis is dealt with in a later section of this article.

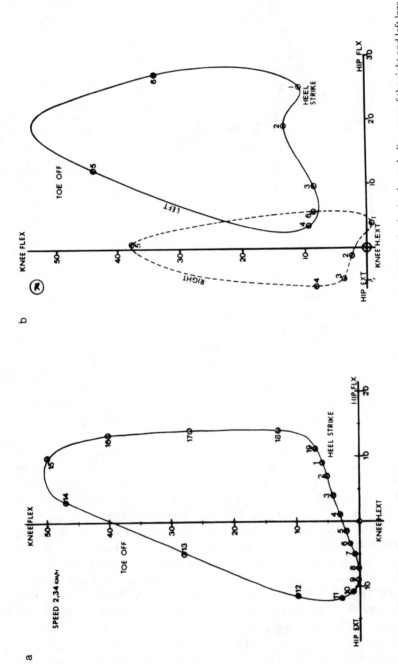

Figure 5—a. Angle-angle diagram of knee vs hip during gait of a normal subject walking at 2.34 km/hr. b. Angle-angle diagrams of the right and left legs for a patient with osteoarthritis of the right hip. Such plots are fairly sensitive to changes in either the knee or hip angle curves.

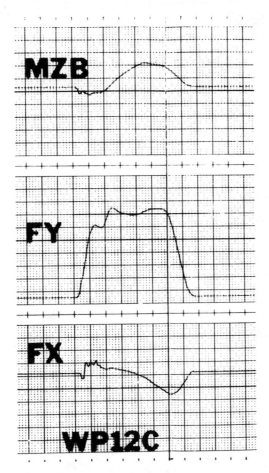

Figure 6—Pen recordings of the vertical (Fy) and horizontal (Fx) ground reaction forces of a patient with a knee arthroplasty. Curves indicate an abnormal pattern, but yield no information as to the abnormal muscle activity that is causing that pattern.

Assessment of Orthotic Devices

This section concentrates on the use of biomechanical analyses in the comparison of two orthotic devices: a spiral ankle foot orthoses (AFO) vs the more conventional rigid AFO. The two devices were assessed in the gait of a 7-year-old girl with a flail foot. Kinematic analyses of ankle, knee and hip angles over the stride are plotted in Figure 7. During the stance phase, the three joints had somewhat similar patterns; the major differences were during the swing phase, especially at the knee and ankle. The spiral AFO resulted in a more normal knee flexion while the ankle joint pattern was more normal for the rigid AFO. Based on these kinematic patterns, no conclusion can be drawn as to which device was

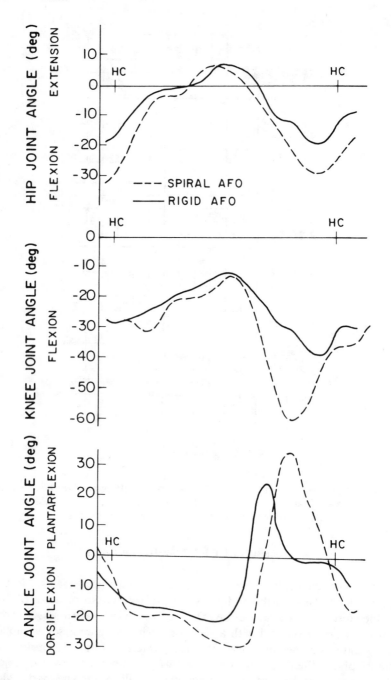

Figure 7—Kinematic pattern of hip, knee and ankle angles of a young child with a flail foot walking with two different types of braces. The rigid and spiral polypropylene orthoses had slightly different patterns during the stance phase and somewhat different during the swing phase but differences were such as to prevent a judgment as to which brace was better.

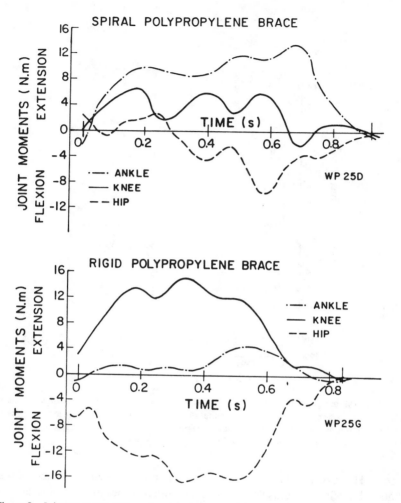

Figure 8—Joint moments of force for the same two trials presented in Figure 7. The spiral AFO gave a much higher plantar flexor moment than did the rigid AFO at the ankle; the compensation moments at the knee and hip for the spiral AFO were much less than those for the rigid orthoses, indicating that the spiral AFO was superior.

better. A look at the kinetic analyses of joint moments of force, however, reveals distinct differences over the stance period (see Figure 8). The ankle moment for the spiral AFO showed that the device was generating a good extensor moment (averaging 10 N.m) as the leg rotated over the flat foot. The rigid AFO, on the other hand, produced a very small moment which was below 2 N.m for the first half of stance and rising to only 4 N.m during push-off. The compensation for this low ankle moment can be seen at the other two joints. The knee had a strong extensor moment for the majority of stance, and the hip a strong flexor mo-

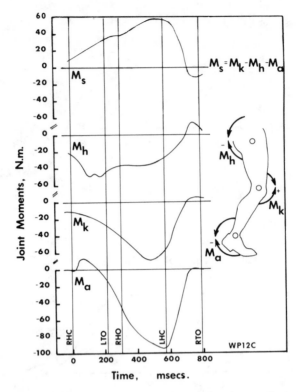

Figure 9—Moments of force calculated for the knee arthroplasty patient whose ground reaction forces appear in Figure 6. Abnormal moments are present at the knee which has a strong flexor moment during most of the stance phase, with a major extensor compensation at the hip to prevent collapse of the knee during the stance phase.

ment. However, when the child wore the spiral brace the hip and knee had relatively low muscle activity, presumably because the brace she wore was giving adequate ankle support as already described. Therefore, it could be concluded from this kinetic comparison that the spiral AFO was somewhat better than the rigid AFO.

Diagnosis of Gait Pathologies

Diagnosis is the term that describes the cause of a pathological pattern, and therefore must focus on the forces, energies or work related to the cause of movement. The ideal analysis would yield the forces or moments generated by each major muscle, an impossibility at this time. What is possible is a joint moment of force analysis, and such an analysis is presented for a knee arthroplasty patient (see Figure 9). This 73-year-old patient walked at a good cadence with a fairly stiff leg during stance (maximum knee flexion = 3°) and produced the ground reaction forces as presented earlier in Figure 6. The cause of the stiff-legged gait was not, as might be suspected, an abnormally high knee extensor moment. Rather, she had strong knee flexor activity for the entire stance time. What produced the almost-locked knee was a strong burst of hip extensor activity immediately after heel contact and continuing during almost all of stance. The diagnosis of this patient's gait concluded that she was unloading (flexing) her pathological joint and preventing knee collapse with her hip extensors. Such a diagnosis was not available from the joint kinematics nor from the ground reaction force curves.

Discussion and Conclusions

No definitive answers are available at this time as to which of the many variables give the best diagnostic information or are most sensitive to the rehabilitation process. However, evidence has been presented to show that certain kinematic variables are suitable for monitoring purposes only but are inadequate as a diagnostic measure. Also, it is evident that the baseline data for normals are still fairly incomplete. We must strive, however, to assemble sufficient data and ensure that it is normalized so that direct comparisons can be made among laboratories, or between a given patient and the normal population. Finally, when diagnostic assessments are being attempted it would be useful to get at the kinetics or EMG variables that reflect the cause rather than the effect of the abnormal movement.

Acknowledgment

The author wishes to acknowledge the support of the Medical Research Council (Grant MT 4343) and the technical assistance of Mr. Paul Guy.

References

BATTYE, C.K., and Joseph, J. 1966. An investigation by telemetering of the activity of some muscles in walking. Med. Biol. Engng. 4:125-135.

DUBO, H., et al. 1976. Electromyographic-temporal analysis of normal gait. Arch. Phys. Med. & Rehab. 57:415-420.

FOLEY, C.D., Quanbury, A.O., and Steinke, T. 1979. Kinematics of normal child locomotion—A statistical study based on TV data. J. Biomech. 12:1-6.

GRIEVE, D.W., and Gear, R.J. 1966. The relationships between length of stride, step frequency, time of swing and speed of walking for children and adults. Ergonomics 9:379-399.

LONGHURST, S. 1980. Variability of EMG during slow walking. Human Locomotion I, Proc. Spec. Conf. of Cdn. Soc. Biomech. London, Canada, pp. 10-11.

MILNER, M., Basmajian, J.V., and Quanbury, A.O. 1971. Multifactorial analysis of walking by electromyography and computer. Am. J. Phys. Med. 50:235-258.

MURRAY, M.P., Drought, A.B., and Kory, R.C. 1964. Walking patterns of normal men. J. Bone Jt. Surg. 46-A:335-360.

MURRAY, M.P., Kory, R.C., and Sepic, S.F. 1970. Walking patterns of normal women. Arch. Phys. Med. & Rehab. 51:637-650.

PEDOTTI, A. 1977. A study of motor coordination and neuromuscular activities in human locomotion. Biol. Cybernetics 26:53-62.

STATHAM, M.S., and Murray, M.P. 1971. Early walking patterns of normal children. Clin. Orthop. & Rel. Res. 79:8-24.

SUTHERLAND, D.H. 1966. An electromyographic study of the plantar flexors of the ankle in normal walking on the level. J. Bone Jt. Surg. 48-A:66-71.

SUTHERLAND, D.H., Olshen, R., Cooper, L., and Woo, S.L. 1980. The development of mature gait. J. Bone Jt. Surg. 62-A:336-353.

WINTER, D.A., et al. 1974. Kinematics of normal locomotion—A statistical study based on TV data. J. Biomech. 7:479-486.

WINTER, D.A. 1980. Overall principle of lower limb support during stance phase of gait. J. Biomech. 13:923-927.

Methodology and Technical Aids for Substitution of Upper Human Extremities Functions — Where Are We Going?

Adam Morecki

Technical University of Warsaw, Warsaw, Poland

This article presents some ideas concerning the possibilities for modern biomechanics in connection with rehabilitation engineering and its limitations. According to statistical data, about 10 to 12% of the population in developed countries are disabled and half of these individuals need the assistance of special devices for substituting or supporting the lost functions in manipulation and locomotion activities. The experiences of the last 20 years show that it is not enough to transfer new technology to hospitals and clinics, but it is necessary to create proper conditions for its adaptation and use.

Before starting with the design of technical aids intended as substitutes for lost functions of the human being, it is necessary to recognize the possibilities of the basic systems of human motion.

Functional electrical stimulation methods are a group of very interesting aids for the rehabilitation of spinal cord injuries. Some selected problems concerning the live manipulators will be discussed as an example of human prehension movement. In the case of spinal cord injuries at the C5 to C6 level, some kinematic functions of the upper extremities are possible, but not prehension. Some results of application of the so-called prosthesis of the human control system — functional electric methods for performing the prehension movements — are presented.

Recently, independent from the prosthetics-orthotics direction, new trends in the design of devices for supporting the lost functions of the upper extremities of humans are being developed. There are different types of manipulators, which enable patients to contact the world while at home, and permit them to carry out some work activities. One possible solution — the so-called head manipulator with chin control — is an example.

Locomotion activity is one of the most important movement activities in humans. Due to various injuries, the mechanism of locomotion can be partially or fully damaged. Much work has been done in the area of design of assistive devices for locomotion, for example, exoskeletons with active control and electrical wheelchairs. During the rehabilitation process, in supporting or restoring the gait capabilities a criterion for diagnosis is needed.

In the second part of this article a dynamic model of human locomotion is presented and a proper criterion is proposed.

Modelling and Identification Problems of Basic Systems of Human Movement Activity

Very interesting possibilities for rehabilitation engineering are opened by the so-called cybernetic machines—artificial devices, which are designated for partial substitution of energetic, physiological and intellectual human functions. In special cases of diseases such a machine can replace some human parts, for example, an extremity.

The substitution of energetic functions means the replacement of physical work. The substitution of physiological functions means replacing a patient's internal or external organs and fitting him or her, for example, with an artificial kidney or extremity. The substitution of intellectual functions means adapting properties of a cybernetic machine to the surroundings. Technical realization of such a machine can be obtained usually in the form of a bionic system, for the proper degree of anthropomorphism.

Under cybernetic mechanisms, one can classify those parts of a cybernetic machine which replace some human motor activity, like manipulation, locomotion, blood flow, respiration and so on. Such a cybernetic mechanism in bionic realization constitutes a special tool, which increases the possibilities of the extremities or organs of a human and supports or substitutes for them in either action or function.

Figure 1 shows a general cybernetic model of human movement activity proposed by Morecki, et al. (1981). The model shown in Figure 1 consists of three basic systems: control, movement, and supply. The fourth system represents the environment with which the entire system interacts. General feedbacks and interactions are indicated in the diagram with arrows.

From the point of view of a modern theory of automatic control, the whole system can be characterized as a hierarchic, multilevel, and centralized system having limited autonomy at determined levels.

The control system consists of two basic parts: somatic and vegetative. The system of higher intellectual activity, which controls such functions as determining, memory, emotions, etc., is the superior part connected

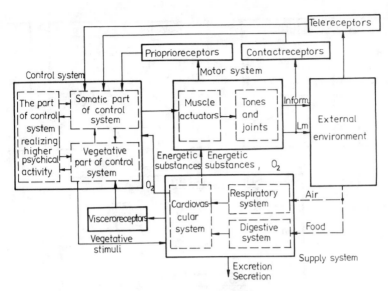

Figure 1 — General cybernetic model of human movement activity.

with the mechanical elements. Movement stimuli are transmitted to each particular muscle actuator which produces the movement in corresponding joints. The outputs of the system are at the same time the inputs from the external environment. Information about the state of the environment is transmitted into the somatic system by teleoperation and contact-reception loops and information about the state of the motor system by the proprioreception loop. The role of the supply system is to take food and air from the environment, transform the food into energetic substances, and absorb oxygen from the air. These substances are distributed by the cardiovascular system. By visceroreception the supply system transmits information to the vegetative part of the control system, from which the information flows back through the respective feedback loops. Energetic substances transmitted both to the motor and the control systems are energy carriers.

The proposed model of human motor activity enables one to conduct different identification and operation analyses. These analyses can be carried out, for example, in the following fields:

1. Informational and energetic states of the systems. In the case of energetic actions of the system (by maximum values of nerve impulses) the connection between motor and supply systems is secured and the output of the system is an energetic one. In the case of informational activity of the system the connection between motor and control systems is secured and the output of the whole system is one of information. Such a system creates optimal motor control (technique of motion) as well as expression of thoughts and feelings (writing, speech, facial expression).

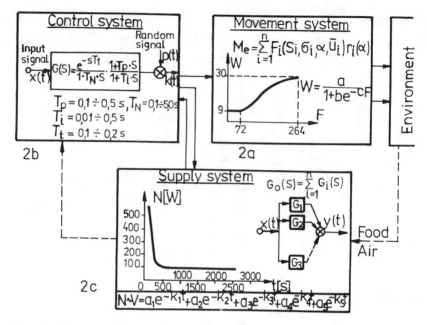

Figure 2—Mathematical description of the cybernetic model of human movement activity.

2. Topological description of a structure, that is, the number of functions of muscle actuators (drives) and their distribution in relation to the number of joints (biokinematic pairs). It should be noted that the motor and control systems for some defined movements are comparatively well examined.

3. Establishment of control laws as subordinate to motor activity, especially for the upper and lower extremities.

4. Estimation of muscle actuators to cooperate in the execution of different movement combinations under static and dynamic conditions.

Figure 2 shows the attempt of a mathematical description of the motion, control, and supply systems of movement activity of a human. The moving system can be described by three relationships, namely (Figure 2a): a) the logistic curve, b) interrelated functions, and c) level of cooperation of muscle actuators operated in a joint with 1°, 2°, and 3° of freedom.

A model of the control systems which would satisfy experimental requirements is shown in Figure 2b. The model, as a system of automatic control, includes both systems for movement and control. In this system an eye is placed as the input.

One can estimate the suppiy system by the value of the output developed during control of movement. The relationship of the output versus time is given in Figure 2c. This curve was determined experimentally (Morecki, 1980a; Morecki, et al., 1981a).

The results can be used by designers and biomedical engineers who have an interest in technical aids for rehabilitation purposes. On the other hand, the complexity of the systems imposes limitations on the designs of these aids.

Reasons for Motor Dysfunction and Analysis of Retained Motor Functions

Motor dysfunction can be caused either by trauma, or past as well as current sickness (e.g., poliomyelitis, apoplexy, polineuropatia, and myopathia). Another reason, which practically eliminates the use of technical means for restoring the motor functions, is muscular spasticity. The latter occurs as a result of high-level damage to the motor control system, or, in other words, damage to the central nervous system.

All of these causes have one feature in common: regardless of the level of primary damage, they cause total disintegration of motor functions and lead to various disturbances and constraints on the functions of the inflicted motor organ. The primary changes are followed by secondary ones, which deteriorate the passive properties of the extremity (for example, they affect the extremity's neutral position, change its range of motion, etc.).

In a case of extremity paralysis, the patient's motor organ has to be analyzed in order to reveal all the activities and motions which may be used in order to compensate for lost functions, or in the control of an external supporting apparatus.

The deficit of motor function cannot be considered apart from the extremity's sensory function. Existing sensory function (such as tactile feeling in the hand) calls for application of technical equipment in restoring prehension. Lack of this function creates very unfavorable conditions for application of any equipment, since the parts of the device exert pressure on the patient's flesh and may irritate the skin and cause mechanical damage. Another important factor to be considered in the patient's rehabilitation is the compensatory function, which consists of taking advantage of other functional parts of the motor organ.

The fitting of a patient with an upper extremity prosthesis, orthoses, implanted stimulators or a medical manipulator, should be preceded by a thorough analysis (Morecki et al., 1974). Before a supporting device is applied, the following tasks should be completed: 1) thorough analysis of the patient's motions and control sites, 2) determination of the needs of a patient, in relation to his or her current possibilities, 3) determination of the patient's emotional and psychological states and predisposition to work with the apparatus, 4) consideration of the possibilities for the patient to resume a professional life in terms of health, profession, family status, financial situation, and the like.

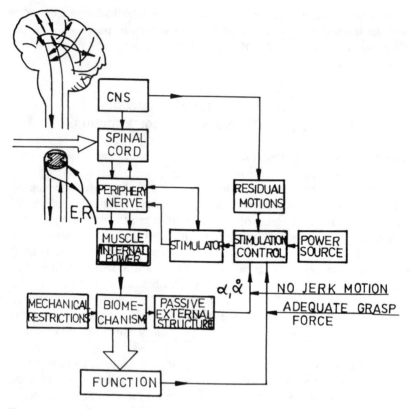

Figure 3 — Diagram of a stimulation orthostesis.

Live Manipulators — Stimulation Orthostheses

Damage to the spinal cord, regardless of the lesion level, results in vast changes in motor activity of the organism. The lesion site becomes a temporary or permanent breaking point of ascending and descending signal pathways.

In such a case, the brain cannot receive information being sent from parts of the body served by the spinal cord and nerves below the level of the lesion. This lack of information makes it impossible for the brain to form an adequate answer to what is really happening in the peripheral zone. On the other hand, the commands sent by the brain do not reach the peripheral zone because the descending pathways of information are blocked by the lesion.

In such a situation, that which happens in the peripheral zone cannot be controlled by the brain, and may be created by the part of the spinal cord which extends below the lesion. Such activity can be manifested by uncontrolled spasticity, or by aimless and undirected muscular response to various stimuli coming from the peripheral zone. The patients with such damage are indifferent to such environmental stimuli as high and low temperatures, pain, and touch. They are also unable to perform a purposeful motion since the flow of necessary brain impulses through the spinal cord and peripheral nerves to the muscles is blocked.

Figure 3 shows diagramatically the nature of the damage. The receptor (R) and effector (E) are intact. Nevertheless, the stimuli received by the receptors reach only the spinal cord below the lesion, and cannot be transmitted to the brain level. The effector, also intact, cannot fulfill its task because of the lack of nerve impulses from the brain, which normally controls its activity.

Spinal centers above and below the place of trauma — analogically to the peripheral nerves — are not damaged; both the nerve and the muscle preserve their properties of reacting to various kinds of stimuli, which means that they retain their excitability.

In this situation a special method for forcing the grasping function was proposed. In this method, the links of the extremity are driven by the retained muscles powered from their own biochemical sources of energy. Low-energy external sources of electrical energy used in this method only play a role of control for the system (Pasniczek et al., 1977; Morecki et al., 1978; Morecki et al., 1980b; Morecki et al., 1981; Weiss et al., 1981).

When proposing the method, the following are the more important acceptable assumptions.

1. When the connection of the peripheral centers with the central nervous system has been functionally blocked as a result of trauma or so-called 'spinal shock' (while the connection remains uninjured), it is the aim of this method to reproduce the connection by means of 'wiping clear ways' through constant 'bombardment' with stimulating impulses in order to obtain some, even if only partial, neurological improvement.

2. The development of a suitable stimulation system making it possible to force the function of the paralyzed hand by means of implanted stimulators to stimulate nerves, which supply the motor fibers in the extremity.

3. The assurance of constant stimulation training according to a program decided upon for the supplying nerves, taking into account the prevention of secondary changes and muscle atrophy.

The apparatus would ensure that the patient with paralyzed upper extremities has a possibility of grasping various objects and displacing them in the desired direction. Taking into account existing disturbances of sensation in the paralyzed extremity and the fact that the patient lacks information through the central nervous system about the surface and the mass of the object that should be displaced and/or lifted, this information should first be obtained by technical means and then the force of stimulating impulses should be suitably and automatically chosen.

Results of Introductory Investigations

During investigations carried out in the years 1971 to 1976, the influence of electrical stimulation of peripheral nerves on functional improvement of the paralyzed hand of tetraplegic patients, in cases of spinal cord injuries at the C5 to C7 level, was verified under clinical conditions (Morecki et al., 1978; Pasniczek et al., 1977).

The median nerve (medianus) was stimulated with the use of implanted stimulators in order to obtain a grasping function of the hand. In several cases, the radial nerve (radialis) was also stimulated with a view toward increasing hand opening. The stimulators were implanted under the skin of the patient's forearm and the electrodes were attached to the epineurum of the nerve.

Fourteen tetraplegic patients were examined, in whom 26 stimulators were implanted. Patients were chosen for experiments in whom there was no neurological improvement as a result of treatments with traditional methods.

The stimulation was performed each day for 6 months following the operation. The daily program (three to five stimulations each lasting five min, five sec of stimulation, 30 sec relaxation for five min, and after that a rest interval for 55 min) was performed in a cycle lasting from two to four hours. The duration of stimulation was 0.7 msec and the frequency 45 Hz.

Using this stimulation method, in 11 examined cases, an increase in the force developed by the oponens pollicis was obtained as well as an increase of time in which the maximum moment of the force during stimulation was maintained.

During the examination an interesting effect was observed, which may have great practical importance when functional stimulation of the paralyzed upper extremities is to be conducted. This effect is the possibility for the patient to increase the force exerted by the stimulated muscles.

Based on the results obtained during the preliminary investigations, an apparatus for program control of paralyzed muscles in order to obtain useful movements of the paralyzed extremity of the tetraplegic patient was designed, and suitable experiments were carried out from 1976 to 1980.

Conception of a Hybrid System
which Forces the Grasping Movement

A system which forces the grasping movement can be achieved by simultaneous cooperation between a patient and a system of stimulators implanted on the nerves, and external apparatus.

The whole system is composed of the following subsystems: 1) palm forming (orthotic device), 2) implanted stimulators placed on the nerves, 3) starting and ending devices (input), and 4) information transducers, including a position-measuring transducer, force transducer and sensor transducer.

A position-measuring transducer-precision potentiometer assures the possibility of controlling the position of the fingers during opening and closing of the hand. The force transducer, by maintaining contact with the object, makes it possible to develop the force necessary for picking up or shifting the given object.

This system insures direct control of the duration of the stimulating signal so as to secure a firm grip. In turn, the programming system controls the work of an individual subsystem in such a manner that after a patient starts a movement, all the remaining operations occur automatically at suitable time sequences.

Stimulators, implanted on the radial and medial nerve of the paralyzed extremity, have been used to stimulate selected muscle groups. Suitably programmed cooperation of such pairs of stimulators creates favorable conditions for performing functionally useful motion (Morecki et al., 1978; Morecki et al., 1980b; Weiss et al., 1981).

The Basic Tasks of the Grasping System

Figure 4 presents the sequence of activity in grasping. The whole process consists of three phases and activities A and B. In Phase 1, the device is turned on and is ready for operation. The flexors and extensors are not stimulated. In Phase 2, the hand is extended (the fingers are straightened). The stimulation of the flexors is turned off. The stimulation of the extensors is turned on and the third phase consists of a sequence of two activities:

1. Activity A: After turning the stimulation on and increasing with time the stimuli sent to the flexors, the hand grasps an object. The system of velocity control is turned on to coordinate the finger flexion.

2. Activity B: The object is lifted. As a result of the stimulation of the flexors by means of the force reflecting transducer, the grasping force is adjusted to the weight of the lifted object.

The device is controlled by means of one signal source. Switching on successive phases of motion is accomplished through application of impulses evoked by the patient's head movement.

Clinical Testing of the Hybrid Device

Tests were carried out on two patients with spinal cord injury at the C6

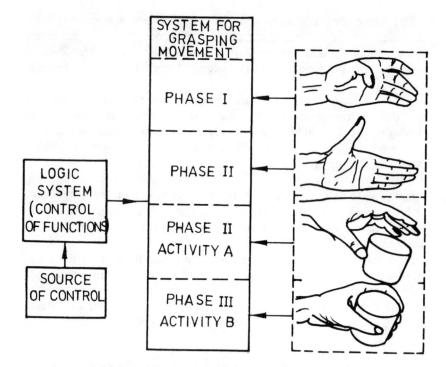

Figure 4—The sequence of activities in the grasping movement.

level by implanting two stimulators on the median and radial nerves. An orthotic with the position transducer installed in the metacarpophalangeal articulation was mounted on the right extremity (see Figure 5). Touch and displacement transducers were attached to the fingers by a bandage. A touch transducer was placed in the region of the proximal interphalangeal articulation of the second finger and that of displacement near the distal interphalangeal articulation of the third finger. Control of the device was achieved by means of a switch activated by a person performing the experiments.

The tension level of the radial nerve stimulation, responsible for finger extension, was chosen in such a manner as to obtain full opening of the hand. The range of tension or duration of stimulation impulses to the median nerve were so defined as to have the force developed by the muscles of finger flexion almost maximum, and simultaneously keep the stimulation pain as low as possible.

The preliminary investigations were conducted using a system in which the grasp force was controlled by changes in the voltage amplitude of the stimulation impulses, the duration and frequency of which were set at 0.2 msec and 50 kHz, respectively. The record of the angle of finger closing versus time (see Figures 6a and 6b) shows that the motion is performed in

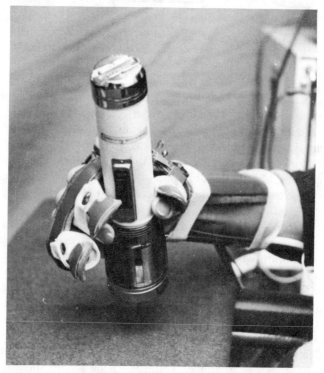

Figure 5—Lifting the object.

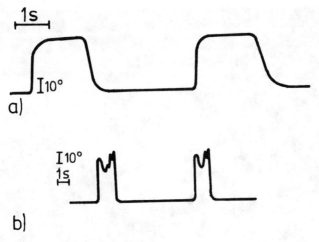

Figure 6—Angle of finger closing versus time.

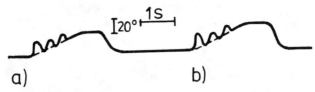

Figure 7—Finger flexion controlled by stimulation of finger extensors.

a very short time and is independent of the rate of change of voltage. The feedback loop has only a small, if any, effect, because of the large inertia of the muscles and time delay of signal transmission in the neuro-muscular system (Morecki et al., 1981b).

Further investigations have shown that a very short duration of motion was caused by the first impulse of about 50 μsec, which resulted in almost full closing of the fingers (more than 80% of the entire motion range). The estimated time delay was about 80 msec.

Another concept investigated was that in which the controlled finger flexion was coupled with stimulation of finger extensors in order to slow down the motion when a predetermined angular velocity of fingers was reached (see Figure 7).

Also, the characteristics of another method of finger flexion control were investigated. In this method the impulse time, associated with the state of muscular excitation, was used to control flexion of the fingers. The amplitude of voltage was set so that 50 μsec impulses caused no motion. The impulse time could be stretched to 220 μsec (the control performed by a change in applied voltage). Using this method in an open control system, a record of finger closing angle versus time was made (see

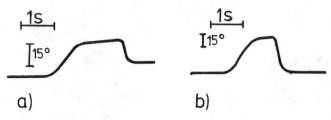

Figure 8—Angle of finger closing versus time controlled by impulse time.

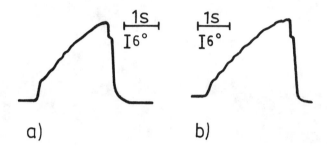

Figure 9—Angle of finger closing versus time with feedback loop control.

Figure 8). It is evident from the graph that motion is performed without abrupt changes in acceleration. The deviations of the graph of actual motion from the theoretical relationship represented by a straight line are less than ± 10%.

Figure 9 shows the record of the angle of closing of the hand versus time for the system fitted with a feedback loop. In this case, the performed motion is fairly smooth.

The possibility of voluntary support of muscle contraction during efferent overthreshold stimulation (or at the threshold border) creates the possibilities for reeducation of tetraplegic functions without the need for creating a control system by several stimulators placed on several nerves. However, the question arises: whether the artificial stimulation of the median nerve influences the possibility for active strengthening of muscle contractions conditioned by stimulation. This system is still under investigation in the Clinic of Rehabilitation at the Warsaw School of Medicine.

Head Manipulator for
Bilateral Shoulder Disarticulated Children

The purpose of the development of a head manipulator was to create the possibility for bilateral shoulder disarticulated children to play and work

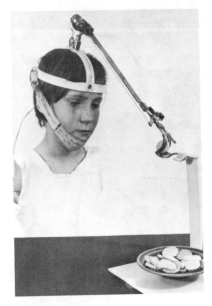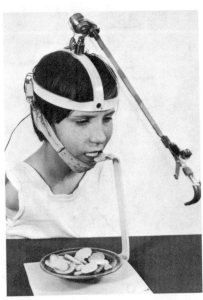

Figure 10 — Manipulator in action.

Figure 11 — Feeding using a manipulator and the spoon grip.

on a table surface rather than on the floor or a bed surface.

Observing a patient with a bilateral arm amputation using gross body motion for manipulatory functions, as in writing on a table surface with the help of the prosthesis, one can see that the lumbar and [thoracic areas of the spine move] the upper body while the prosthesis only transfers this motion to the pencil. This activity is performed by a large group of muscles which moves the large mass. Such an activity has extremely low efficiency and the patients get tired quickly. On the other hand, restoring the manipulatory function by means of current technical aids is today a very difficult task. It seems that the only reasonable way for optimum utilization of the existing motor functions of patients is to equip them with simple devices.

Writing is one of the most important activities for children. Unfortunately, without special devices patients are required to keep a pencil in their mouth and in this case the distance between the eyes and the pencil is not adequate. It requires extra focusing of the eye on the written text. It was decided to support this kind of activity by fixing the gripping mechanism on the patient's head at the distance of 250 mm from the eyes (see Figures 10 and 11).

The electrically operated hook is mounted on the end of the outrigger bar which is connected through two passive joints with the hand fixed on the patient's head. The stabilization of the headband is achieved by carefully fitting it to the shape of the head and adding the jaw strap. The tension of the jaw strap is controlled by the forward shifting of the jaw. The slight tension on the strap controls, by means of a four-position micro-switch unit, provides for the opening and closing of the hook. The children could spoon-feed themselves (see Figure 11). Three children, aged 3, 10, and 16 years, were fitted with this manipulator with very good results. They played and worked with the help of a manipulator for several hours each day (Ober et al., 1977).

The manipulators provide the children with the capability of writing, drawing, and playing on a table surface for many hours without fatigue. The manipulator may also be used by quadriplegic patients in electric wheelchairs.

Criterion for the Analysis of Human Gait

At the Warsaw Technical University, during the past several years, an intensive study of human locomotion has been undertaken. A mechanical model of the human body with 13° of freedom has been used for this purpose (Morecki et al., 1975). At the beginning of this investigation, human motion during running was recorded by a photographic chrono-cyclographic method. The three components of ground reaction force were measured by a single force plate. The model was verified by these experiments and good results were obtained using off-line data analysis (Morecki et al., 1975; Morecki et al., 1981a).

Recently, the mechanical model of the human body described by Morecki et al. (1975) has been modified to include a double support phase in the gait. In the previously used model, the system of coordinates was attached to the ankle joint, which required excessive numerical sensitivity during computations. For this reason, the coordinates have been moved to the hip joint.

The aim of the investigation was to determine the static and dynamic moments acting at all joints, as functions of time or relative angles, and to propose a proper criterion for diagnosis.

The modified planar model with 13° of freedom and equations of motion are given in Morecki et al. (1981c). The equations are complemented with positions of centers of gravity and values of mass movements of inertia.

Moreover, it is necessary to know the change of joint angles ϕ_i, their first and second derivatives, and two linear coordinates for the displacement of the hip joint axis (x_{ref}, y_{ref}). It is also necessary to know two components of the ground reaction forces of the left leg (F_{x4}, F_{y4}) and moment of F_{y4} force about the axis of the metatarsus joint.

After substituting these data into equations one obtains: 1) reactions F_{x1}, F_{y1} and M_1, for the right leg, and 2) moments of forces versus time M_i (t) and relative angles M_i (Θ_k).

The purpose of experimental verification was to compare values of the measured quantities, $F_{x1_{exp.}}$, $F_{y1_{exp.}}$, with the calculated ones. In the experiment, two force platforms were used. For small differences between the compared values, it was assumed the model was adequate and the calculated values of the moments (M_i) were close to reality.

The procedure followed in solving the formulated problem is as follows:

1. After shooting a film of a given motion, one can obtain from the film Cartesian coordinates of chosen joints (x_i, y_i) where $i = 1, 2, \ldots, 7$). Based on these coordinates, joint angles ϕ_i are calculated.

2. Using the method of adjustment calculus, velocities and accelerations are computed. In the same step, x_{ref} and y_{ref} are calculated. In this case a symmetry of motion is assumed, which means that calculations are conducted for six angles (ϕ_1, ϕ_2, ϕ_3, ϕ_7, ϕ_8, ϕ_9) and the remaining angles are determined on the basis of these calculations.

3. With use of the two force platforms, the vertical and the horizontal components of the ground reaction force are measured for the left and the right foot, and then the moments M_1 and M_4 are computed.

First, the numerical values for F_{x4}, F_{y4}, and M_4 are substituted into the equations and then F_{x1}, F_{y1}, and M_1 are calculated. Forces F_{x1} and F_{y1} are measured and moment M_1 is calculated with the purpose of verifying the moments (M_i) developed by muscles in the individual joint.

The computations are conducted with the use of a set of computer programs labeled HLS (Human Locomotion System). The programs are written in FORTRAN IV language and are intended for a CYBER 70 computer system.

The instrumentation used in this system includes: 1) PENTAZET-35 movie camera; 2) K-202 minicomputer with CPO-2 digital image processor (Kulpa and Dernalowicz, 1978); 3) CYBER-70/7216 model computer system; 4) HP-25 programmed calculator; 5) two force platforms; 6) recorder for recording the acceleration signals (from accelerometers) and the ground reaction components (from the force platforms); and 7) stereocomparator (Kulpa and Dernalowicz, 1978; Jaworek and Morecki, 1981).

The procedure runs the programs in the following sequence: 1) Cartesian Coordinates, 2) Converter, 3) Adjustment Calculus, 4) Force-Point-Moment, 5) Moments in Human Joints, and 6) Comparator.

The program 'Cartesian Coordinates' yields rectangular Cartesian coordinates for seven chosen human joints. Program 'Converter' converts information about coordinates of the joint axes from the paper tape to perforated cards. Angular coordinates, velocities and accelerations are determined by means of the 'Adjustment Calculus'. The components of the ground reaction, and moments of those forces are calculated with use of the 'Force-Point-Moment' program (see Figure 12).

The values of components of the ground reaction for the assumed

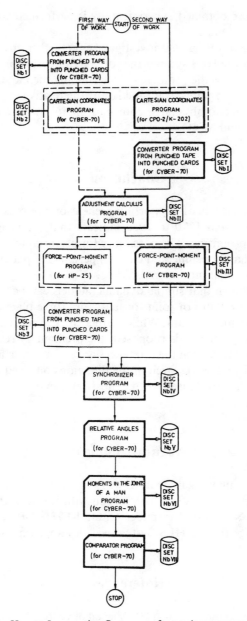

Figure 12 — HLS — Human Locomotion System — software (programs package).

model are computed with use of the program 'Moments in Human Joints' and the final results are obtained from running the programs: 'Cartesian Coordinates', 'Adjustment Calculus', and 'Force-Point-Moment'. The comparison of the results of computations made for the model with those obtained experimentally is performed with use of the program 'Comparator'.

The objective of the current investigations is further improvement and development of the HLS system, the software in particular. Great attention is being given to the problem of determining the moments of forces developed in human joints as a function of the relative angles. The HLS system enables one to obtain the course of the following quantities: F_{x1}, F_{y1}, M_1, $M_i(t)$, and $M_i (\Theta_k)$.

Conclusion

In this article, methods of supporting prehension are presented and the results obtained from 1970 to 1980 indicate that the proposed hybrid method is very promising. The possibility for performing smooth motion of the fingers (hand opening and closing) permits the patient to perform useful and functional movements.

It has been shown that simple dynamic models can be used to obtain off- or on-line estimates of joint torques from force plate data together with appropriate motion data. While further work is needed to determine the accuracy of the methods proposed, the preliminary results reported here and detailed further in the references are encouraging in this regard. We believe that further research in the directions outlined could lead to improved clinical procedures for diagnosis and management of neuro-musculoskeletal deficiencies.

Acknowledgment

This research on human gait analysis was supported in part by the Polish Academy of Sciences (PAS) in connection with Key Sub-Problem 10.4, and jointly by NSF and PAS under Grant J-F7F061-P. The computer program package for the HLS System was developed by K. Jaworek.

References

JAWOREK, K., and Morecki, A. 1981. Method of verification of kinematic and dynamic properties of a biped locomotion model. 8th International Conference on Dynamics of Machines, Warsaw, April 6-10.

KULPA, Z., and Dernalowicz, J. 1978. Digital image analysis system CPO-2/K-202: General description. Proceedings of 6th Polish-Italian Symposium on Pattern Recognition for Biocybernetic Applications, Porto Ischia Felice, Italy.

MORECKI, A., et al. 1974. Manipulators for supporting and substituting lost functions of human extremities. Proceedings on Theory and Practice of Robots and Manipulators, Vol. 1, Udine, pp. 214-230.

MORECKI, A., Busko, Z., Kedzior, K. and Olszewski, J. 1975. Biomechanical modelling of human motion. Proceedings of 6th World Congress on the TMM, Vol. 4, Newcastle upon Tyne, England, Sept. 8-13, pp. 793-798.

MORECKI, A., Borowski, H., Gasztold, H., Kiwerski, J., and Paśniczek, R. 1978. On two systems for tetraplegics. Colloques IRIA, International Conference on Telemanipulators for the Physically Hancicapped. Institute de Recherche d'Informatique, Rocquencourt, Sept. 4-6, pp. 237-251.

MORECKI, A. 1980a. Biomechanics of motion. International Centre for Mechanical Sciences and Lectures, No.263. Springer Verlag, Wien, New York.

MORECKI, A., Weiss, M., Kiwerski, J., and Paśniczek, R. 1980b. A new method for forcing lost grasping functions of extremities by use of an orthotic manipulator combined with implanted stimulators of nerves. International Conference on Medical Devices and Sports Equipment, August 18-20, 1980, Century 2, Emerging Technology Conferences, August 10-21.

MORECKI, A., Ekiel, J., and Fidelus, K. 1981a-1982a. Cybernetic systems of limb movements in man, animals and robots. E. Horwood, International Publications in Science and Technology, (in preparation), Catalogue.

MORECKI, A., Borowski, H., Kiwerski, J., and Paśniczek, R. 1981b. Substituting of upper human extremities functions – where are we going? Medical Informatics Europe 81 Proceedings, Toulouse, France. Springer Verlag, Berlin, pp. 823-830.

MORECKI, A., Olszewski, J., Jaworek, K., McGhee, R.B., Koozekanani, S.H., and Burnett, C.N. 1981c. Automatic computer analysis of human gait. In: A. Morecki and K. Fidelus (eds.), Biomechanics VII-B, pp. 133-140, PWN, Warsaw, and University Park Press, Baltimore.

MORECKI, A., Koozekanani, S.H., McGhee, R.B., Johnson, E., Jaworek, K., Olszewski, J., and Rahmani, S. 1981d. Reduced order dynamic models for computer analysis of human gait. Preprints, 4th CISM-IFTOMM "Ro.man.sy-81" Symposium, Warsaw, Sept. 1981. (in press)

OBER, J.K., Leonhard, J., and Ogórkiewicz, A. 1977. On the head manipulator for bilateral shoulder disarticulated children. Proceedings on Theory and Practice of Robots & Manipulators, PWN-Elsevier, pp. 206-210.

PAŚNICZEK, R., Kiwerski, J., Wirski, J., and Borowski, H. 1977. Some problems of implant stimulations applied to grasp movements. Proceedings of the 4th International Symposium, Advances in External Control of Human Extremities, Belgrade, pp. 584-602.

WEISS, M., Kiwerski, J., Paśniczek, R., and Morecki, A. 1981. An electronic hybrid device for control of hand functions by electrical stimulation methods. In: A. Morecki and K. Fidelus (eds.), Biomechanics VIIA, pp. 397-404, PWN, Poland, and University Park Press, Baltimore.

A.
Evaluation

Evaluation of EMG Activities Measured with Surface Electrodes During Abduction of the Shoulder in Patients with Spinal Muscular Atrophy and Facioscapulohumeral Dystrophy

R.J. Boukes, K.L. Boon, G.J. Docter, and J. Ensink
Vrije Universiteit, Amsterdam, Holland

A characteristic symptom of neuromuscular disorders, especially in progressive muscular dystrophies and spinal muscular atrophies, is the progressive loss of muscular strength. Whether and to what extent this loss of muscular strength can be used as a proper measure in obtaining an objective impression of the course of the disease is still a matter of discussion. Qualitatively the strength of muscles can be determined by manual muscle testing techniques (Lovett, 1915; Kendall and Kendall, 1938; Daniels, 1956). Forces can be measured in a quantitative way by means of dynamometers (measurement of the maximal voluntary contraction force, MVC). In patients, this method has been used by Fowler and Gardner (1967), Hosking, Dubowitz, et al. (1976), and de Lateur and Giaconi (1979). It has been our experience, however, that MVC measurements in patients with a neuromuscular disorder are very difficult to interpret, because of the considerable variability in the values of MVC.

In most cases, it was hardly possible to get an impression of the course of disease by this method (see Figure 1a). Measurement of the abduction angle of the shoulder yields more reproducible results (see Figure 1b) and seems to be a better test for this purpose (Boukes et al., 1981). We were interested whether EMG analysis can contribute to our understanding of the relation between loss of function and the deterioration of individual muscles in these kinds of disorders. In the present study attention is focused on surface EMG, which are the interesting parameters. It would be valuable to know to what extent these parameter values deviate from the values in healthy subjects. Three groups were examined:

1. patients with spinal muscular atrophies (M. Wohlfart-Kugelberg-Welander = MWKW).

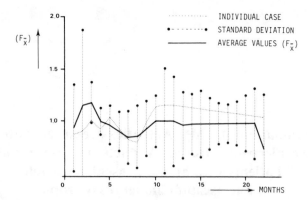

Figure 1a — Relative mean values for torques ($F_{\bar{x}}$) of MVC during isometric elbow flexion of the right arm in patients with a neuromuscular disorder (n = 11). The course of the "relative mean strength," $F_{\bar{x}}$ obtained in 11 patients (uninterrupted line) and the course obtained in one individual case (dotted line). $F_{\bar{x}}$ is obtained by the following procedure: each patient performed a test during which a maximal strength (MVC) has to be exerted three times during three seconds. The maximum of these three trials is taken as *the* maximum. The maxima thus obtained in 11 patients are averaged. This leads to an average value at successive time events. All mean values (for each 'time point') are also averaged and this value is called F_{ref}. $F_{\bar{x}}$ has been obtained by dividing the average values at each time point by F_{ref}. In addition to these values for each time point the standard deviation is indicated. In the same figure also the values for an individual case are shown (without standard deviation).

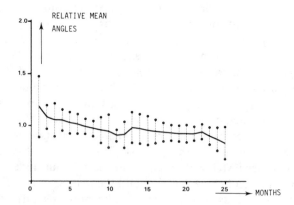

Figure 1b — Relative mean values for angles of abduction (right shoulder) in patients with a neuromuscular disorder (n = 11). The same procedure as described in Figure 1a is followed for the values of the abduction angle. Notice that in contrast to the values for the strength measurements a significant decrease can be established (student-t test, α = .95) with relative low values for the standard deviations.

2. patients with facioscapulohumeral dystrophy (FSHD).

3. healthy subjects.

With respect to the first two categories the following remarks can be made. First, it must be realized that highly progressive congenital neuromuscular disorders such as M. Werdnig-Hoffmann and M. Duchenne occur only in young children. With these children, problems can arise with respect to the understanding of the tasks which have to be performed. Adults, in this case patients with MWKW and FSHD, have a better understanding of the instructions and are generally more cooperative. This is the reason why the highly progressive disorders are not included in this study. In the MWKW and FSHD types of neuromuscular disorders usually proximal muscles are involved and these muscles can be studied relatively easily with surface electrodes. A special point of interest is the so-called "stationary" period appearing in the course of the FSHD type of muscular dystrophy. The term 'stationary' means that during longer periods (e.g., three years) the disease tends to stabilize. Contrary to the MWKW type of atrophy, in the FSH type of dystrophy it is expected that during this time the EMG parameters remain constant.

Methods and Material

Bipolar surface electrodes (type Beckman, $\varnothing = 16$ mm) were positioned on the bulkiest parts of the m. biceps brachii, the m. deltoideus (clavicular, acromial, and spinal part, respectively); the m. trapezius (ascending, horizontal, and descending part, respectively); the m. teres major and in some cases also on the m. serratus anterior and the m. infraspinatus. The electrodes were always placed parallel to the fiber arrangement. The distance between the two electrodes was approximately 11 mm (from border to border). The EMG signals were processed by means of an eight-channel amplifier in-house developed (input impedance, 10 Giga Ohm $/\!/$ 10 pF; adjustable gain, 500 to 10,000; bandwidth, 10 to 500 Hz; cut-off, 12 dB/oct.). The angle between arm and gravity was measured by means of an electrogoniometer (Peat, 1976). The error of the goniometer was less than $0.95°$. The subject was instructed to perform a tracking task by abducting his arm, viz. at step-wise intervals of approximately $5°$ (each position was maintained for at least two sec and was indicated on a visual display). After a few trials the subject knew how to perform this test, called 'intermittent isometric' abduction.

All patients were able to perform the test for angles between $20°$ and $50°$, although it was not always possible to abduct the arm in a purely coronal plane. A maximal deviation of $10°$ towards the scapular plane was permitted. In the healthy subjects, the same procedure was followed. In addition, abduction was also performed with an external load of 10 and 20 N, respectively. The goniometer was adjusted to a special

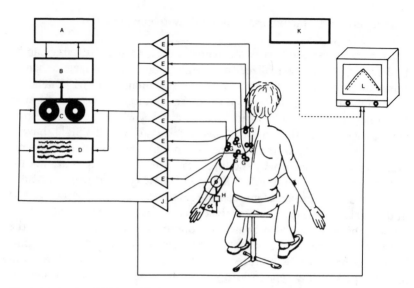

Figure 2 – A scheme of the general measuring arrangement. The patient is sitting in a chair, the arms are kept in an extended position. Both arms have to be elevated symmetrically. EMG recordings are stored on magnetic tape (Hewlett-Packard, 3514 A). The signals are processed by means of a PDP-11 computer. The subject observes a visual display on which the abduction angle (as a function of time) together with a target position are shown. The subject is instructed to follow the target. In this way a feedback is obtained. A = computer terminal, B = PDP-11 computer, C = tape recorder, D = X-T recorder, E = 8-channel EMG amplifier, G = bipolar surface electrodes, H = goniometer, α = abduction angle, J = goniometer amplifier, K = pattern generator, L = monitor.

hexcelite[R]-cast that forced the externally rotated arm into a straightened position. The EMG signals together with the 'angle signal' were recorded on magnetic tape. Afterwards signal processing was performed by means of a computer (12 bit AD conversion, sample rate 1000 Hz.). Figure 2 shows a scheme of the complete experimental arrangement.

The patient group consisted of 12 patients with WKW type of spinal muscular atrophy (MWKW), 10 males and two females, aged 21 to 38, and four patients with FSH type of muscular dystrophy, three males and one female, aged 21 to 51. The group of healthy subjects consisted of nine males and one female, aged 20 to 35.

EMG Parameters

In healthy subjects, the myoelectrical signals appeared as rather noisy signals when a muscle contracted. These signals were analyzed by means of statistical techniques. In healthy subjects, the amplitude showed a Gaussian distribution. Since the mean value was always zero (amplifications were AC coupled) these amplitude characteristics can be described completely by means of the variance or standard deviation (square root of the variance). The wave shape characteristics of the EMG signal can be described by means of the power density spectrum (pds) curve. From this curve the following parameters are more or less arbitrarily derived: (1) center frequency (median of the pds), and (2) F_{pmax}: frequency at which the maximum of the spectrum occurs. Besides these variables the variable, peaks, was also investigated. Since the mean peak-to-peak amplitude has been used as a variable by various investigators (Visser and Zilvold, 1979; Fusfeld, 1978), this value was also examined.

Results

A very important finding is the fact that in patients with MWKW the surface-EMG can reveal some striking patterns at first sight. Figures 3a to d show the EMG pattern of m. deltoideus, acromial part, during 'intermittent isometric' abduction of the shoulder at three levels: (a) phase I, beginning of abduction; (b) phase II, intermediate position; and (c) phase III, maximal abduction.

Whereas, in healthy subjects a noisy pattern is usually observed, recordings were found that were quite similar to needle EMG recordings (see Figures 3c, d, and e). Whether they were myopathic or not (Daube, 1978), large repetitive peaks were observed that were apparently related to single (giant) active motor units (Lenman and Ritchie, 1969). These specific patterns were observed in 8 of 12 patients with MWKW, especially at the m. biceps brachii, the m. teres major and the m. deltoideus. In patients with FSHD these motor unit patterns were not found. In one patient with FSHD a remarkable pattern was found in phase III (see Figure 3b). In this case, the interference pattern showed an asymmetric shape with the base line shifted upwards. The degree of interference (phase III) in the EMG pattern was different in both patient groups as compared with the group of healthy subjects. In the MWKW patient group a rather peaked pattern was observed, whereas, in patients with FSHD, a pattern was found with significantly smaller peaks as compared to the patterns observed in healthy subjects. In Figure 3e an EMG pattern is shown of a MWKW patient scoring 0 to 1 on the MRC scale (Medical Research Council, 1943). Notice that an interference pattern cannot be obtained in this patient who was exerting a maximal voluntary contraction.

In Figure 4a to e, some examples are given of results obtained with statistical techniques (signal analysis with the aid of the computer). The relationship between peak-to-peak amplitude and angle of abduction during an intermittent isometric abduction is shown for the m. deltoideus and m. biceps in Figures 4a and 4b.

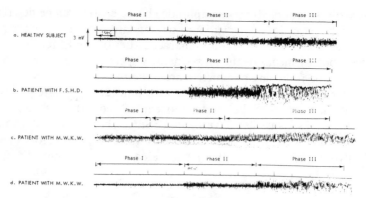

Figures 3a to d—EMG activities during abduction at different levels. Phase I: beginning of abduction; Phase II: intermediate abduction; Phase III: maximal abduction.

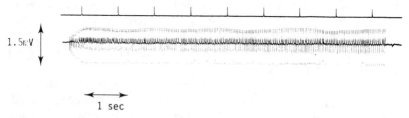

Figure 3e—EMG pattern of a MWKW patient scoring 1 on the MRC scale.

In contrast to what has been observed in the m. deltoideus, it appears that in a patient with M.W.K.W. the mean peak-to-peak amplitude in the m. biceps reaches a much higher level than occurs in healthy subjects. During abduction the m. biceps brachii and m. teres major are hardly active in healthy subjects. On elevating the arm the m. biceps brachii shows only a slight increase of activity, while the m. teres major remains almost inactive in healthy subjects. In the patients both muscles show remarkable and increasing activity, similar to the activities of the other muscles investigated.

Figures 4c and d show the differences of the center frequency level between a patient and a healthy subject at the m. biceps brachii and the m. trapezius (ascending part).

Discussion

In this study, special attention was paid to surface EMG, until now a method hardly used in clinical practice. Only some investigators used surface EMG as a diagnostic tool (Lovelace, 1970). For routine examinations, the technique still most commonly used is the application of

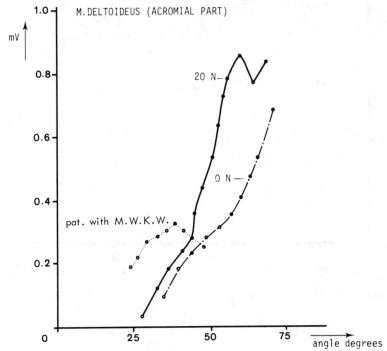

Figure 4a — Mean peak-peak amplitudes vs. angle units.

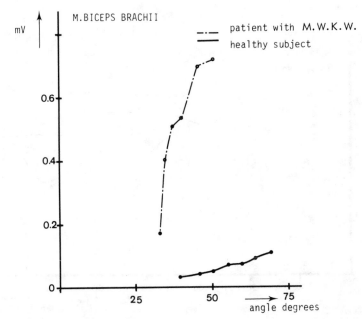

Figure 4b — Mean peak-peak amplitude vs. angle units.

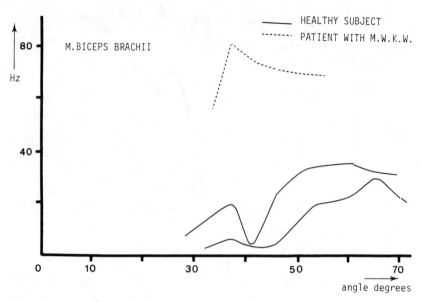

Figure 4c — Center frequency vs. angle units.

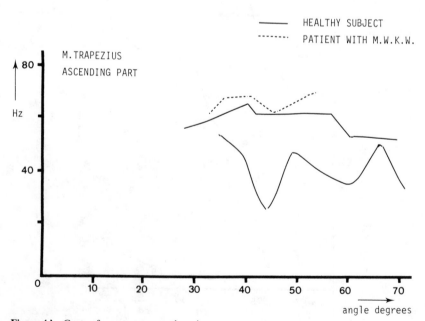

Figure 4d — Center frequency vs. angle units.

needle-EMG (Agarwal and Gottlieb, 1975). There are a number of arguments that favor the application of surface EMG as compared to needle EMG. These are (a) the technique is noninvasive, (b) for sequential examinations the cooperation of the subjects is much better, since they fear needle EMG, (c) the pick-up area of myoelectrical activities (the 'electrode window') is much larger as compared to the muscle area 'seen' by needle EMG (parts of muscles are observed instead of motor units) and (d) routine investigations can be carried out by non-clinicians. One of the obvious disadvantages is the relative lack of information about typical findings from the surface EMG in relation to particular disorders. In the application of surface EMG, usually only the RMS value (so-called IEMG) is chosen as a parameter of interest. Lindström et al. (1970) showed in more general terms the importance of the power density spectrum (pds). Using the parameters center frequency and mean peak-to-peak amplitude, it appears from these experiments that there are significant differences between patients and healthy subjects in the behavior of individual muscles (see Figures 4a to d). It is evident that in this way a more detailed description can be obtained as compared to strength measurements of a complex of muscle groups.

Future research will deal with the changes of the above mentioned parameters in order to better understand the decrease in the range of motion as found in these patients.

Acknowledgment

This research project was supported in part by a grant from the Social Security Council (Arbeids Ongeschiktheids Fonds), Zoetermeer, the Netherlands.

References

AGARWAL, G.C., and Gottlieb, G.L. 1975. An analysis of the electromyogram by Fourier, simulation and experimental techniques. IEEE Trans. B.M.E. 22-3:225.

BOUKES, R.J. 1981. Force and surface EMG measurements in patients with neuromuscular disorders. Unpublished thesis, Amsterdam.

DAUBE, J.R. 1978. The description of motor unit potentials in electromyography. Neurology, 28-7:623.

DANIELS, L., Williams, M., and Worthingham, C. 1956. Muscle Testing: Techniques of Manual Examination, 2nd ed., W.B. Saunders, Philadelphia.

FOWLER, W.M., and Gardner, G.W. 1967. Quantitative strength measurements in muscular dystrophy. Arch. Phys. Med. 12:629-644.

FUSFELD, R.D. 1978. Instrument for quantitative analysis of the electromyogram. Med. & Biol. Eng. & Comput. 16:290.

HOSKING, G.P., and Dubowitz, V., et al. 1976. Measurements of muscle strength and performance in children with normal and diseased muscle. Arch. Dis. in Childhood, 51:51.

KENDALL, H.O., and Kendall, F.P. 1948. Normal flexibility according to age groups. J. Bone and Joint Surg. 33A:690-694.

LATEUR, S. de, and Giaconi, R.M. 1979. Effect on maximal strength of submaximal exercise in Duchenne muscular dystrophy. Am. J. Phys. Med. 58:26-36.

LENMAN, J.A.R. 1974. The integration and analysis of the electromyogram and related techniques. In: J.N. Walton (ed.), Disorders of Voluntary Muscle, 3rd ed. Churchill Livingstone, London.

LINDSTRÖM, L., Magnussen, R., and Petersén, I. 1970. Muscular fatigue and action potentials conduction velocity changes studied with frequency analysis of EMG signals. Electromyography 10:341.

LOVETT, R.W., and Martin, E.G. 1915. A method of testing muscular strength in infantile paralysis. Journal of the American Medical Association 65:1512.

PEAT, M. 1976. An electrogoniometer for the measurement of single plane movements. J. Biochem. 9:423.

VISSER, S.L., and Zilvold, G. Electromyographische Befunde zur Beurteilung der funktionellen Elektrostimulation bei spastische Hemiparese.

WILLISON, R.G. 1971. Quantitative Electromyography. In: Electrodiagnosis and Electromyography (3rd ed. New Haven).

Evaluation of the Supporting Force of the Leg
for Cerebral Palsy Subjects

T. Yamashita, S. Shitama, and T. Taniguchi
Kyushu Institute of Technology, Kitakyushu, Japan

T. Matsuo and C. Saeki
Shinko-En Children Hospital, Kasuya, Japan

The force plate which measures the force acting between the foot and the ground during standing or walking has been effectively used for a long time. The present study was jointly started in 1979 by an engineering team from Kyushu Institute of Tech (KIT) and a medical team from Shinko-En Children Hospital (SCH) to apply the force plate method to cerebral palsy (CP) subjects specifically to evaluate improvements as a result of surgical operation. The fluctuation of the supporting force in standing and in the transient phase of starting to walk have been studied.

The steadiness of standing posture has been rather well studied (Kapteyn, 1972; Stribley et al., 1974; Murray et al., 1975) in the normal subject, but few studies have been conducted using disabled subjects. The transient characteristics of walking have not been studied extensively except by Yamashita et al. (1978), who studied the characteristics of the normal adult.

A few CP subjects were studied twice, that is, before and after an operation, to evaluate the degree of improvement by the operation. These characteristics will be compared with those of the normal with respect to both the steadiness of standing and the stability of the transient phase of walking.

Methods

Two force plates of the same type, constructed in our laboratory, were used to measure the force vertically acting on each leg. The measuring

Table 1

List of Subjects

Subj	Sex	Yrs	Wt (kg)	Experimental date		Note
D.S.	m	6	16.7	'80. 1.25	–	CP
K.H.	m	6	14.6	'80. 1.25	'80. 5. 9	CP
F.N.	m	11	23.9	'80. 1.25	'80. 5. 9	CP
S.M.	m	9	21.7	'80. 3.12	–	CP
T.H.	f	11	35.9	'80. 5. 9	'80. 8.29	CP
Y.S.	f	19	67.0	'80. 8.29	–	CP
E.Y.	f	8	21.9	'80. 1.25	–	Norm
S.S.	m	23	63.4	'80. 7.11	–	Norm

principle of the force plate has been described elsewhere together with some of its applications (Yamashita and Katoh, 1976). The magnitude and the point of application of the vertical supporting force of each leg, as well as the resultant force, can be obtained as a function of time by processing experimental data with a computer in our procedure.

Eight subjects including a normal child and an adult were studied as shown in Table 1. Cerebral palsy subjects who were studied twice had a surgical operation after the first evaluation. The second evaluation was carried out after about three months following the operation. Some CP subjects could walk without aids although at slower speeds than the normal subjects, but the severely disabled ones needed a cane or walker.

Two kinds of experiments were done: 1) standing quietly on the force plates for a given period, and 2) walking over the force plates along the auxiliary walkway. In the standing experiment, the subjects were required to take a natural posture or to support the body weight mainly on one leg according to a given signal. As to the stance width, no special control was exercised, then each subject took the natural width of his or her own stance so that each foot was placed on the respective force plate. In the walking experiment a free walking was studied, that is, no special requirements were imposed.

The force data were automatically digitized by an A/D converter. Six channels (each force plate had three channels to detect the components of the vertical force) were sequentially sampled at proper speeds: in the standing experiment the speed was 0.12 sec; in the walking experiment 0.024 sec. The locations of the footprint which were needed to define the point of application in the space coordinate were measured manually and fed into the computer. All calculations were done by the computer.

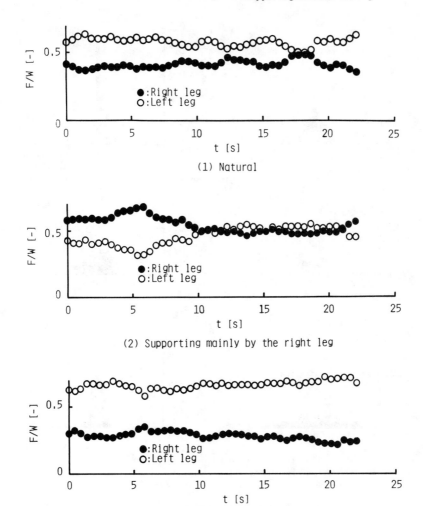

Figure 1—Supporting force of each leg for the different postures of CP (subject T.H.). Supporting force F is divided by the body weight W of the subject.

Results and Discussion

The supporting force of the leg and the point of application were analyzed to find the differences between the normal and the CP subjects. Characteristics in quiet standing are described first, and the characteristics of walking follow.

Standing

The supporting force of each leg showed that compared with the normal

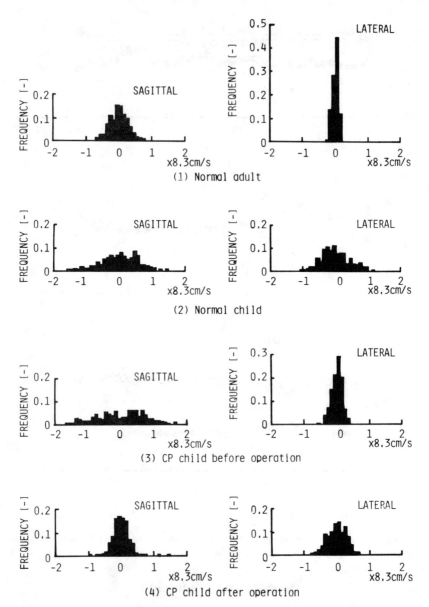

Figure 2—Distribution of the fluctuating speed for the resultant force. Frequency was normalized by the total number of the sample, 300.

subject, the disabled subjects supported the body unsteadily or asymmetrically. Supporting characteristics of a CP in different standing postures, which were ordered orally before the measurement to maintain, are shown in Figure 1. As shown in this diagram the body is supported mainly by the left leg in the natural posture. Although the subject

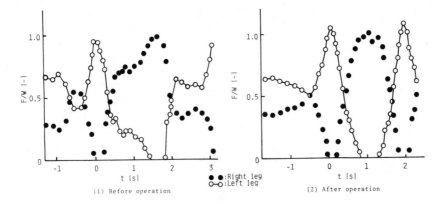

Figure 3—Improved supporting characteristics of CP by an operation.

was required to support the body weight mainly by the right leg in the second situation, the support of the left leg gradually took over from the right leg.

As for the supporting force of each leg of the same subjects, no significant difference was observed between before and after a surgical operation in this standing experiment. But on this point further investigations need to be done for many additional cases.

The point of application of the supporting force fluctuates even when one attempts to stand quietly. The fluctuation in the child was greater than the normal adult. In some cases, the fluctuation of CP was improved and was less after an operation.

The speeds at which the point of application fluctuated were approximated by dividing the difference of positions of the point of application at successive sampling instants by the sampling interval. The speed distributions for various subjects are shown in Figure 2, where the frequency is expressed in a normalized form by the total number of samples, 300. A decrease in speed suggests a higher capability of maintaining static equilibrium in standing. The following features can be drawn from this figure: 1) the normal adult showed the greatest balancing capability; 2) an improvement as a result of the operation on the CP is clearly shown in the sagittal direction.

Walking

Both the supporting force of each leg and its point of application were obtained as a function of time during the walking experiment. The normal child exhibited characteristics similar to those of the adult. The case in which an operation on the CP improved the characteristics is described in the next paragraph.

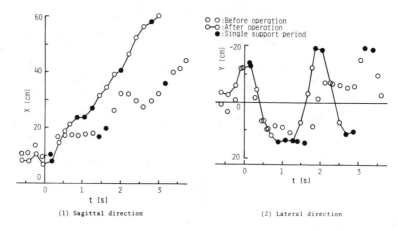

(1) Sagittal direction   (2) Lateral direction

Figure 4—Point of application for the resultant force as shown in Figure 3.

Figure 3 shows the supporting force of each leg before and after an operation. The figure clearly shows weight transfer onto the supporting leg is carried out in a shorter time and more smoothly after the operation. Before the operation the weight transfer was delayed. This results in a lack of the stability as the support leg is changed. These characteristics were, also, exhibited in the movement of the point of application of the resultant force as shown in Figure 4.

The characteristics of the CP described in Figures 2, 3, and 4 were all obtained from the same subject. These results suggest that the operation was effective and improved the balancing capability. These facts have confirmed the subjective observations of medical professionals.

The other two CP subjects studied twice were not discussed with regard to the difference between before and after an operation because one subject could not stand independently even after the operation and the other showed only a slight difference. Further studies for various cases are needed to determine other effective measures for improving balancing capability during standing and walking.

Conclusion

Characteristics of the support in both standing and walking were studied for normal and disabled CP subjects by using two force plates. The method allowed the measurement of the supporting force of each leg and the resultant force of both, together with the points of associated force application. Some CP subjects were studied twice to ascertain the differences between before and after a surgical operation.

Measures which are able to show the difference between the normal and the CP subjects as well as the effectiveness of an operation are important in this type of study. In this study, simple measures were used: 1) in standing, speeds at which the point of application of the resultant force fluctuated were analyzed; 2) in walking, the supporting force of each leg and the point of application of the resultant force were studied. The results evaluated by these measures have confirmed quantitatively the subjective observations by the medical professionals.

References

KAPTEYN, T.S. 1972. Data processing of posturographic curve. Agressologie 13(B):29-34.

MURRAY, M.P., Seireg, A.A., and Sepic, S.B. 1975. Normal postural stability and steadiness: quantitative assessment. J. of Bone and Joint Surgery 57(A):510-516.

STRIBLEY, R.F., Albers, J.W., Tourtellotte, W.W., and Cockrell, J.L. 1974. A quantitative study of stance in normal subjects. Arch. Phys. Med. Rehabil. 55(2):74-80.

YAMASHITA, T., and Katoh, R. 1976. Moving pattern of point of application of vertical resultant force during level walking. J. of Biomechanics 9(2):93-97.

YAMASHITA, T., Ishida, A., Kurisu, Y., and Yamaguchi, M. 1978. Analysis of dynamic characteristics during starting phase of walking. Biomechanisms 4. Univ. Tokyo Press: 182-192 (in Japanese).

Locomotive Mechanics
of Normal Adults and Amputees

Kenji Suzuki, M. Takahama, Y. Mizutani, and M. Arai
Akita University School of Medicine, Akita, Japan

A. Iwai
Miyagi Educational College, Sendai, Japan

The bipedal locomotion of human gait contains a fundamental gait factor—the mechanism of body weight transfer in the double support period. The purposes of this study were to measure the characteristic points of the ground reaction forces during the double support period and to identify the mechanism of body weight transfer of the prosthetic gait.

Methods and Materials

Twenty amputees, age range 16 to 57 years, 10 with unilateral above-knee amputations and 10 with unilateral below-knee amputations, and 25 normal adults, age range 22 to 38 years, were examined.

The following measurements of the characteristic points were made from the ground reaction forces of the gait for normals and amputees by the use of two force plates, an addition amplifier and a data processor (see Figures 1 and 2):

1. Pk = the maximum value of the downward acceleration of the center of gravity (COG) of the body and Po = transfer point of the COG from the backward to the forward limb were obtained from the result of the vertical components.

2. F = the maximum value of forward acceleration of the COG; M = the zero point of the acceleration and A = the maximum value of the backward acceleration of the COG, were obtained from the result of the fore-aft components.

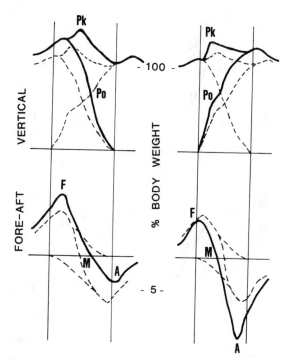

Figure 1—Typical tracing displaying the resultant of the ground reaction forces during the double support period for an above-knee amputee.

Normal adults walked at various cadences (80, 100 and 120 steps/min) and used various step lengths (45, 55 and 65 cm). Amputees walked at a cadence of 80 steps/min and took a step length of 45 cm and were compared to the measurements of normals.

The body weight transfers both from the sound limb to the artificial limb and from the artificial limb to the sound limb were examined in each amputee.

A t-test was used to assess the significance of the differences between the measurements of amputees and normal adults.

Results

Normal Adults

The values of the characteristic points of Pk, Po, F, and A became

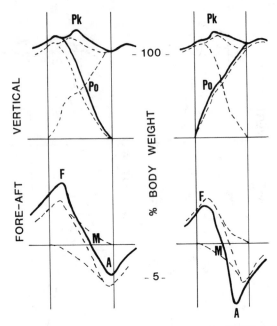

TRANSFER FROM S. TO A.   TRANSFER FROM A. TO S.

A : ARTIFICIAL LIMB   - - - NORMAL

S : SOUND LIMB   —— AMPUTEE

Figure 2 — Typical tracing displaying the resultant of the ground reaction forces during the double support period for a below-knee amputee.

greater and the timing of them became faster with the increase in the cadence and step length. Moreover, the vertical acceleration was particularly influenced by the cadence and the forward acceleration was mainly influenced by the step length.

Pk always preceded the point of Po in % walking cycle. The faster the walking speed, the shorter the Pk-Po interval and the F-A interval.

There was no difference of the timing between Po and M at the various walking speeds. The ground reaction forces were balanced in the timing of Po and M and during the transfer of the COG of the body from the backward to the forward limb.

Amputees

The means and standard deviation of the measurements of the characteristic points are shown in Table 1, where the significance of the differences between the measurements of amputees and normal adults is indicated (Table 1).

Table 1

Comparison of Characteristic Points in Double Support Period
(Cadence: 80, Step Length: 45 cm)

Characteristic Points		Normals	Above-Knee Amputees		Below-Knee Amputees	
			S to A	A to S	S to A	A to S
Pk reaction force		116.2±5.5	128.5±9.0**	121.2±5.7*	120.8±5.3*	120.0±5.4*
Po	(% body	55.2±3.5	62.5±3.1**	58.7±4.5*	58.3±3.1*	59.8±3.6*
F	weight)	5.9±1.4	8.9±2.7**	6.6±2.4*	9.5±1.9**	6.3±2.9
A		6.1±1.2	3.3±2.2**	12.6±2.4**	7.0±2.7	8.3±2.0**
Pk		51.1±1.1	58.3±2.2**	48.2±2.8**	56.2±3.4**	51.4±3.4
Po	time	54.3±1.9	61.0±1.4**	48.3±2.5**	57.9±3.6**	51.9±2.6**
F	(% walking	48.1±2.6	56.4±1.6**	44.6±2.8**	53.2±4.9**	48.0±3.7
M	cycle)	54.3±2.4	64.1±1.6**	47.3±2.7**	59.5±3.6**	51.9±2.8*
A		62.1±2.6	67.6±1.0**	52.7±3.2**	65.9±3.1**	57.7±2.6**

* $P < 0.05$ A: artificial limb
** $P < 0.01$ S: sound limb

Above-knee Amputees. In the transfer from the sound limb to the artificial limb, the values of Pk and Po increased significantly and were different from those for normals and below-knee amputees. The value of F was greater, but the value of A was only about half that of the mean for normals. All of the timing of these characteristic points were markedly delayed and the timing of Po preceded the timing of M.

In the transfer from the artificial limb to the sound limb, the values of Pk and Po were also increased and the value of F was almost the same, although the value of A was increased about twice as much as the mean of normals. All of the timing of these characteristic points became faster than the means of normals and below-knee amputees.

Below-knee Amputees. In the transfer from the sound limb to the artificial limb, the values of Pk and Po were increased less than the means of above-knee amputees. The value of F was increased, but the value of A was almost the same as the mean of normals. All of the timing of the characteristic points were moderately delayed.

In the transfer from the artificial limb to the sound limb, the values of Pk and Po were increased more than mean of normals. The value of F was almost the same as the means of normals, but the value of A was increased a little. All of the timing of the characteristic points were faster than the means of normals.

Discussion

In the transfer from the sound limb to the artificial limb for above-knee amputees (Figure 1), the great increase of the value of Pk means that excessive downward displacement of the COG of the body occurs just after heel-strike of the forward-placed artificial limb.

In this transfer, the body weight is loaded slowly to the forward-placed artificial limb and is reduced slowly from the backward-placed sound limb in order to compensate for the disability of the artificial limb. Then, the value of Po increases over the normal range. The gait of the above-knee amputees is characterized by the fact that they are not able to flex the prosthetic knee during the double support period. The COG of the body in this transfer must be elevated abnormally because of the inability to actively flex the prosthetic knee.

The COG of the body is accelerated forward to the maximum and decreases slowly by the compensation of the sound limb and the insufficient restraint that reduces it to about half of the normal value.

In the transfer from the artificial limb to the sound limb (see Figure 1), the body weight is loaded rapidly on the forward-placed sound limb and is reduced rapidly from the backward-placed artificial limb. The value of Po is also increased over the normal range by the compensation of the forward-placed sound limb to propel the COG of the body.

Propulsion of the artificial limb is reduced by the inability to actively plantar flex the prosthetic ankle and actively flex the prosthetic knee. On the contrary, restraint of the sound limb is increased markedly in order to transfer the COG of the body safely through compensation by the sound limb.

In the transfer from the sound limb to the artificial limb (see Figure 2) by below-knee amputees, the ground reaction forces are similar to those of normals and the gait is characterized by the fact that they are able to flex the knee actively during the double support period.

The COG of the body in this transfer is elevated subnormally and smoothly because of the active knee flexion of the forward-placed artificial limb and then the value of Po is moderately increased by the compensation of the sound limb.

Propulsion of F is increased, but restraint of A is decreased slightly because of the insufficiency of the prosthetic ankle of the forward-placed artificial limb.

In the transfer from the artificial limb to the sound limb (see Figure 2), the COG of the body is displaced nearly the same as for normals. The value of Po is increased because of the compensation of the forward-placed sound limb. Propulsion of F is decreased by the inability to actively plantar flex the prosthetic ankle, and the restraint of A is increased by the compensation of the sound limb, compared to those for normals.

Conclusions

In this study:

1. The value and timing of the characteristic points, obtained from the ground reaction forces, were quite variable because of the limitations of the prosthetic knee and ankle and the compensation of the sound limb.

2. Insufficient knee-ankle interaction in the prosthetic gait was noted in the double support period mainly when the center of gravity of the body was transferred from the sound limb to the artificial limb.

3. The increased value of Po, over the normal limits, indicates the extent of the compensation of the sound limb in the prosthetic gait.

References

EBERHART, H.D., et al. 1954. The locomotor mechanism of the amputee. In: P.E. Klopsteg and O.D. Wilson (eds.), Human Limbs and Their Substitutes, pp. 472-480, New York.

SAUNDERS, J.B., et al. 1953. The major determinants in normal and pathological gait. J. Bone and Joint Surg. 35(A):543-558.

SUZUKI, K. 1972. Force plate study on the artificial limb gait. J. Jpn. Orthop. Ass. 46:503-516.

Quantitative Back Muscle Electromyography in Idiopathic Scoliosis

Carl Zetterberg, Roland Björk,
Gunnar Andersson, and Roland Örtengren
University of Göteborg, Sahlgren Hospital,
Göteborg, Sweden

The human spine is quite unstable without muscles. Some stability is provided by the ligaments and discs, some by skeletal structures, but without active muscular control the spine would collapse. The normal spine is straight in the antero-posterior aspect, curved in the lateral. A curve in the frontal plane is called scoliosis. Scoliosis is a common deformity in a number of neurological and muscular diseases, for the reasons already given. In the majority of cases the etiology of scoliosis is unknown. Adolescent idiopathic scoliosis comprises about 70% of all cases diagnosed, and of these 80% are female (Brooks et al., 1975).

To study the role of the trunk muscles in idiopathic scoliosis, myoelectric investigations were conducted over a number of years. These studies have shown in increased activity on the convex side of the curve near its apex (Riddle and Roaf, 1955; Henssge, 1962; Zuk, 1962; Hoogmartens and Stuyck, 1978; Güth and Abbink, 1980). In some patients there is no difference between the convex and the concave side, however, a fact which can have importance for the liability of the curve to progress (LeFebvre, 1961; Hoogmartens and Basmajian, 1976; Alexander and Season, 1978).

Histological examinations have revealed an asymmetry in the distribution of fiber types in para-spinal muscles in scoliotic patients. On the convex side there is an increase in fatigue resistant type I fibers and on the concave side fewer fatigue resistant type II fibers (Fidler et al., 1974; Spencer and Eccles, 1976; Spencer and Zorab, 1976).

Whether myoelectric and morphologic changes are primary or secondary to scoliosis is unknown. It is clear from the studies, however, that asymmetries in muscle function in scoliotic patients can be studied using electromyography and that the results can be of prognostic value; i.e., aid in the evaluation of whether the scoliosis is stable or progressing.

The purpose of this investigation was to study the function of paraspinal muscles in girls with and without scoliosis by quantitative electromyographic evaluation of activity levels and fatigue response.

Material

The investigation included two groups of subjects. The first, called the pilot group, consisted of five girls with a thoracic scoliosis of 20 to 45°, and a lumbar of 13 to 29° (Cobb, 1948). Their mean age was 14.7 years (range 14.6 to 14.9 years). All thoracic curves were convex to the right.

The main study group consisted of 36 girls with adolescent idiopathic scoliosis. All had thoracic curves convex to the right, with a vertex at the T7 to T9 levels. The mean age was 13.4 years (range 11.2 to 15.1 years). The primary thoracic curve had angles from 5 to 70°, and the compensatory lumbar curve ranged from 4 to 47°. All patients were right handed and otherwise healthy. A group of 13 age-matched, healthy girls were used as controls. Their mean age was 13.2 years (range 10.1 to 15.8 years). Scoliosis was excluded by clinical examination and Moiré topography.

Methods

In the pilot group, several aspects of spinal loading were studied. By means of a specially designed yoke, an axial, vertical load was applied with the subject in the normal standing position. Different loads could be put on the yoke and total loads between 100 and 400 N were used. No muscle power had to be used to balance the yoke, which rested on the shoulders.

To study horizontal loading of the trunk in standing postures, a frame was used which permitted the pelvis to be fixed by straps. A harness was then placed on the trunk of the subjects and loads applied from the harness over a pulley at right angles to the spine at the T6 to T7 level (see Figure 1). Loading was done in four directions: forward, backward, and to both sides. In all four directions the subjects were asked to actively pull against a strain-gauge. In addition, tests were conducted of resisting the pulling of a weight attached to a cord.

In the main group the load application was by means of only the trunk. In the prone position, the girls were asked to raise their trunks and to maintain that position for 2 min (see Figure 2).

The myoelectric signals were picked up by means of four bipolar surface electrodes, placed symmetrically about 2.5 cm from the midline on

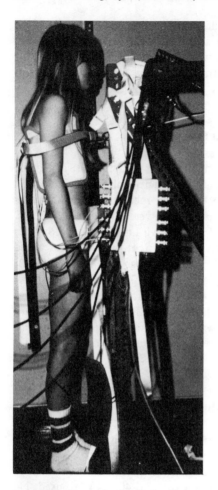

Figure 1 — A scoliotic patient standing in the loading frame with the harness on (the pelvic straps not tightened).

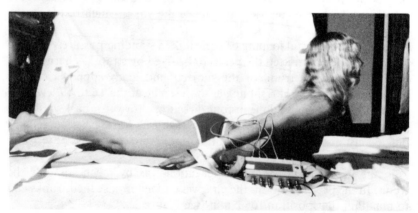

Figure 2 — The prone position. The patient is arching her back during EMG registration.

both sides of the back at T8 and L3 levels. The signals were amplified, and fed to an analog-to-digital converter of a computer for storage and subsequent analysis. The sampling rate was set to 2048 Hz per channel. Recordings were made over the 2-min study period. The quality of the signal was checked on an oscilloscope screen visually and also during analysis, using an automatic artifact control program. For amplitude analysis, the rms value of the signal was detected during the first eight sec of the recording period. This period was chosen to minimize possible amplitude alterations occurring during sustained muscle contraction.

Based on the rms values, an amplitude quotient (Q) was computed for each spinal level and each subject.

$$Q = (V_{convex} - V_{concave}) / (V_{convex} + V_{concave}) \qquad (1)$$

This quotient was used in comparisons of the degree of curvature of the scoliosis among subjects, and in the main group.

Muscle fatigue is known to produce a shift of the myoelectric signal power spectrum toward lower frequencies. To indicate such changes the signal recordings were analyzed according to the fatigue index method (Lindström et al., 1977), in which the logarithmically-scaled center frequency values of the power spectra are calculated. The center frequency values are plotted versus time giving fatigue curves, and such a curve—estimated by means of linear regression analysis—is the fatigue index which is used to quantify the fatiguing effect of muscle contractions. In the present material, the curves often displayed sudden changes after about 30 sec. Therefore, the regression analysis was performed over a period of 30 sec only. The slope was tested for significance using Student's t-test at the 5% level.

Results

The Pilot Group

When different loads were applied to the yoke, the myoelectric signal was small and intermittent, changing from side to side in the back. Activity was more often present on the convex side. No development of muscle fatigue was observed.

When the trunk was loaded in the loading frame it was difficult to precisely control the posture. Difficulties in defining postures and load led to abandoning pulling against a strain gauge. When resisting a weight applied to produce a flexion moment on the trunk, the rms values at the T8 level were, on the average, higher on the convex side, while no difference was found at the L3 level. In lateral flexion during resisting, a higher rms value always occurred on the ipsilateral side in the thoracic

Table 1

**The rms Value in μV of the Four Recordings in Loading Forwards,
to the Right and to the Left (Pilot Study)**

Load direction	Electrode level	Left	Right
Forwards	T8	39.3 ± 22.6	76.0 ± 50.3*
	L3	165.5 ± 121.5*	176.75 ± 82.0
To the right	T8	23.3 ± 15.5	108.2 ± 106.2*
	L3	156.0 ± 145.8*	41.4 ± 20.7
To the left	T8	144.8 ± 103.5	32.8 ± 19.4*
	L3	59.6 ± 41.5*	143.2 ± 211.0

*convex side
n = 5

region and on the contralateral side in the lumbar region (see Table 1).

The result of the fatigue analysis showed a varied degree of fatigue development. In lateral pulling, however, there was a tendency to fatigue faster on the convex side both at thoracic and lumbar levels, irrespective of the direction of the pull.

The Main Group

Arching of the back in the prone position was the posture found to be easiest to control, and the one which yielded the most reproducible muscle fatigue. The results of the amplitude analysis in the 36 girls with thoracic scoliosis are shown in Figure 3. The girls are grouped according to their scoliotic angle. In the thoracic region, the myoelectric signal amplitude was generally higher on the convex side of the curve. The difference was found to increase significantly with increasing scoliotic angle ($r = 0.36$, $p < 0.05$). At the lumbar level, a dominance of convex activity was also seen, which, however, was not significantly related to the angle of the scoliosis.

The results of the fatigue index analysis are shown in Table 2. No difference in the development of fatigue between the convex and concave sides could be seen, either at the thoracic or at the lumbar levels. Also no difference could be seen between the scoliotic group and the control group.

The Control Group

The healthy girls had an amplitude quotient of −0.03 (SEM 0.23) at the thoracic level and 0.04 (SEM 0.18) at the lumbar level, i.e., no difference

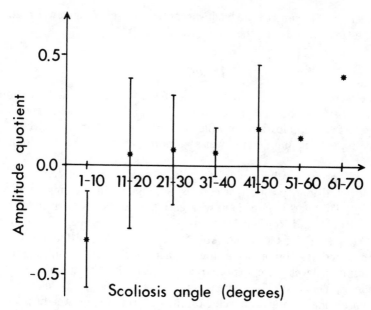

Figure 3—The amplitude quotient plotted versus scoliosis angle thoracically. The regression line is significant, (p < 0.05) N = 36.

Table 2

The Fatigue Index, in dB, for Subjects Having a Scoliosis Convex to the Right in the Thoracic Region

Electrode level	Left	Right
T8	18.6 ± 10.1	16.8 ± 9.6*
L3	13.9 ± 5.4*	11.7 ± 3.5

*convex side
The probability of fatigue > 0.95, N = 22

The Fatigue Index, in dB, for the Control Group

Electrode level	Left	Right
T8	16.9 ± 7.6	18.8 ± 9.4
L3	14.5 ± 5.4	14.2 ± 5.3

The probability of fatigue > 0.95, N = 10

between the two sides of the back was noted. Muscle fatigue was regularly seen on both sides of the back (see Table 2).

Discussion

There were two major characteristics of the loading method. It should permit a stable body posture, and at the same time cause back muscle fatigue. Load application with the yoke gave only slight myoelectric activity in the back muscles and was abandoned as the fatigue development was poor.

Load application by means of the harness was more effective and, in the forward load position, a convex dominance in the rms value was present. This was in agreement with earlier reports of similar load situations (Riddle and Roaf, 1955; Henssge, 1962; Zuk, 1962; Hoogmartens and Stuyck, 1978; Gúth and Abbink, 1980). In lateral loading the ipsilateral activity dominance at the thoracic level and the contralateral at the lumbar level were expected as a result of the loading method (Andersson et al., 1977). There was no significant fatigue development in the paraspinal thoracic muscles of the subjects in the weight-resisting part of the pilot study. Therefore, for the purpose of studying fatigue, the weight-resisting postures were considered less suitable. Also, in the lateral loading positions there was often a significant change in the trunk posture.

Arching of the trunk in the prone position gave the best fatigue development in the thoracic part of the erector spinae muscles. As fatigue regularly developed in both the thoracic and lumbar parts of the erector spinae it was decided to choose that position for loading of the trunk in the main study. The results showed a convex amplitude dominance, both for the primary thoracic curve and also for the lumbar curve. This was in agreement with previously cited literature.

It was also possible to demonstrate a higher degree of convex amplitude dominance with the larger scoliosis angle (see Figure 3). One explanation for this can be that the muscles on the convex side in scoliosis have to perform greater work for biomechanical reasons.

In the main material, the fatigue index values were similar on the convex and concave side, both in the thoracic and lumbar regions, and corresponded well to the results in the control subjects, i.e., there is no side difference in the ability of the erector spinae muscles to withstand a fatiguing contraction in scoliosis. An explanation of the symmetrical fatigue development—in contrast to the amplitude difference—can be that the muscles in scoliosis have adapted to their new working situation with greater demands on the convex side muscles.

It was not possible from this material to evaluate if progressive scoliotic patients had a different EMG pattern compared to nonprogressive. A follow-up over several years is in progress.

Acknowledgments

Thanks are due to Anders Arvidsson, Sven Friberg, and Lars Lindström who provided parts of the programs included in the analysis package.

Financial support was given by the Göteborg Medical Society, the Greta and Einar Asker Foundation, and by the Britta and Felix Neuberg Foundation.

References

ALEXANDER, M.A., and Season, E.H. 1978. Idiopathic scoliosis: An electromyographic study. Arch. Phys. Med. Rehab. 59:314-315.

ANDERSSON, G.B.J., Örtengren, R., and Herberts, P. 1977. Quantitative electromyographic studies of back muscle activity related to posture and loading. Ortho. Clin. North America 8:85-96.

BROOKS, H.L., Azen, S.P., Gerberg, E., Brooks, R., and Chan, L. 1975. Scoliosis: A prospective epidemiological study. J. Bone Joint Surg. 58(A):968-972.

COBB, J.R. 1948. Outline for the study of scoliosis. In: Instructional Course Lectures, The American Academy of Orthop. Surgeons. 5:261-275. JW Edwards, Ann Arbor.

FIDLER, M.W., Jowett, R.L., and Troup, J.D.G. 1974. Histochemical study of the function of multifidus in scoliosis. In: P.A. Zorab (ed.), Scoliosis and Muscle.

GÜTH, B., and Abbink, F. 1980. Vergleichende electromyographische und kinesiologische Untersuchungen an Kongeitaler un idiopathischen Skoliosen (Comparisons of electromyographic and kinesiological investigations of cases of diopathic scoliosis). Z. Orthop. 118:165-172.

HENSSGE, J. 1962. Electromyographische Befunde der Rückenmusculatur nach Poliomyelitis und bei idiopatischen Skoliosen (Electromyographical findings of the back musculature after poliomyelitis and with idiopathic scoliosis). Z. Orthop. 96:324-334.

HOOGMARTENS, M.J., and Basmajian, J.V. 1976. Postural tone in the deep spinal muscles of idiopathic scoliosis patients and their siblings. Electromyogr. Clin. Neurophysiol. 16:93-114.

HOOGMARTENS, M.J., and Stuyck, J. 1978. First results with vibration electromyography as a differential measurement of postural tone in the left and right spinal muscles. Agressiology 18:341-343.

LE FEBVRE, J. 1961. Electromyographic data in idiopathic scoliosis. Arch. Phys. Med. and Rehab. 42:710-711.

LINDSTRÖM, L., Kadefors, R., and Petersén, I. 1977. An electromyographic index for localized muscle fatigue. J. Appl. Physiol.: Respirat. Environ. Exercise Physiol. 43(4):750-754.

RIDDLE, H.F.V., and Roaf, R. 1955. Muscle imbalance in the causation of scoliosis. The Lancet, June 18:1245-1247.

SPENCER, G.S.G., and Eccles, M.J. 1976. Spinal muscle in scoliosis. Part II: The proportion and size of Type I and Type 2 skeletal muscle fibres measured using a computer-controlled microscope. J. Neurol. Sciences 30:143-155.

SPENCER, G.S.G., and Zorab, P.A. 1976. Spinal muscle in scoliosis. Part I. Histology and Histochemistry. J. Neurol. Sciences 30:137-142.

ZUK, T. 1962. The role of spinal and abdominal muscles in the pathogenesis of scoliosis. J. Bone Joint. Surg. 44(B):102-105.

B.
Miscellaneous

On the Biomechanics of Slipping Accidents

Lennart Strandberg and Håkan Lanshammar
National Board of Occupational Safety and Health
Solna, Sweden

According to official statistics of Sweden, more people are killed in falls than in motor vehicle accidents. In 1975 accidents caused about 3900 fatalities. There were 1700 (43%) classified as 'caused by falls' and 1200 (31%) by motor vehicles. In the same year, occupational injuries resulted in about 3.1 million absentee-days. Five percent were due to occupational diseases, 18% were the result of accidents with vehicles, ships or aircraft, and 26% of the days were recorded as falling accidents.

These statistical classification schemes group all events from each accident under a single cause. A subgroup of falls, slipping, appeared in less than 5% of the occupational accidents in 1975. With an entirely different scheme in the present occupational injury information system (ISA), slipping has been registered in more than *11%* of all occupational accidents during 1979.

Selective outputs from ISA may point to the circumstances where slipping is most frequent. However, proper measurement techniques and criteria are needed for the objective selection of satisfactory slip-resistant shoes and walking surfaces. None of more than 60 different slip-resistance meters and methods found in the literature (Strandberg, in preparation) has achieved international acceptance. Many of the methods are based on the classic laws of friction, although most of the laws have been falsified by tribology research — particularly for viscoelastic materials (such as rubber) and lubricated conditions (Moore, 1972). Rabinowicz (1956) concludes that "there is really no such thing as a static coefficient of friction for most materials."

In current tribometry the kinetic coefficient of friction (COF) is considered dependent on sliding velocity and normal force and its time of application. So, if COF measurements are to correlate with real slip-resistance, they must be based on accurate evaluations of motion and forces from situations in which humans have slipped.

While the biomechanics of walking on non-slippery surfaces is quite well documented in the available literature, experimental data on slipping of walking subjects are sparse. With 36 subject/shoe combinations Perkins (1978) used a 30 to 100 Hz stroboscope, yielding multiple-image photographs for evaluation of shoe motions, and paper records from a lubricated Kistler piezoelectric force plate for ground reaction force data. Bring (1978) presented kinematic data from high-speed cinematography on eight intentional and eight unintentional skids. His data on skid forces originated, however, from two other experiments where no kinematic data were recorded.

Method

The conclusion of Perkins (1978) that "the static friction coefficient is the most relevant for slip-resistance testing" may be due to his method's detection threshold for sliding velocities. Differentiation of shoe position data means noise amplification (Lanshammar, 1980). So if short duration skids are to be detected and quantified in the COF-peak sliding velocity interval (i.e., 0.02 to 0.5 m/sec according to Conant and Liska, 1960), a comparatively high sampling rate and carefully chosen evaluation methods are necessary. Therefore, the recently developed (Gustafsson and Lanshammar, 1977) gait analysis system, called ENOCH, was implemented for laboratory investigations of human walking and skidding. These investigations are part of a continuing project, synthesizing tribology and biomechanics in order to develop suitable apparatus and adequate methods of measuring slip-resistance (Strandberg and Lanshammar, 1981).

The ENOCH system consists of a Hewlett-Packard computer with substantial software, connected to a Kistler piezoelectric force plate yielding ground reaction force data, and to a Selspot optoelectronic device for 315 Hz coordinate logging of shoe and body markers (six light-emitting diodes on each side).

The present study was based on computer records from 124 walking experiments on an even and level track covered with stainless steel. The force plate, in the center of the track, was lubricated (with soap solution) or cleaned between various experiments without the subject's knowledge. Injuries were prevented with a rail-suspended safety harness. The experiments and intermediate activities were tracked with a time lapse video recorder.

One female and three male subjects with heights ranging from 1.51 to 1.91 m, two sole-heel materials, and two shoe types were used. Cadence was controlled by a metronome at 90, 100, 110 or 120 steps/min. The subjects were asked to walk normally without looking at the force plate surface.

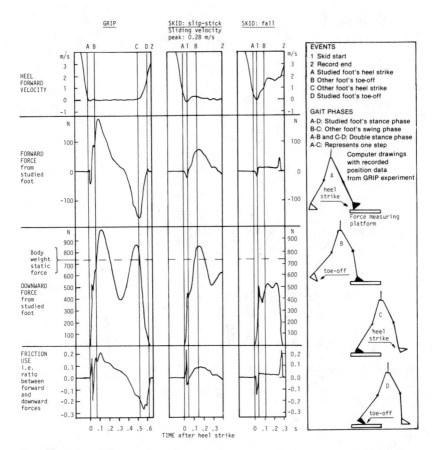

Figure 1 — Selected diagrams from three experiments with the same subject (HL in Table 1).

Force plate misses and gait abnormalities were sorted out with video recordings and gait analysis software. On the basis of error estimates (Lanshammar and Strandberg, 1980) sliding motions were considered only if the sliding *velocity* exceeded 0.05 m/sec (Lanshammar and Strandberg, 1983). The threshold value for detection of sliding *distance* was approximately 4 mm.

Results and Conclusions

Falling or loss of balance occurred only if the shoe slid forward in the beginning of the stance phase. In 39 experiments, called 'skids', such critical shoe-heel motions were detected. The remaining 85 experiments have been called 'grips'. However, many of the grips exhibited backward sliding just before toe-off, which never resulted in loss of balance.

Table 1

Skidding Time, Sliding Velocity and Distance, and Absolute Value of 'Friction Use'
(Ratio Between Forward and Downward Shoe Forces)
Mean Value ± Standard Deviation for Different Skid Types

SKID TYPE Measured quantity no. in subgroup	SLIP-STICKS				FALLS with different subjects				ALL
	4 mini	9 midi	3 maxi	16 total	5 AL	10 B0	8 HL	23 total	39 skids
Skid start* time after heel strike (msec)	48±39	55±15	44±13	51±22	61±25	45±18	37±14	46±20	48±21
Skidding time at peak velocity (msec)	33±17	52±20	52±34	47±22	Record ended mostly before peak.				
Peak velocity of forward slip (mm/sec)	230±40	490±220	560±500	440±280	Above walking speed (1-2 m/sec).				
Forward sliding distance (mm)	12±4	51±47	(86±37)§	(48±45)	§Record ended before skid stop.				
Friction Use abs. value at skid start	.12±.07	.10±.07	.04±.03	.09±.07	.04±.02	.13±.05	.04±.04	.08±.06	.09±.06
Ditto 50 msec after skid start	.15±.02	.14±.04	.06±.02	.13±.05	.02±.02	.12±.06	.04±.02	.07±.06	.09±.06
Ditto 100 msec after skid start	.13±.03	.15±.06	.08±.04	.13±.05	.02±.01	.11±.05	.03±.02	.06±.05	.09±.06

*Skid start is defined by the first minimum in heel forward velocity if sliding continues after heel strike.

In the skids, the average critical sliding motion started 50 msec (48 msec mean value ± 21 msec standard deviation) after heel strike, when the vertical floor reaction force was 60% (64 ± 16) of body weight, and the shoe's angle of contact was 6° (5.5 ± 5.9). These data cast serious doubt on slip-resistance measurements where some kind of *static* COF is supposed to be gauged after a long and uncontrolled contact time, with normal forces being a small fraction of body weight, or with test specimens that do not resemble the rear edge of a shoe heel.

Twenty-three skids resulted in a 'fall' and were restrained by the safety harness. In the remaining 16 skids, called 'slip-sticks', walking continued with the balance undisturbed or regained. Some qualitative differences between grips, slip-sticks and falls are summarized in Figure 1. According to their perceived severity, the slip-sticks were divided into three categories: 1) The subject was unaware of the sliding motion in four cases called 'mini-slips'; 2) No apparent gait pattern disturbances were observed in nine other cases called 'midi-slips'; 3) Large compensatory swing-leg and arm motions occurred in the remaining three cases called 'maxi-slips'.

Statistics on the skids have been collected in Table 1, indicating that the risk-oriented but partly subjective skid classifications outlined correlate well with objectively measured quantities. The transition from mini- to maxi-slips can be observed as decreasing friction use values, as well as increasing sliding distances and peak velocities. Note that the peak velocity values are related to the COF-peak interval [Conant and Liska (1960) cited under the method outline]. The fall-safe upper limit of sliding velocity, however, as well as the kinetic COF demand (see Friction Use in Table 1), depends on walking speed, step length, and anthropometric parameters. Thus, COF limit values should be assessed from human gait characteristics.

Although the experiments presented in this article are far from being representative and unbiased, they clearly demonstrate the potential of integrating biomechanics and tribology into slip-resistance research.

Acknowledgments

This study is based on experiments supported with grants from the Swedish Work Environment Fund. Since 1980 the project proceeded with grants from the funds for special work environment projects at the Swedish National Board of Occupational Safety and Health. Some of the material in this article has been reproduced from the Journal of Occupational Accidents with kind permission from Elsevier Scientific Publishing Company.

References

BRING, C. 1978. Provning av halksäkerhet. (Testing of slipperiness.) Ph.D. dissertation, Royal Institute of Technology, Stockholm.

CONANT, F.H., and Liska, J.W. 1960. Friction studies on rubberlike materials. Rubber Chem. & Technol. 33:1218-1258.

GUSTAFSSON, L., and Lanshammar, H. 1977. ENOCH — an integrated system for measurement and analysis of human gait. Ph.D. dissertation, Institute of Technology, Uppsala University, Sweden.

LANSHAMMAR, H. 1980. Precision limits on derivatives obtained from measurement data. In: K. Fidelus and A. Morecki (eds.), Biomechanics VII, pp. 586-592, Polish Publishers' House and University Park Press, Baltimore, Maryland.

LANSHAMMAR, H., and Strandberg, L. 1980. Halkmekanik — Etapp 1. (Slipping accident mechanics — Stage 1.) National Board of Occupational Safety and Health, Solna, Sweden.

LANSHAMMAR, H., and Strandberg, L. 1983. Horizontal floor reaction forces and heel movements during the initial stance phase. In: H. Matsui and K. Kobayashi (eds.), Biomechanics VIII-B, pp. 1123-1128, Human Kinetics Publishers, Champaign, Illinois.

MOORE, D.F. 1972. The Friction and Lubrication of Elastomers. Pergamon Press, Oxford.

PERKINS, P.J. 1978. Measurement of slip between the shoe and ground during walking. In: C. Anderson and J. Senne (eds.), Walkway Surfaces: Measurement of Slip Resistance, pp. 71-87. ASTM, STP 649, Philadelphia, PA.

RABINOWICZ, E. 1956. Stick and slip. Scientific American 194:5, pp. 109-118.

STRANDBERG, L., and Lanshammar, H. 1981. The dynamics of slipping accidents. J. Occupational Accidents 3:153-162.

Automated Acquisition and Analysis of Data in Clinical Gait Evaluations

Erich Schneider, Edmund Y.S. Chao, R. Keith Laughman, Joan E. Bechtold, and Byron P. Cahill

Mayo Clinic/Mayo Foundation, Rochester, Minnesota, U.S.A.

Gait evaluations, if used as a routine clinical service, have to meet high standards. Referring physicians expect fast and complete information, patients hope for a short evaluation, and biomechanicians wish to evaluate a considerable number of normals. Keeping up with these requests and the large amount of data gathered during gait evaluations requires versatile and reliable instrumentation as well as a high degree of automation.

During a period of eight years, an automated system for gait analysis has been successfully developed in this laboratory involving joint angular motion of the lower extremities, foot-floor contact patterns, ground reaction forces and other temporal distance factors including step length and step width (Chao and Hoffman, 1977; Chao et al., 1980). Further progress has recently been made by replacing the film analysis system with a new computer-interfaced video system, currently used for evaluations such as foot function during walking.

Automation in the area of data analysis is much more difficult, especially for the comparison of signal shapes and the distinction between normal and abnormal gait cycles prior to averaging. Fourier analysis and signal correlation are well established procedures in electronic signal processing, but their use in gait analysis is surprisingly rare. Fourier analysis has been applied to accelerographic signals (Smidt et al., 1971), body kinematics (Zarrugh and Radcliffe, 1979), and ground reaction forces (Jacobs et al., 1972). It has been used for the determination of frequency contents of gait signals (Winter et al., 1974) as well. Signal correlation has been used by Marsh et al. (1980) in EMG analysis to investigate the interdependence of different muscles. In this laboratory, Fourier analysis has been used for efficient storage, reconstruction and

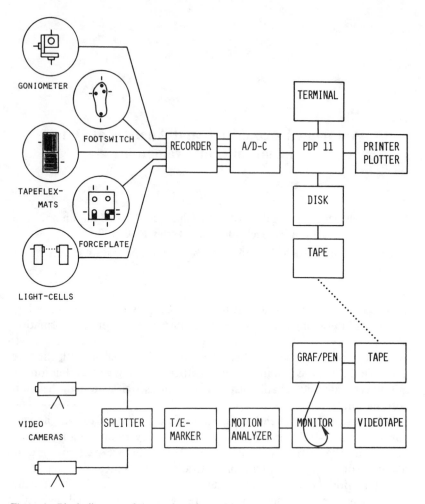

Figure 1—Block diagram of the equipment used in the gait laboratory.

averaging of gait signals as well as for quantification of signal shapes. Signal correlation is utilized for objective differentiation between normal and abnormal gait cycles prior to averaging.

The objectives of this study were 1) to describe the instrumentation of the gait laboratory including the video system, 2) to illustrate the application of Fourier analysis for storage, averaging and quantification of signal shapes with the example of ground reaction forces, and 3) to demonstrate the use of signal correlation for the distinction between normal and abnormal gait cycles.

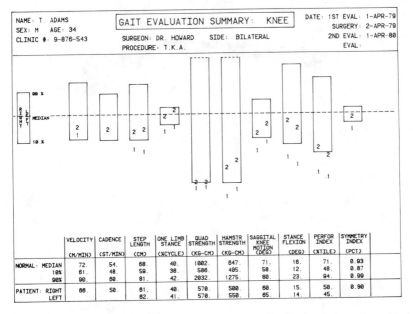

Figure 2—Patient report form for the results of a knee evaluation in tabular and graphical form, including values of a normal population (rectangles) for comparison. Numbers '1' and '2' describe subsequent evaluations.

Instrumentation

An overview of the equipment in the gait laboratory is given in Figure 1. It consists of an instrumented walkway equipped with two piezoelectric force plates to measure the ground reaction forces as well as two mats with embedded stainless steel strips for the determination of step length and step width. Four foot switches, taped on each shoe of the patient, measure the time spent on specific areas of the foot. Electrogoniometers determine the three-dimensional, bilateral relative motion of hip, knee and ankle joints. Infrared light cells determine the average walking speed and initiate sampling of the data by the PDP 11/34 computer. A multi-channel strip chart recorder is used for continuous monitoring during the evaluation. Results are computed and presented within 30 min after an evaluation. An example of a patient report form containing the pertinent parameters of a knee evaluation are shown in Figure 2.

The new video system, as shown on the bottom of Figure 1, incorporates two cameras, one usually placed underneath the transparent force plate to determine the location of the foot, the other one used for patient monitoring. The two signals are combined in a splitter/inserter, and time and event marks are added. The video signal of the whole evaluation is stored on a video tape recorder, essential parts of it on a

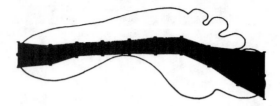

Figure 3—Mean and band of standard deviation of the center of pressure distribution for barefoot walking in thirty normals.

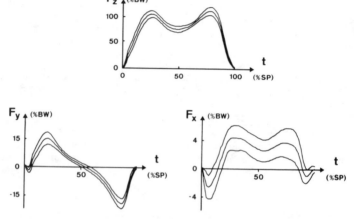

Figure 4—Average and band of standard deviation of the vertical (F_z), fore-aft (F_y) and medial-lateral (F_x) ground reaction forces of 26 normals walking at their own speed. (Forces in % bodyweight, time in % stance phase length.)

video sheet recorder. This important piece of equipment is similar to a video disc allowing frame by frame analysis. Video information is digitized directly from the screen using a sonic pen digitizer with microphones mounted on the video monitor. Data are transferred to the computer by tape and combined with other gait data to obtain an overall analysis. Figure 3 shows the normal pattern of the center of pressure distribution for barefoot walking, a result obtained with this video system.

Fourier Analysis

Fourier has shown that a periodic signal can be described by a sum of cosine and sine terms, the Fourier coefficients A_n and B_n. Details of this theory and solutions for computers can be found in Ralston and Wilf (1960) and Brigham (1974). Once Fourier analysis is performed on the sampled signals, only the coefficients need to be stored, resulting in a

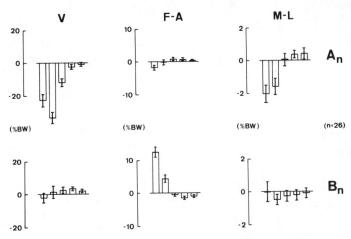

Figure 5 — Average and standard deviation of the first five Fourier cosine (A_n) and sine (B_n) coefficients of the vertical (V), fore-aft (F-A) and medial-lateral (M-L) ground reaction forces of 26 normals walking at their own speed. (Forces in % bodyweight, constant term not shown.)

significant reduction of computer storage space. The number of essential terms must be determined very carefully for each signal according to the characteristics of the waveform. An objective method to do this is currently being developed. The original signal can easily be reconstructed using the Fourier series approximation. The determination of an average pattern, as shown in Figure 4 for the ground reaction forces of normal walking, usually requires extensive normalizing of every curve by body weight and stance phase period to the same number of samples, before corresponding points of each curve can be averaged. Due to the linearity property of Fourier analysis (Brigham, 1974), the same average force pattern can be obtained by averaging the Fourier coefficients of matching harmonics and reconstruction of the curve.

Since Fourier coefficients describe a curve in an unequivocal way, they can be used for quantitative comparison of different shapes. Figure 5 shows the Fourier coefficients describing the average force curves of Figure 4. The vertical force is mainly described by A_n (cosine) coefficients. This is due to the fact that the vertical force resembles an even function, $f(t) = f(-t)$. A change of this 'even' shape would increase the magnitudes of the B_n coefficients. The constant term, by definition the average of the signal, yields the total vertical impulse (no negative contribution). The magnitudes and positions of the two peaks are mainly described by the first three harmonics. The contributions of higher coefficients to the averaged pattern are relatively small, but they become more significant for individual tracings depending on their specific waveform. The fore-aft force is dominated by the B_n (sine) coefficients. This

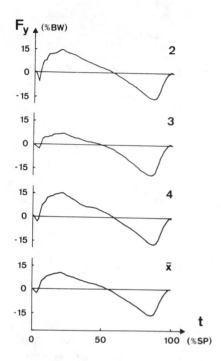

Figure 6 — Fore-aft ground reaction forces of subject E.K., showing the recordings Nos. 2, 3, and 4, and the average of a total of seven recordings. (Forces in % bodyweight, time in % stance phase length.)

is due mainly to the odd nature of this curve, $f(t) = -f(-t)$. A change of this 'odd' shape would increase the magnitudes of the A_n terms. The constant term equals the net (average) forward or backward impulse. The main contributors of this shape are certainly the first two B_n coefficients. The medial-lateral force has the smallest absolute amplitude and the largest variation in individual tracings, reflected in a broad band of the standard deviation. The constant term here reflects the net medial impulse. The first two A_n coefficients are the principal determinants of this signal.

Correlation

One of the main obstacles to automated gait analysis is the difficult decision between normal and abnormal curves. Correlation, which is basically a multiplication and integration of two signals (Brigham, 1974), is an appropriate way to solve this problem. At the present time, this technique is used prior to the averaging of single cycles during multiple evaluation runs for the establishment of the average pattern of each individual subject. The result of this procedure at zero shift is reflected in a correlation coefficient, +1 being identical curves, 0 showing no similarity at all between the curves. Figure 6 shows three separate and the overall

Table 1

Correlation Coefficients of the Fore-Aft Forces of Subject E.K.
(Individual Cycle Nos. 2, 3, 4, and Individual Average)
and the Normal Pattern

	individual cycles			average $\bar{x}$	normals $\bar{n}$
	2	3	4		
cycle 2	1.0	0.855	0.991	0.983	0.976
cycle 3		1.0	0.806	0.931	0.914
cycle 4			1.0	0.965	0.964
average				1.0	0.990
normals					1.0

averaged tracings of the fore-aft force of the same subject. Table 1 presents the correlations between all the signals and the average normal pattern of Figure 4. A number of conclusions can be drawn from this example: 1) The correlation between cycles 2 and 3 is much lower than the correlation between cycles 2 and 4, indicating that cycle 2 is similar to cycle 4, but different from cycle 3. Due to the very good repeatability of gait signals, correlations between individual cycles are high in fore-aft forces, usually above 0.9. 2) The higher correlation between cycle 2 and the normal pattern than between cycle 3 and normal, reinforces the difference of cycle 3 from normal. 3) The high correlation of the average of this subject with the normal pattern after exclusion of signal 3 reflects the normal behavior of this subject. Of course, the ideal ranges of the correlation coefficient to differentiate abnormals from normals must be analyzed through the entire population.

Summary

An automated system for gait analysis including joint angular motion, foot-floor contact pattern, ground reaction forces, and other temporal distance factors has been developed, and a new video system has been added to the overall capacity of our evaluation facilities. Fourier coefficients are used for storage, reconstruction and comparison as well as for quantification of signal shapes. Signal correlation is used to facilitate the automated differentiation between normal and abnormal cycles within individual tracings or between subjects and normal patterns of populations.

Acknowledgment

This study has been supported by NIH Grants AM 18029 and CA 23751.

References

BRIGHAM, E.O. 1974. The Fast Fourier Transform. Prentice Hall, Englewood Cliffs, New Jersey.

CHAO, E.Y.S., and Hoffman, R.R. 1977. Instrumented measurement of human joint motion. ISA Transactions 17(1):13-19.

CHAO, E.Y., Laughman, R.K., and Stauffer, R.N. 1980. Biomechanical gait evaluation of pre and postoperative total knee replacement patients. Arch. Orthop. Traumat. Surg. 97:309-317.

JACOBS, N.A., Skorecki, J., and Charnley, J. 1972. Analysis of the vertical component of force in normal and pathological gait. J. Biomechanics 5:11-34.

MARSH, E.M., de Bruin, H., Brandstater, M.E., Clarke, B.M., and Gowland, C. 1980. The use of a coherence function in processing EMG gait patterns. Proceedings of the Canadian Society of Biomechanics: Human Locomotion I, 18-19.

RALSTON, A., and Wilf, H. 1960. Mathematical methods for digital computers. J. Wiley and Sons, New York.

SMIDT, G.L., Arora, J.S., and Johnston, R.C. 1971. Accelerographic analysis of several types of walking. Am. J. Phys. Med. 50(6):285-300.

WINTER, D.A., Sidwall, H.G., and Hobson, D.A. 1974. Measurement and reduction of noise in kinematics of locomotion. J. Biomechanics 7:157-159.

ZARRUGH, M.Y., and Radcliffe, C.W. 1979. Computer generation of human gait kinematics. J. Biomechanics 12:99-111.

IV.
ELECTRO-
MYOGRAPHY

Variability of the Electromyogram
Prior to a Rapid Voluntary Movement in Man

Katsumi Mita, Hisashi Aoki,
Kentaro Mimatsu, and Kyonosuke Yabe
Institute for Developmental Research, Aichi, Japan

Electromyographic activity abruptly changes from a tonic discharge to a phasic discharge in a rapid voluntary movement following a slight sustained contraction (see Figure 1a). Ikai (1955) reported that a silent period appeared prior to the onset of the phasic discharge (see Figure 1b). This phenomenon is referred to as the 'premotion silent period' in the present study. Yabe and Murachi (1975) and Yabe, Mita, and Aoki (1976) assumed that the premotion silent period was due to an inhibition originating in the upper centers of the central nervous system. It is difficult to estimate the mechanism of the silent period using raw electromyograms (EMG), because the raw EMG has a complex temporal pattern which is considered to be a random signal. The present study was designed to examine the statistical properties of the EMG prior to the onset of a rapid voluntary movement and to investigate the physiological mechanism of the premotion silent period on the basis of the statistical properties.

Methods

Experimental Procedure

The experiment was performed on three healthy adults ranging in age from 27 to 39 years. An isometric plantar flexion in response to a visual stimulus was chosen as the rapid voluntary movement. The experimental trials were repeated about 200 times on each subject. Each subject was seated comfortably in a chair with the hips flexed at 90°, knee flexed at 60°, and ankle at 0°. Following an oral signal, 'ready', the subject was

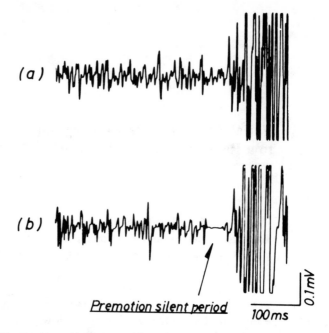

Figure 1—Electromyographic patterns without the premotion silent period (a), and with the premotion silent period (b) recorded from the triceps surae muscle during a plantar flexion of the right foot.

asked to make a slight voluntary contraction and to maintain the sustained contraction corresponding to a level of 10% of the maximum strength during 2 to 5 sec. After this preparatory phase, plantar flexion was performed as quickly as possible whenever a light stimulus (xenon lamp) was activated. To help the subject maintain a stable tonic discharge during the preparatory phase, an indicator which displayed the integrated EMG (the rectified and low-pass filtered EMG) was used. Time constant of the low-pass filter was 0.5 sec. The action potentials of the triceps surae and anterior tibial muscles of the right foot were led off by surface electrodes (10 mm dia.) placed on the long axis of the muscle about 30 mm apart. The EMG responses were recorded on a pen writing polygraph and magnetic tape.

Data Processing

The EMG was analyzed with a NEAC-3200 mini-computer. The raw EMGs recorded on analog magnetic tape were sampled at a rate of 2000 samples/sec and digitized with 12-bit accuracy using analog-to-digital converter. Digitized EMG data were stored on digital magnetic tape. Afterwards, the full-wave rectified EMG was computed from the raw

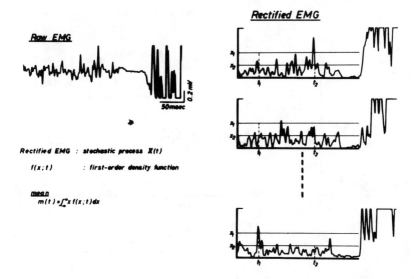

Figure 2 – Computational processing procedure for the statistical analysis of the rectified EMG.

EMGs, and the statistical properties of the rectified EMG were examined by the following procedures.

Assume that a family of the rectified EMGs obtained from a subject, $X(t)$ is a stochastic process. Then, each rectified EMG record, $x_i(t)$ (i = 1---n) can be viewed as an outcome of the process $X(t)$ (see Figure 2) (Papoulis, 1965). To examine the statistical properties of the process $X(t)$, an amplitude probability density of the rectified EMG during 200 msec before the onset of the phasic discharge was computed. The mean (ensemble average) of the rectified EMG was derived from the amplitude density. The processing was performed on the EMG obtained from the triceps surae muscle with and without the silent period, respectively.

Results

Figure 3a illustrates the characteristic mean activity of the rectified EMG with a silent period from subject TKR. The abscissa indicates a time course and the origin of the abscissa corresponds to the onset of the phasic discharge. The ordinate shows the mean activity of the rectified EMG. The histogram below the abscissa shows a frequency of the duration of the silent period. The mean activity remained stable throughout the period between − 200 msec and − 100 msec. Then, a time average for this period was calculated and is illustrated by a thin straight horizontal line on the trace of the mean value. Judging from the time average, a depression of the mean activity before the phasic discharge was observed

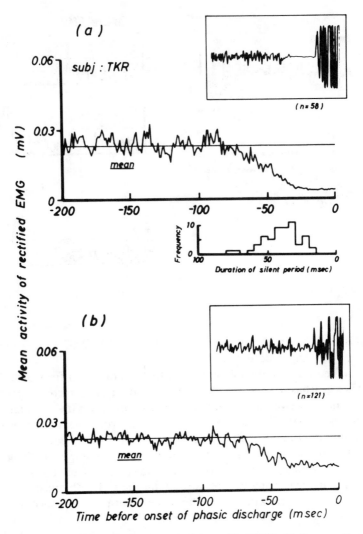

Figure 3 — Mean activity (computed average) of the rectified EMG with the premotion silent period (a) and without the premotion silent period (b) before the onset of the phasic discharge. The histogram below the abscissa of the trace (a) indicates the frequency of the premotion silent period duration.

for all subjects, and the beginning of the depression was distributed between −100 msec and −90 msec. As for the duration of the silent period, the maximum value of the duration ranged from 53 msec to 79 msec. Thus, it was noted that the depression of the rectified EMG began preceding the appearance of the silent period.

Figure 3b shows a mean activity of the rectified EMG without the silent period. The thin straight line indicates a time average between

−200 msec and −100 msec. It was also found that there was the depression before the phasic discharge not only with the silent period but also without it. The rate of the depression, however, was not remarkably different in comparison to the upper trace with the silent period.

Discussion

An electromyographic activity switch from tonic discharge to phasic discharge was recorded in a reaction movement. There was a peculiar phenomenon in which the electrical activity disappeared just prior to the onset of the phasic discharge. The silent period does not always appear, however, and interindividual variance in the frequency of the silent period appearance is large (Yabe and Murachi, 1975). On the other hand, since the electrical activity detected by surface electrodes consists of many action potentials of the active motor units, the electromyographic record has the properties of a random signal. Furthermore, the statistical properties of the EMG activity with the silent period are not thought to be 'invariant to a shift' of the time origin. Then, the EMG activity might be described as a non-stationary process. To investigate the mechanism of the silent period, a statistical method was used for these reasons.

It was found that whether the silent period appeared or not, the mean activity of the rectified EMG decreased preceding the phasic discharge. The depression of the mean activity with the silent period began before the appearance of the silent period. Also the same characteristics were noted in the EMG activity without the silent period. Therefore, it may be concluded that a brief suppression of the muscle activity is always followed by the start of a reaction movement.

Previous reports have stated that the amplitude of the H-reflex with a reaction movement has a tendency to increase prior to the phasic discharge and the increase is preceded by a decreasing period (Requin, 1969; Piellot-Deseilligny et al., 1971). The researchers suggested that the depression of the H-reflex might result from a presynaptic inhibition acting on the afferent Ia pathway.

In the analogy of the variabilities of the H-reflex and the EMG activity in a reaction movement, it is suggested that the excitability of the motoneurons in the spinal cord are reduced temporarily prior to the initiation of the reaction movement. Such a variability of the excitability in the spinal cord may produce the momentary depression of the muscle activity. As to the relationship between the silent period and the depression of the rectified EMG, it is estimated that the depressing influence is so pronounced that disappearance of electrical activity is observed.

References

IKAI, M. 1955. Inhibition as an accompaniment of rapid voluntary act. J. Physiol. Soc. Jap. 17:292-298. (In Japanese)

PAPOULIS, A. 1965. Probability, Random Variable, and Stochastic Process. McGraw Hill, New York.

PIELLOT-DESEILLIGNY, E., Lacert, P., and Cathala, H.P. 1971. Amplitude et variabilité des réflexes monosynaptiques avant un mouvement voluntaire. (Amplitude and variability of monosynaptic reflexes with a voluntary movement.) Physiol. Behav. 7:495-508.

REQUIN, J. 1969. Some data on neurophysiological processes involved in the preparatory motor activity to reaction time performance. Acta Psychol. 309:358-367.

YABE, K., and Murachi, S. 1975. Role of the silent period preceding the rapid voluntary movement. J. Physiol. Soc. Jap. 37:91-98. (In Japanese)

YABE, K., Mita, K., and Aoki, H. 1976. Premotion silence observed in the contralateral limb. In: E. Asmussen and K. Jørgensen (eds.), Biomechanics VI-A, pp. 75-81, University Park Press, Baltimore.

Electromyographic Study of the Bifunctional Leg Muscles during the Learning Process in Infant Walking

Tsutomu Okamoto
Kansai Medical School, Osaka, Japan

Yukihiro Goto
Osaka City University, Osaka, Japan

Hirotake Maruyama
Seibo Junior College, Kyoto, Japan

Nobuyuki Kazai
Bukkyo University, Kyoto, Japan

Hiroshi Nakagawa
Osaka University of Economics, Osaka, Japan

Hideo Oka
Osaka Kyoiku University, Osaka, Japan

Minayori Kumamoto
Kyoto University, Kyoto, Japan

In normal adult gait, the knee and hip joints are simultaneously extended before the heel contact, and the electrical discharges of the bifunctional Biceps femoris (Bf) and the Semimembranosus tend to decrease or disappear in many cases when the bifunctional Rectus femoris (Rf) increases in electrical activity just before the heel contact (Kumamoto, 1979). This discharge pattern of the bifunctional muscles shows that the resultant force at the sole is limited by the knee joint, and when the upper body flexes at the hip and the resultant force is limited by the hip joint, the discharge pattern is reversed (Kumamoto et al., 1980). Whereas in child gait, the discharge patterns of the bifunctional muscles show mostly the reversed pattern and co-contraction at the heel contact (Okamoto, 1973).

In this paper, changes in the discharge patterns of the antagonistic bifunctional muscles (Bf and Rf) just before the heel contact were examined during the learning process of infant walking.

Method

Three male infants at 7, 9, and 12 months after birth were employed. They were followed up from the very initial period to about 1 month before learning to walk. Electromyograms (EMGs) were recorded at approximately 1-week intervals. The EMGs of two of the three infants had been recorded at approximately 2- or 4-week intervals for more than 1 year. One of the two was followed-up at approximately 1-month intervals until about 3 years of age. In addition, the EMGs of 10 children ranging in age from 1.2 to 8 years were also recorded. In order to confirm the discharge pattern of the bifunctional muscles in normal adult gait, 40 adults ranging in age from 18 to 45 years were employed. Some of the adult subjects were asked to walk in a squatting posture with the upper body erect and slightly flexed. All subjects tested were requested to walk at an adequate normal walking speed for their respective ages. Some of the younger children were tested also while they were running.

The EMGs were recorded from the six leg muscles, Tibialis anterior (Ta), Gastrocnemius lateral head (Gl), Vastus medialis (Vm), Rectus femoris (Rf), Biceps femoris (Bf) and Gluteus maximus (Gm), using surface electrodes 10 mm in diameter. The EMG recordings were made with an 18-channel pen-writing electroencephalograph (60 mm/sec) and a 12 channel electromagnetic oscillograph (200 mm/sec). A 16 mm motion picture (at 32 frames/sec) of the heel and toe contacts with the ground and angular changes of the ankle, knee, and hip joints were recorded simultaneously with the EMGs.

Results

In all subjects tested, the EMG patterns of the ankle joint muscles (Ta and Gl) and of the monofunctional extensors (Vm and Gm) at the knee and hip joints were similar to the EMG patterns reported by Okamoto (1973), and there were no signs of pathological EMG pattern. In 65% of the adults tested, the discharge of the Bf tended to decrease or disappear while that of the Rf increased just before the heel contact (adult pattern) and, in almost all of the rest, the co-contraction of the Bf and Rf appeared. Seldom was there a reversed pattern where the Bf showed sustained discharges and the Rf tended to decrease or disappear just before the heel contact. In the child subjects, the Bf and Rf showed mostly the reversed pattern and partially the co-contraction in the subjects under 3 years old. The adult pattern was observable in subjects over 3 years of age.

Therefore, during the process of learning to walk in the infant subjects, the changes in the discharge patterns of the bifunctional Bf and Rf were examined in terms of the adult, co-contraction, and reversed patterns.

In the very initial period, almost all discharge patterns of the Bf and Rf showed the co-contraction pattern. At about 1 week after learning to walk, the patterns were mostly the co-contraction but only partially the adult and reversed patterns. Thereafter, the co-contraction pattern tended to decrease and the adult and reversed patterns tended to increase until about 1 month after learning to walk. At this time the contribution of the three patterns was roughly one third each. After that time, the co-contraction pattern and the reversed pattern, respectively, tended to decrease and increase continuously, but the adult pattern started to decrease until about 2 months after learning to walk. At about 2 months after learning to walk the reversed pattern was about 80%, the co-contraction pattern was found partially, but the adult pattern was hardly seen. From 2 months after learning to walk to the end of 2 years of age, there was no obvious change in the three discharge patterns. At 2.9 years of age, the adult pattern reappeared to some extent.

When the adults in a squatting posture walked at normal speed with the body erect, the bifunctional muscles showed the adult discharge pattern just before the heel contact, but with the upper body flexed the reversed pattern occurred. On the other hand, when the younger children in a half squatting posture ran with the body sway forward, the discharge patterns of the bifunctional muscles showed the strong co-contraction just before the heel contact.

Discussion

The remarkable changes observed in the discharge patterns of the bifunctional muscles from the initial period to 2 months after learning to walk suggested that there might be changes in the walking posture and speed. The co-contraction pattern observed in the very initial period might be induced by the peculiar walking style of the infant in a deep squatting posture with the body sway forward, and with quick stepping to prevent falling before the heel contact. This was confirmed by the co-contraction pattern observed in the conditioned experiments where the younger children in half-squatting posture were running with the body sway forward.

The results obtained in the conditioned gait experiment in the adults indicated that the adult and the reversed patterns partially observed after 1 week might be induced when the infant in a deep squatting posture occasionally held the trunk erect (adult pattern) or walked slowly instead of quick stepping with the body sway forward (reversed pattern). The co-contraction pattern decreased further probably because the infant in a deep squatting posture began to cease quick stepping. This was confirmed by the fact that the discharge of the monofunctional knee extensor, Vm, started to disappear at the latter part of the swing phase. In this period, the infant occasionally walked with the trunk erect, resulting in roughly equal appearance of the adult and reversed patterns. One month after learning to walk, the foot base narrowed and the deep squatting posture began to shift to the half squatting posture, in which the walking became less unstable. This might be related to the decrease in the adult pattern and further decrease in the co-contraction pattern. Thus, the stable walking in the infant in a half squatting posture with the upper body sway forward resulted in the increase of the reversed pattern up to 80% (Okamoto, 1973; Kazai et al., 1976). These tendencies continued until about 2 months. Such stable infant walking was continued until the end of 2 years of age (Okamoto, 1973), and this corresponded to the fact that the predominant reversed pattern appeared during the same period. From this period, when the infant began to acquire the real adult walking with the upper body erect, the bifunctional muscles as well as the other leg muscles tested, the Ta, Gl, and Gm, began to show the adult pattern.

Thus, the changes in the discharge patterns of the bifunctional muscles were closely related to the changes in the infant walking style during the learning process.

References

KAZAI, N., Okamoto, T., and Kumamoto, M. 1976. Electromyographic study of supported walking of the infants. In: P.V. Komi (ed.), Biomechanics V-A, pp. 311-318. University Park Press, Baltimore.

KUMAMOTO, M. 1979. Antagonistic inhibition in double-joint leg muscles during normal gait cycle. Proc. 4th Cong. I.S.E.K., 164-165.

KUMAMOTO, M., Oka, H., Kameyama, O., Okamoto, T., Yoshizawa, M., and Horn, L. 1980. Possible existence of antagonistic inhibition in double-joint leg muscles during normal gait cycle. In: A. Morecki, K. Fidelus, K. Kedzior and A. Wit (eds.), Biomechanics VII-B, pp. 157-162.

OKAMOTO, T. 1973. Electromyographic study of the learning process of walking in 1- and 2-year-old infants. In: S. Cerquiglini, A. Venerando and J. Wartenweiler (eds.), Biomechanics III, pp. 328-333. Karger, Basel.

Properties of Human Motor Unit Potentials Detected by Surface Electrodes

Shigeru Morimoto, Yoshiki Umazume, and Makoto Masuda

The Jikei University School of Medicine, Tokyo, Japan

During a study of shivering in humans, a clearly discriminable train of spikes was detected unexpectedly with ordinary silver disc electrodes fastened to the skin surface overlying the muscle. Further investigation showed that these discriminable spikes were also detected during voluntary activation of the muscle. In the present study, an attempt was made to observe the properties of these spikes obtained by the surface electrodes.

Methods

The subjects of this study were healthy male adults. In the major part of the experiment, the subject was the principal author, S.M.

The subject sat on a high stool which allowed free movement of the lower legs. The tension of knee extension was detected by a strain gauge (Nihon-Kohden, Tokyo, RTB-100K) connected to the ankle. The output of the strain gauge was fed into a carrier amplifier (Nihon-Kohden, Tokyo, RP-3).

For detecting the electrical activity, surface and intramuscular electrodes were used. Two silver discs with a diameter of 10 mm each were fixed by elastic adhesive plaster, at a center-to-center separation of 20 mm, to a given position on the skin surface overlying the *m. vastus medialis* in line with the general direction of the muscle fibers. Contact was completed with electrode jelly. The inserted electrode was prepared in the laboratory following Kurata's method (1974). Copper wires with a diameter of 0.1 mm each and insulated with polyurethane were joined. The bipolar electrode thus prepared was inserted into a 1/3 hypodermic-injection needle and bent at 2 mm from the tip. After the electrode was inserted in the muscle with the aid of the needle, the latter was withdrawn. The ground electrode consisted of a silver plate, 30 mm in diameter, placed on the skin surface of the knee joint where there was no underlying muscle tissue. Signals from both surface and inserted electrodes were amplified by conventional differential amplifiers (Nihon-Kohoden, Tokyo, RB-2) with a time constant of 0.03 sec. The input impedance of each amplifier was 5 MΩ.

Mechanical and electrical signals were simultaneously displayed on a cathode ray oscilloscope (Nihon-Kohden, Tokyo, VC-9) and photographed with a continuously recording camera (Nihon-Kohden, Tokyo, PC-2B).

In one of the experiments, the temperature of the muscle was measured with a thermistor (Shibaura Electronics, Tokyo, MGP-III) closed at the tip injected into the muscle with a 1 mm needle.

Results

Relationship Between the Spikes
Recorded by the Surface and Inserted Electrodes

An attempt was made to find out if the spike detected by the inserted electrode synchronized with the one indicated by the surface electrode from the m. vastus medialis. After fixing the surface electrode in a position in which the largest amplitude of the spikes was detected, the intramuscular electrode was pushed into the muscle via the skin at a point 10 mm from the proximal side of one of the disc electrodes. An attempt was made to locate the spikes synchronized with the surface spikes by moving the inserted electrode to and fro in the direction of the wire axis. Only when the tip of the electrode was located in the muscle closely under the fascia were the synchronized potentials detected. Spikes detected in the deeper part of the muscle were not synchronized with those detected on the surface. Figure 1 shows the typical synchronized potentials recorded from both surface and inserted electrodes. When the subject extended the knee joint gradually, spikes appeared at a tension of 5 N or more. Both trains of spikes were synchronized with each other. Other large synchronized potentials appeared at a tension of about 20 N and over. It was difficult to discriminate the synchronized potentials at a tension of 50 N and over because of the interference of spikes from other sources.

Conduction Velocity of the Spikes

The time difference between both spikes was a function of the distance along the muscle axis between the surface and the inserted electrodes. The time difference was determined by measuring the period between the negative upward peak of the spikes (see inset of Figure 2). The distance between the surface electrodes and the inserted electrode was defined as the distance between the inserted electrodes and the mid point of two disc electrodes on the skin surface. Figure 2 shows a rectilinear relationship between the two parameters. From this kind of measurement, the conduction velocity of the spike potential could be calculated: the results were, for example, 3.3, 3.5, 3.6 and 3.7 m/sec. Figure 2 also shows that the spike potential is conducted in both directions. The intersection of the two lines indicates the starting point of excitation, i.e., 'end-plate'. The end-plate was located at a point one-third of the observed length from the distal end in this particular case. Because of the finite size of the surface electrodes and the broad distribution of the end plates, the intersection occurred in the negative region of the ordinate.

Shape of Spikes at the Vicinity of End-Plate

Figure 3 shows spike potentials detected by the surface electrodes. In trace b, one of the disc electrodes, connected to the negative input of the differential amplifier, was put in a position corresponding to the intersection of the two lines in Figure 2. This steeply rising phase can be interpreted as follows: considering the volume conductor, the shape of the rising curve of an ordinary spike potential is smooth because of the current flow into the excited part in front of the electrode. On the other hand, if the electrode is put directly on the end-plate, there is no current flow into the electrodes before the excitation starts. Trace a and c were recorded at the distal and proximal parts from the end-plate, respectively. In this case, the proximal disc electrodes of each pair of electrodes were connected to the negative, and distal discs to the positive, inputs of the differential amplifiers. The inversion of the phase of two potentials can be interpreted as a result of the conduction of the excitation in the opposite directions from the end-plate.

Reproducibility and Generalization of the Spike

To confirm the reproducibility of the experiment, the threshold value was examined, the tension at which the spike was recruited, and the wave forms recorded at the same point on the skin but on different days. Figure 4 shows the wave forms which were obtained from three experiments performed on different days. The threshold values obtained from each experiment performed on different days were 24.0 ± 1.9 N (n = 24), 24.4 ± 2.5 N (n = 22) and 22.1 ± 2.0 N (n = 19). Wave forms were highly reproducible.

Finally, to generalize our findings, attempts were made to determine

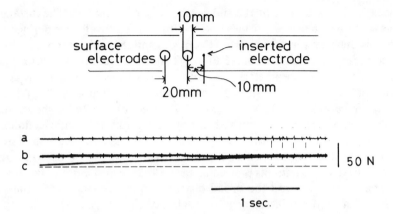

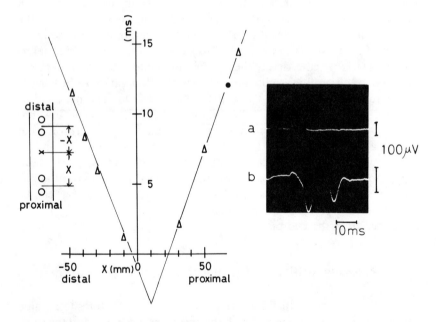

Figure 1—Synchronized potentials detected by surface and inserted electrodes. Trace a: signal from the surface electrode. Trace b: signal from the inserted electrode. Trace c: tension of knee extension. Broken line shows zero level of tension.

Figure 2—Relationship between time difference and electrode distance (X). Ordinate: the time difference in msec of responses detected by the surface and inserted electrodes. Abscissa: the distance (X) in mm between the inserted electrode (x) and the surface electrodes. The zero point on the abscissa corresponds to the position of the inserted electrode. The lines were drawn by a method of least squares fit. Conduction velocities were calculated to be 3.7 m/sec (distal) and 3.6 m/sec (proximal). Inset is the record of both spikes. This record corresponds to the filled circle on the plot. Trace a: the signal from the inserted electrode. Trace b: signal from the surface electrode. Upward deflection means negative voltage.

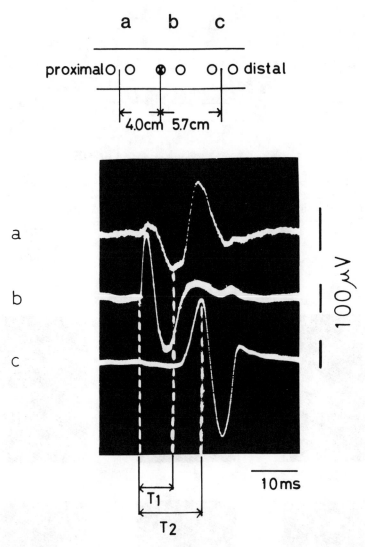

Figure 3 – Spike potential detected simultaneously from surface electrodes at three different positions along the direction of conduction. x shows the position of the end-plate. The conduction velocity calculated using the distance between a and b (4.0 cm) or b and c (5.7 cm) and T_1 or T_2 on this record is 3.7 m/sec. Upward deflection means negative voltage.

the surface spikes as mentioned previously in other muscles and other subjects. They were found in *m. vastus lateralis, m. tibialis anterior, m. flexor digitorum superficialis* and *m. flexor pollicis brevis* in one subject, S.M., and in *m. vastus medialis* in five other subjects. Figure 5 shows some examples of the records of these findings.

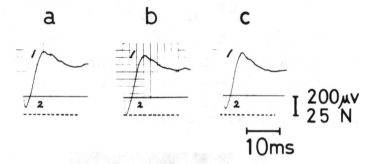

Figure 4 — Spike potentials recorded using the internal trigger mode of an oscilloscope in three independent experiments. In each photo: Trace 1: spike potential. Trace 2: tension. The broken line shows the zero level of tension. a, b and c were recorded on March 4, 6 and 7, 1978, respectively. Downward deflection means negative voltage.

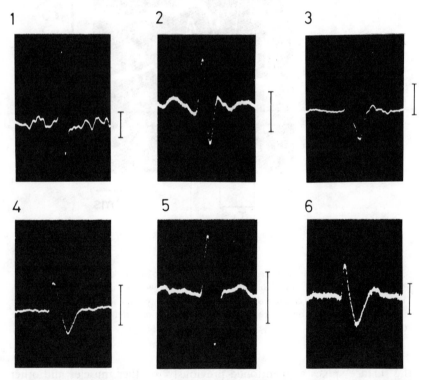

Figure 5 — Examples of surface spikes recorded from different muscles and subjects. 1) m. tibialis anterior in S.M.; 2) m. biceps brachii in S.M.; 3) m. flexor pollicis brevis in S.M.; 4) m. vastus medialis in Y.O.; 5) m. vastus medialis in H.K.; 6) m vastus medialis in Y.U. Calibration bars indicate 100 μV and 10 msec, respectively. Upward deflection means negative voltage.

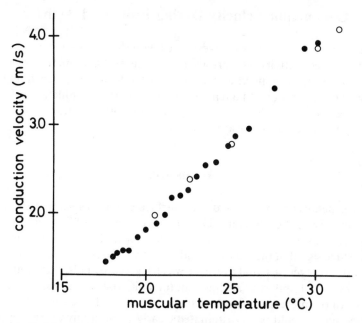

Figure 6 — Relationship between conduction velocity and muscular temperature. • = data points during cooling phase; ○ = data points during warming phase.

Conduction Velocity at Different Temperatures

The effect of temperature on the conduction velocity of the spike was also investigated. In this experiment, two pairs of surface electrodes, 38 mm apart, were fixed on the skin. The conduction velocity was calculated from the time difference between the point of the negative peak on both spikes and the distance between the two pairs of electrodes. The position of a pair of electrodes was defined as the mid-point of two discs. The temperature was changed in the muscle by the following means: the skin surface overlying the muscle was covered with ice wrapped in a plastic sheet after fixing the two pairs of electrodes. The temperature of the muscle was measured by a thermistor inserted into the muscle just under the fascia at the midpoint between the two pairs of electrodes. To avoid the effect of heat conduction, the needle containing the thermistor was wrapped in adiabatic material. The temperature decreased gradually from 31 to 17°C in 30 min after the ice was applied. After the ice was removed the temperature rose again to 31°C in 20 min. During this process the spikes were recorded. Figure 6 shows the relationship between the conduction velocity and the temperature. The two variables showed a linear relationship and the rate was about 0.2 m/sec • deg. This rate was maintained during both the cooling and recovery phases.

Conduction Velocity During Prolonged Activity

The conduction velocity was investigated with the surface technique during prolonged activity at various tension levels. Conduction velocity changed during prolonged activity in a different manner depending on the level of the tension. At a tension of 60 N and less, conduction velocity decreased during prolonged activity. On the other hand, conduction velocity increased at a tension of 80 N.

Discussion

It has generally been accepted that surface electrodes sample the electrical activity of the whole muscle. Recently, Yemm (1977) reported that the activity of a few motor units was visible as a discrete wave form on the surface signal in the masseter and temporal muscles. According to his method, however, a signal averager was triggered by pulses derived from a succession of spikes of a single motor unit from the needle electrode signal. In the present experiment, on the other hand, the spikes from the surface signal could be discriminated clearly without any averaging procedure and synchronized with spikes of a single motor unit from the inserted electrode signal. The surface responses gave information similar to the responses detected by an inserted electrode.

There are many reports on measurements of conduction velocity in human muscle. Activating muscle fibers in m. biceps brachii voluntarily, Buchtal et al. (1955) found a velocity of 4.7 m/sec. The surface electrode samples the electrical signal from the wide part of the muscle. Because of the property of the surface electrode, the surface spike is the sum of the spikes of each single fiber. The conduction velocity found in the present investigation was the mean value of each of the muscle fibers belonging to the same motor unit. By considering the difference between the muscle and room temperature, the present results, of around 3.5 m/sec, are in agreement with the finding of Buchtal and his colleagues (1955).

The voltage of the intracellular action potential in the frog sartorius muscle increased with rising temperature in the range of 0-30°C (Macfarlane and Meares, 1958). The length constant decreased with temperature in the frog sartorius muscle (Castillo and Machne, 1953). But in the giant squid axon, Sjodin and Mullins (1958) found a decrease in the threshold with increasing temperature when stimulated with 1 msec duration pulse. There was a positive linear relationship between the muscular temperature and the conduction velocity in the range 17-31°C in the present experiment. The cause of this phenomenon would be the changes in membrane characteristics.

It has been shown that the intracellular action potential in the isolated muscle fibers decreased during repeated stimulation (Schanne et al.,

1962; Persson, 1963; Lüttgau, 1965). Schanne et al. (1962) showed that the internal resistance increased in the frog sartorius muscle during prolonged stimulation. The present results show that the various changes of conduction velocity during prolonged activity were dependent on the level of tension. The factors which influence the conduction velocity during prolonged activity are still unknown. But it seems that the changes in membrane characteristics, composition of extracellular fluid, and muscular temperature would influence the conduction velocity.

References

BUCHTAL, F., Guld, C., and Rosenfalck, P. 1955. Innervation zone and propagation velocity in human muscle. Acta Physiol. Scand., 35:174-190.

DEL CASTILLO, J., and Machne, X. 1953. Effect of temperature on the passive electrical properties of the muscle fiber membrane. J. Physiol. 120:431-434.

KURATA, H. 1974. Characterization of the motor unit in voluntary contraction. A relationship between recruitment order and the gradient of tension increase. J. Phys. Fitness Jpn., 23:125-133.

LÜTTGAU, H.C. 1965. The effect of metabolic inhibitors on the fatigue of the action potential in single muscle fibers. J. Physiol. (Lond) 178:45-67.

MACFARLANE, W.V., and Meares, J.P. 1958. Intracellular recording of action and after potentials of frog muscle between 0-45°C. J. Physiol. 142:97-109.

PERSSON, A. 1963. The negative after potential of frog skeletal muscle fibers. Acta Physiol. Scand. 58:Suppl. 205.

SCHANNE, O., Kern, R., and Shäfer, B. 1962. Zum Problem des Innewiderstandes biologisher Zellen. (The problem of inner resistances of biological cells.) Naturwissenschaften. 49:161.

SJODIN, R.A., and Mullins, L.J. 1958. Oscillatory behavior of the squid axon membrane potential. J. Gen. Physiol. 42:39-47.

YEMM, R. 1977. The representation of motor unit action potentials on skin surface electromyograms of the masseter and temporal muscles in man. Arch. Oral Biol., 22:201-205.

Analysis of Myoelectric Signals
during Dynamic and Isometric Contractions

M. Muro
Tokyo College of Pharmacy, Tokyo, Japan

N. Nagata
Tokyo Metropolitan University, Tokyo, Japan

T. Moritani
Texas A & M University, U.S.A.

Recent histochemical and biochemical studies (Peter, Barnard, Edgerton, Gillespie, and Stempel, 1972; Burke and Edgerton, 1975; Prince, Hikida, and Hagerman, 1976) have shown that skeletal muscles can be classified into three fiber types (e.g., SO, FOG, and FG) which may correspond to slow twitch fatigue resistant (S), fast twitch fatigue resistant (FR), and fast twitch fatigable (FF) types of motor units (MUs). These studies have generally suggested that different MUs are expected to be recruited in the reverse order of their fatigability, i.e., in the order of S, FR, and FF MUs. Therefore, it seems likely that FR and FF MUs would be activated at higher force levels and/or during rapid movements (Grimby and Hannerz, 1968; Burke and Edgerton, 1975). With respect to EMG power spectral changes in connection with different force output levels, no generally accepted patterns seem to exist. For example, Fay, Jones, and Porter (1976) reported a power increase in the lower frequency band (0 to 50 Hz) up to ⅓ of maximum voluntary contraction (MVC) and a substantial increase in the higher band (0 to 100 Hz) for ⅔ of MVC while observing a corresponding decrease of energies in the high frequency band. O'Dannell, Rapp, Berkhout, and Adey (1973) found that the percentage of power at lower frequencies (26 to 49 Hz) was higher for a moderate contraction than during a nearly maximum contraction and a continuous increase in the high frequency band (> 60 Hz) with increasing contraction was observed. It was the purpose of this investigation to

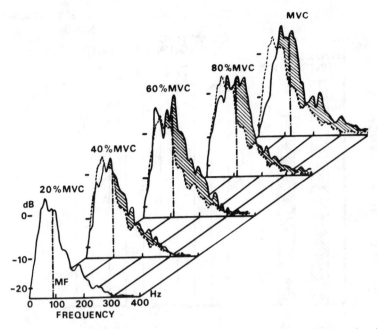

Figure 1 — A typical set of data obtained during isometric contractions at different fractions of MVC (sub: T.M., biceps brachii). Note that the power spectrum obtained at 20% of MVC was superimposed with broken line.

elucidate the MU recruitment pattern systematically with respect to various movement conditions (fast and slow dynamic contractions and isometric contractions at different fractions of MVC) by means of surface EMG power spectra.

Material, Methods, and Procedures

On the first visit to our biodynamics laboratory, eight male subjects were tested for MVC (for 2 sec) with the arm flexor muscle group (biceps brachii) according to the method described by Moritani and deVries (1978). This procedure was repeated 10 times with 3-min rest intervals between contractions. After a 30-min rest, each subject performed isometric contractions at 20, 40, 60, and 80% of MVC which were chosen at random, with each contraction lasting 5 sec and repeated 6 times with 2-min rest intervals. The same procedure was also used for the rectus femoris muscle group. On the second visit, the subject performed fast (1.5 ± 0.3 rad/sec) and slow (0.17 ± 0.03 rad/sec) contractions with a load corresponding to about 20% of MVC 20 times with a 30-sec rest period between contractions. A 10-min rest period was given between

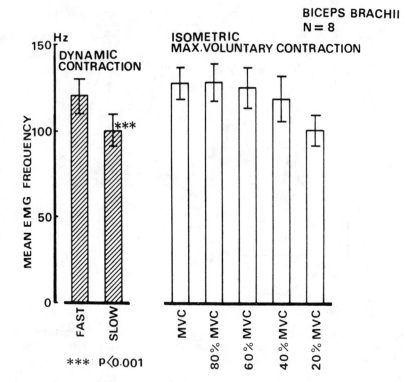

Figure 2—Changes in mean power frequency (MPF) during fast and slow dynamic contractions and isometric contractions (mean ± S.E.) for the biceps brachii (n = 8).

fast and slow contraction trials.

Myoelectric signals were picked up by two miniature size electrodes (2 mm contact diameter, 6 mm inter-electrode distance) attached over the motor point area of the biceps brachii and the rectus femoris muscle. The reference electrode was placed over the volar aspect of the wrist of the non-contracting arm. All electrode placements were preceded by abrasion of the skin surface so that source impedance was reduced to less than 3000 ohms. The EMG power spectral analysis method used in this study has been fully described elsewhere (Moritani, Nagata, and Muro, 1981; Nagata, Muro, Moritani, and Yoshida, 1981).

Results

Figure 1 represents a typical set of data showing the EMG power spectral changes with respect to different fractions of MVC being held. As is seen in this figure, the contribution of higher frequencies to the spectra seems to increase progressively with increasing force. Analysis of variance with

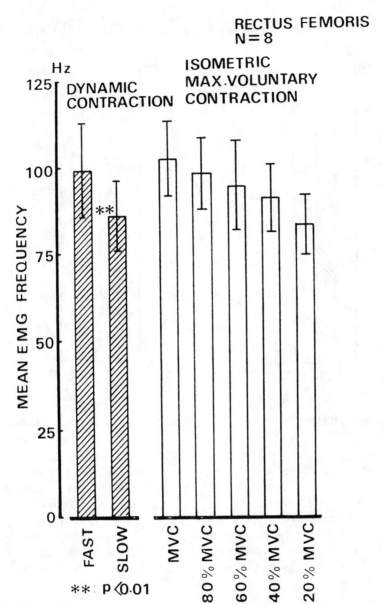

Figure 3 — Changes in mean power frequency (MPF) during fast and slow dynamic contractions and isometric contractions (mean ± S.E.) for the rectus femoris (n = 8).

a repeated measures design revealed that there were significant shifts in mean power frequency (MPF) at different fractions of MVC for both the biceps brachii (F = 25.15, p < 0.001) and the rectus femoris (F = 12.26, p < 0.001).

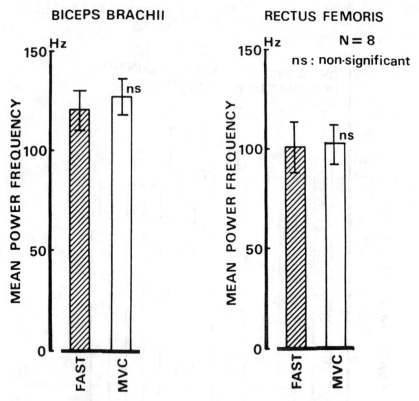

Figure 4—Comparisons in the MPF obtained during fast dynamic contractions and during MVC for both muscle groups.

When comparisons were made for the differences in MPF during fast and slow dynamic contractions, highly significant differences were observed for the biceps brachii (mean diff of 20.6 Hz, p < 0.001) and for the rectus femoris (12.7 Hz, p < 0.01) (see Figures 2 and 3). However, when the MPF obtained during MVC was compared with that obtained during a fast dynamic contraction, no statistically significant difference could be observed in the biceps brachii (t = 0.76, p > 0.05) and in the rectus femoris (t = 0.25, p > 0.05).

Discussion

Available human studies on the relative contributions of MU recruitment and firing frequency have suggested that the largest contribution of MU recruitment may occur at relatively low force levels while the contribution of increased firing frequency becomes more important at higher force levels (Milner-Brown, Stein, and Yemm, 1973; Gydikov and

Kosarov, 1974). Furthermore, it has been shown that low threshold MUs (tonic MUs) recruited at low contraction strengths as a rule had smaller potential amplitude, slower rise time, and lower firing frequency than high threshold MUs (phasic MUs) (Tanji and Kato, 1973b; Gydikov and Kosarov, 1974; Stephens and Ushwood, 1975). Phasic MUs have also been shown to have higher nerve conduction velocities (Freund, Budingen, and Dietz, 1975) and shorter duration of after-hyperpolarization (Eccles, Eccles, and Lundberg, 1958), thus possessing the capacity for higher firing frequencies. Our data seem to be in good agreement with those previous findings on the basis that there were significant changes in MPF during isometric contraction at various fractions of MVC, suggesting the existence of a significant degree of MU recruitment shift.

At higher levels of MVC, e.g., above 60% of MVC, the difference in the MPFs did not achieve any statistical significance. In all cases, however, we observed shifts of MPF toward higher frequencies which agree with the previous findings (Gydikov and Kosarov, 1974; Komi and Viitasalo, 1976). Using the rectus femoris muscle, Komi and Viitasalo (1976) suggested that at approximately 60% of MVC practically all of the tonic MUs have attained their maximal firing frequency and that the further increase in muscle force was due to increases in recruitment and firing frequency of phasic MUs. Therefore, it may be possible that a small increase of a very high frequency zone in the power spectrum observed at higher MVC levels, although not large enough to shift the MPF to any significant extent, could certainly contribute to the exerted force, since high threshold phasic MUs develop contractile force which might be many times that of low threshold MUs (Burke and Edgerton, 1975; Walmsey, Hodgson, and Burke, 1978).

As to the recruitment pattern of different MUs during dynamic contractions with low force outputs, Tanji and Kato (1973a) reported that during ramp contractions the peak value of firing frequency was higher with greater speeds. This finding was substantiated by the work of Freund et al. (1975) who found a positive correlation between the nerve conduction velocity and the threshold force at which the MUs were recruited. Therefore, it seems likely that even at low force output which can be maintained by innervating tonic MUs, an increase in the speed of contraction would bring about higher threshold MUs. With the facts that phasic MUs have higher firing frequency, shorter after-hyperpolarization period, and faster contraction time (Buchthal and Schmalbruch, 1975; Freund et al., 1975), the significant increases (20.6 and 12.7 Hz for the biceps brachii and rectus femoris, respectively) in the MPF during fast dynamic contraction may have been a result of phasic MU recruitment, giving rise to a shift of the power spectral density to higher frequencies. This may suggest that non-invasive EMG power spectral analysis can provide a sensitive measure of the MU recruitment pattern during rapid movements in which the use of conventional needle electrodes may not be suitable.

References

BUCHTHAL, F., and Schmalbruch, H. 1970. Contraction times and fiber types in intact human muscle. Acta Physiol. Scand. 79:435-452.

BURKE, R.B., and Edgerton, V.R. 1975. Motor unit properties and selective involvement in movement. Exer. Sport. Sci. Rev. 3:31-81.

ECCLES, J.C., Eccles, R.M., and Lundberg, A. 1958. The action potentials of the alpha motoneurons supplying fast and slow muscle. J. Physiol. 142:275-291.

FAY, D.F., Jones, N.B., and Porter, N. 1976. Spectral analysis of the myoelectric activity of the pelvic floor during voluntary contractions. Electromy. Clin. Neurophysiol. 16:525-551.

FREUND, H.J., Budingen, H.J., and Dietz, V. 1975. Activity of single motor units from human forearm muscles during voluntary isometric contractions. J. Neurophysiol. 38:933-946.

GRIMBY, L., and Hannerz, J. 1968. Recruitment order of motor units on voluntary contraction: changes induced by proprioceptive afferent activity. J. Neurol. Neurosurg. Psychiat. 31:565-573.

GYDIKOV, A., and Kosarov, D. 1974. Some features of different motor units in human biceps brachii. Pflugers Arch. 347:75-88.

KOMI, P.V., and Viitasalo, H.T. 1976. Signal characteristics of EMG at different levels of muscle tension. Acta Physiol. Scand. 96:267-276.

MILNER-BROWN, H.S., Stein, R.B., and Yemm, R. 1973. Changes in firing rate of human motor units during linearly changing voluntary contractions. J. Physiol. 230:371-390.

MORITANI, T., and deVries, H.A. 1978. Reexamination of the relationship between the surface integrated electromyogram (IEMG) and force of isometric contraction. Am. J. Phys. Med. 57:263-277.

MORITANI, T., Nagata, A., and Muro, M. 1981. Electromyographic manifestations of muscular fatigue. Med. Sci. Sport. Exer. 14:198-202.

NAGATA, A., Muro, M., Moritani, T., and Yoshida, T. 1981. Anaerobic threshold determination by blood lactate and myoelectric signals. Jap. J. Physiol. 31:585-597.

O'DANNELL, R.D., Rapp, R., Berkhout, J., and Adey, W.R. 1973. Auto spectral and coherence patterns from two locations in the contracting biceps. Electromy. Clin. Neurophysiol. 13:259-269.

PETER, J.B., Barnard, R.J., Edgerton, V.R., Gillespie, C.A., and Stempel, K.E. 1972. Metabolic profiles of three fiber types of skeletal muscle in guinea pigs and rabbits. Biochemistry 11:2627-2633.

PRINCE, F.P., Hikida, R.S., and Hagerman, F.C. 1976. Human muscle fiber types in power lifters, distance runners and untrained subjects. Pflügers Arch. 363:19-26.

STEPHENS, J.A., and Ushwood, T.P. 1975. The fatigability of human motor units. J. Physiol. 251:37P-38P.

TANJI, J., and Kato, M. 1973a. Firing rate of individual motor units in voluntary contraction of abductor digiti minimi muscle in man. Exp. Neurol. 40:771-783.

TANJI, J., and Kato, M. 1973b. Recruitment of motor units in voluntary contraction of a finger muscle in man. Exp. Neurol. 40:759-770.

WALMSLEY, B., Hodgson, J.A., and Burke, R.E. 1978. The forces produced by medial gastrocnemius and soleus muscles during locomotion in freely moving cats. J. Neurophysiol. 41:1203-1216.

Electrical Inhibition on Bifunctional Muscle during Double Joint Movements

Noriyoshi Yamashita and Minayori Kumamoto
Kyoto University, Kyoto, Japan

Yasuhiko Tokuhara
Teikoku Women's University, Moriguchi, Japan

Fujio Hashimoto
Osaka Electro-Communication University, Osaka, Japan

The pioneer work of the reciprocal motor system was reported by Sherrington in 1906. This was followed by extensive studies of the transmission in 1a inhibitory pathway (Lloyd, 1941; Araki et al., 1960), and the interconnections between the interneurons mediating 1a inhibition and the supraspinal systems of animals (Hongo et al., 1969; Jankowska and Tanaka, 1974). In human joint movement, the reciprocal 1a inhibition has also been observed during voluntary ankle dorsiflexion confirming depression in H-reflex (Tanaka, 1972).

When the serial double joint movements (simultaneous hip and knee extensions) were performed with maximal effort, the activities of the bifunctional muscles were depressed in comparison with those observed in the individual hip or knee extension (Yamashita, 1975). Even in the submaximal conditions, the same tendencies have been reported as observed in the maximal condition (Yamashita and Kumamoto, 1976; Kazai et al., 1978; Tokuhara et al., 1980).

The aim of the present study was to elucidate the real significance of the electrical depression on the bifunctional muscle during the serial double joint movements.

Method

The experimental postures were the same as in our previous studies (Yamashita, 1975; Yamashita and Kumamoto, 1976). Healthy male adults were in a supine position with the hip and knee joints kept at 90°. Total leg extensions with simultaneous hip and knee extensions were performed isometrically along two distinctly different functional force directions. The resultant force exerted at the sole of the foot is not the sum of the forces developed along the functional force direction by the individual hip and knee extension forces but is limited by the weaker force. In the case of the K-line where the line passes through the ankle joint and the point approximately ¼ of the thigh length from the knee joint, the weaker force is developed along the functional force direction by the hip extension force. In the case of the G-line where the line passes through the ankle joint and the center of gravity of the body when in the experimental posture, the weaker force is developed by the knee extension force (Yamashita, 1975).

The evoked potentials in the Rectus femoris (Rf) and the Vastus medialis (Vm) induced by patella ligament tapping (approximately 4 times/sec) were recorded while the forces were produced up to about 30% of the maximal force with tension development speed of about 6 kg/sec. Electromyograms (EMGs) were recorded from the Rf, the Vm, and the Semimembranosus (Sm) with conventional methods utilizing an electroencephalograph (30 mm/sec) and an electromagnetic oscillograph (200 mm/sec). In order to check the intensity of the tapping stimulation, the amplitude of mechanical impacts was recorded via a strain gauge placed on the hammer shaft. The resultant leg extension force exerted at the sole of the foot, the mechanical impacts of the tapping, and the EMGs in the Rf, Vm, and Sm were recorded simultaneously with the evoked potentials in the Rf and Vm.

Results

The spike potentials obtained in the Rf and Vm were assessed whether these were evoked by the tendon tap or not. Examination of the fast speed recordings (200 mm/sec) with the electromagnetic oscillograph revealed that the potentials induced by the tendon tap in the Rf and Vm appeared after some delay from the mechanical impacts. The latencies were about 28 msec in the Rf, and about 22 msec in the Vm. These results indicated that the potentials induced by the tendon tap in the Rf and Vm were not mechanical artifacts, but evoked potentials through a possible monosynaptic arc.

When the total leg extension was performed along the K-line, where the resultant leg extension force was limited by the hip extension force, the amplitude of the evoked potential in the Rf decreased, and the electrical discharge of the Sm increased while the leg extension force gradually increased. However, when the total leg extension was performed along the G-line, where the resultant leg extension force was limited by the knee extension force, the amplitude of the evoked potential in the Rf tended to increase and there was little or no electrical discharge in the Sm. The amplitude of the evoked potential in the Vm increased with the resultant leg extension forces exerted along both K- and G-lines.

Discussion

The amplitude of the evoked potential in the Rf decreased when the resultant leg extension force was limited by the hip extension force. This decrease strongly suggests that an inhibitory input to the α-motoneuron pool innervating the Rf exists, and the inhibitory input may be due to an antagonistic inhibition induced by the increased activity in the hamstring muscles, the Biceps femoris, the Semitendinosus and the Sm.

In the present experimental posture, the bifunctional leg antagonists, the Rf and the hamstrings, have to play roles of both agonist and antagonist, while the hip and knee extensions are performed simultaneously. That is, the Rf acts as the agonist in the knee extension, but acts as the antagonist in the hip extension; on the contrary, the hamstrings act as the agonist in the hip extension, but act as the antagonist in the knee extension. Therefore, the reciprocal motor system of the double joint movements is not the simple agonist-antagonist relationship as is observed in the single joint movement. Thus, more consideration should be given as to which role, agonist or antagonist, prevails on the bifunctional muscle during such double joint movements. In the results obtained in our experiments, when the Rf received the inhibitory input, the resultant leg extension force was limited by the hip extension force. In other words, the Rf had to play the predominant role of the antagonist in this case. If the resultant leg extension force was limited by the knee extension force, the hamstrings possibly received the inhibitory input from the Rf, increasing its activity. Thus, the appearance of the electrical inhibition on the bifunctional Rf or hamstrings playing roles of both agonist and antagonist depends on which joint is limiting the resultant leg extension force.

In other results obtained in the present experiments, although the amplitude of the evoked potential in the Rf decreased when the resultant leg extension force was limited by the hip extension force, that of the Vm did not decrease but increased in the same period. In other words, the Vm, monofunctional knee extensor, was always facilitated independently even though the reciprocal inhibition might occur between the bifunctional antagonistic muscles, the Rf and the hamstrings, during this period. In the detailed animal experiment, the Vasto-crureus, monofunctional knee extensor of the cat, received 1a activation from the hip extensors, the Sm, the adductor femoris, and the anterior biceps (Eccles and Lundberg, 1958). This report suggests that even if there is a general neurophysiological feature in the reciprocal motor system, 1a-excitatory and 1a-inhibitory actions are organized in a more complex fashion. This seems to be supported by the result mentioned above.

References

ARAKI, T., Eccles, J.C., and Ito, M. 1960. Correlation of the inhibitory postsynaptic potential of motoneurones with the latency and time course of inhibition of monosynaptic reflexes. J. Physiol. 154:354-377.

ECCLES, R.M., and Lundberg, A. 1958. Integrative pattern of 1a synaptic actions on motoneurones of hip and knee muscles. J. Physiol. 144:271-298.

HONGO, T., Jankwska, E., and Lundberg, A. 1969. The rubrospinal tract. II. Facilitation of interneuronal transmission in reflex paths motoneurones. Exp. Brain Res. 7:365-391.

JANKOWSKA, E., and Tanaka, R. 1974. Neuronal mechanism of the disynaptic inhibition evoked in primate spinal motoneurones from the corticospinal tract. Brain Res. 75:163-166.

KAZAI, N., Kumamoto, M., Yamashita, N., Maruyama, H., and Tokuhara, Y. 1978. Role of two-joint muscle in joint movements. In: E. Asmussen and K. Jørgensen (eds.), Biomechanics VI-A, pp. 413-418. University Park Press, Baltimore.

LLOYD, D.P.C. 1941. A direct central inhibitory action of dromically conducted impulses. J. Neurophysiol. 4:184-190.

SHERRINGTON, C.S. 1906. The Integrative Action of the Nervous System. Yale University Press, New Haven.

TANAKA, R. 1972. Activation of reciprocal 1a inhibitory pathway during voluntary motor performance in man. Brain Res. 43:649-652.

TOKUHARA, Y., Kazai, N., Maruyama, H., Kumamoto, M., and Yamashita, N. 1980. The influence of antagonistic inhibition in joint movements as deduced from the relation between graded forces applied and the integrated EMGs of two-joint muscles. In: Biomechanics VIIB, pp. 65-70. University Park Press, Baltimore.

YAMASHITA, N. 1975. The mechanism of generation and transmission of forces in leg extension. J. Human Ergol. 4:43-52.

YAMASHITA, N., and Kumamoto, M. 1976. Force generation in leg extension. In: P.V. Komi (ed.), Biomechanics V-B, pp. 41-45. University Park Press, Baltimore.

EMG Activity and Kinematics of Human Stumbling Corrective Reaction During Running

S. Suzuki and S. Watanabe
Kyorin University, Tokyo, Japan

M. Miyazaki
Waseda University, Tokyo, Japan

S. Homma
Chiba University, Chiba, Japan

When mechanical and tactile stimuli are applied to the dorsum of a cat's paw during the animal's locomotion, the so-called stumbling corrective reaction, which influences activity in order to maintain equilibrium and adapts the locomotion to the external conditions, was observed in chronic spinal (Forssberg et al., 1975, 1977), conscious (Forssberg, 1979), and freely moving cats (Wand et al., 1980). To achieve such compensatory reaction in the course of a stereotyped movement, it is essential that some programmed or controlled neuronal mechanisms interact with additional afferent inputs due to external disturbance. It is especially significant during locomotion that kinesthetic inputs impinge directly on the body parts to be moved.

We were interested in observing the resulting movements and reflex patterns when the intact human subject stumbles. For this purpose, stumbling was generated artificially through different and more appropriate physiological conditions than those available in animal experiments. EMG and kinematics were analyzed while the subject was stumbling in various phases of running, and it was found that compensatory reactions are definitely phase dependent.

Procedures

The subjects, with a rope tied around their torsos, were asked to perform

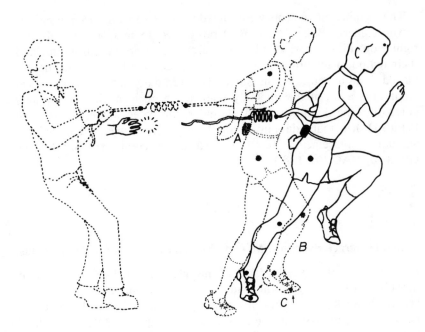

Figure 1 – Experimental conditions: See text for explanation.

strenuous forward running on a hard floor about 5 sec, by overcoming a backward pull of a 10 to 20 kg load applied by the experimenter. Figure 1 provides a schematic illustration of the experimental arrangement. Following maintenance of the running under such conditions, the rope was abruptly and randomly released by the experimenter, relative to the subjects' running phases. When the rope was released, a compensatory movement was induced which may be called a "stumbling corrective reaction" (SCR).

For the EMG recordings, ceramic (Dia Medical System Co. Ltd.) or Ag-AgCl pairs of surface electrodes were placed over the muscle bellies of the lateral gastrocnemius (LG) and tibialis anterior (TA) (see Figure 1.) Electrode wires were connected to a wireless transmitter, which was carried on the subject's back, and the electrical activities were recorded with differential AC amplifiers by telemetry (see Figure 1A). Raw EMG signals were then led to a driving amplifier with a cut-off filter beyond the range of 10 to 3000 Hz at −3 db. The EMGs were then rectified and smoothed. Right foot contact signals were also recorded by telemetry (see Figure 1C). The time course of load tension was recorded by a potentiometer placed on a spring balance which was connected to the rope between the subject and experimenter (see Figure 1D). All signals were monitored on a storage oscilloscope and recorded on FM tape, simultaneously.

The running movements were filmed at 50 f/sec with a 16 mm motor-driven camera (Bolex H 16-SBM) positioned 15 m from the subject's right side. Film records and EMG recordings were synchronized, and their temporal relationship was accurately defined by an event marker placed simultaneously on the film and the FM data recorder. Anatomical landmarks associated with body segments were identified by markers glued onto the skin at the head, shoulder, hip, knee, ankle, and toe (see Figure 1). Kinematic data involving the neck, hip, knee, and ankle joint angular displacements were analyzed by a computer (SPORTIAS GP-2000, NAC Co. Ltd.).

Results

General Characteristics of EMG and Kinematics during Loaded Running

EMG recordings of the LG and TA muscles and the kinematic data of the subject running aginst the load are shown in Figure 2 to the left of the vertical line which indicates the release point. Running in this case was performed at approximately one cycle/sec with approximately 15 kg load (see Figure 2J). The TA and LG muscles then showed considerable coactivation during roughly the first two-thirds of the stance phase (see Figures 2F, H, and I). Concerning this reciprocal activity in the antagonist during the agonist's contraction, it has been suggested that cocontraction is due to the movement that is resisted (Gellhorn, 1947; Levine and Kabat, 1952) or is related to stabilizing postural functions (Seamans, 1959; Stockmeyer, 1967; O'Donovan et al., 1979). Notice that the EMG activation of the TA muscle during the swing phase was almost imperceptible.

Although the timing of LG activity related to foot contact was variable when the parameters (load, cycle) were changed, the LG muscle activity began before or at approximately the same time as the foot touched the ground and increased during the knee extension phase. This EMG activity observed prior to foot contact resembles that preceding the so-called functional stretch reflex involved in landing (Melville-Jones and Watt, 1971a, b). It is probable that this activation is not produced by proprioceptive reflexes (Granit, 1970). In the kinematic record, the yielding phenomenon observed during the stance phase might be correlated to load compensation. The hip and knee angles during the stance phase were completely out of phase with that of the ankle, such that the knee reached peak extension as the ankle reached peak flexion (see Figures 2C, D, and E). This reciprocal displacement of the knee and ankle was common to paw shaking movements of the cat (Smith et al., 1980). The switching of right and left feet during this loaded running had a tendency toward either the right foot making contact as the left foot was off the

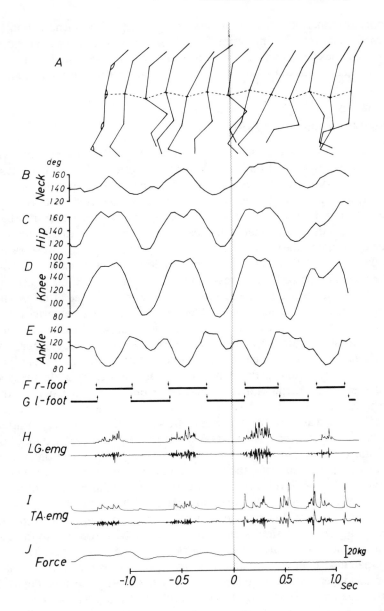

Figure 2—Stumbling corrective reaction as the rope was released within the SWING phase. A: movements consist of stick diagrams taken from 13 consecutive film frames at intervals of 200 msec for four steps. Each hip position is connected by the dotted line. B, C, D, and E: angular movements of the neck, hip, knee and ankle. F and G: the foot contact pattern of the right and left feet. H and I: raw EMG activity, and the rectified and smoothed TA and LG EMGs displayed in synchrony with movements shown in A, B, C, D, and E. J: force displacement against strenuous forward running. Movements analyzed are the same as those shown in Figure 3.

floor, or a period when both feet had contact. This phenomenon indicates that the switching action from flexors to extensors or from extensors to flexors changed very quickly. It is speculated that this phenomenon was due to the necessity for a high degree of load compensation.

Stumbling Corrective Reaction

In Figure 2, the EMG activities and kinematics are shown as the rope was released (vertical line) during right knee extension in the swing phase. The two running cycles, before and after the rope release, are shown. The EMG initiation of TA and LG relative to foot contact during the free running following the rope release, appeared earlier than those of TA and LG during the loaded run before the rope was released (the two-step cycle after rope release was especially significant). This shift forward of the onset of EMG relative to respective joint angles was also seen with increased bicycle pedalling movement (Suzuki et al., 1982). LG activity just after the release increased sharply at around 40 msec after ground contact and reached its maximum at the peak of knee extension. The body configuration is almost straight at this point in the sequential stick diagram (see Figure 2A). After the release, a typical double burst of EMG activity in TA muscle could be observed during the next cycle. The first longer burst during the stance phase was followed by a silent period and then by a sharp, shorter second burst during the first half of the swing phase (see Figure 2I). The initiation of the second burst during knee flexion was close to the timing of foot lift-off. Also, the initiation of the second burst in TA which occurred about 75 msec after the end of LG muscle activity indicates that the second burst in TA initiates dorsiflexion associated with the swing phase which otherwise might be prevented by the EMG activity of the extensor LG during full extension of the lower extremity (Duysens and Pearson, 1980). The ongoing activity in the extensor LG muscle was enhanced during the stance phase following release. A similar enhancement of activity occurred in the TA in its second burst during the swing phase. These EMG activities were associated with an accentuated forward movement as the subject took a step of longer distance (see Figure 3). The kinematic record of Figure 2 shows the knee action ranged from 80 to 165° before release. After the rope was released this range tended to become slightly larger, flexion movement was faster, and yielding of the hip joint during the stance phase was increased (see Figure 2). Hip and neck angular movements shifted to include a greater range of extension after the release (see Figures 2 B and C). The range of angular movement of the ankle joint remained relatively constant (see Figure 2E), between 85 and 130°. Plantar flexion during the swing phase was suppressed and the ankle showed a relatively smoother continuous flexion movement after the release. The

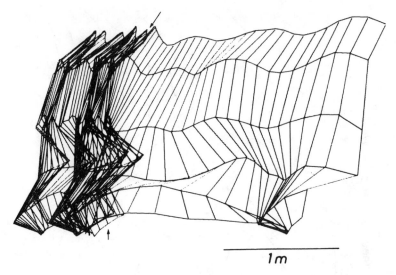

1 m

Figure 3—Single limb movement analysis during stumbling corrective reaction as the rope was released within the swing phase. Stick diagrams are presented at 40 msec intervals. Trajectories of the head, shoulder, hip, knee, ankle and toe are shown by horizontal lines connecting each plotted point, respectively. The timing of the rope release is indicated by arrows (↓ ↑).

period between the lifting of the left foot and contact of the right foot was prolonged after the rope was released (see Figures 2F and G). It is interesting that at this moment of increased forward running velocity, a progressive shortening of the duration of the support phase occurred, whereas the swing phase remained rather constant or longer occasionally (see Figures 2F and G).

Another SCR, triggered by rope release late in the stance phase in the right leg, is shown in Figures 4 and 5. When the rope was released during maximum knee and hip extension, those joints flexed more and faster than before the rope release. Therefore, the lower extremity was lifted higher (see Figure 5), and the forward movement covered a shorter distance as compared with the distance shown in Figure 3. In Figure 4, the EMG activities of the LG muscles had ceased before the rope was released. The TA response in the swing phase was observed to be greater in amplitude than that of an ordinary burst during the swing phase. The onset of the following burst of LG activity during the stance phase shifted slightly ahead of the EMG relative to the knee angle (see Figures 4D and H). When the release occurred at just the beginning of the foot lift-off phase, TA response during the swing phase was also larger in amplitude (not illustrated).

When the rope was released rather slowly at the beginning of the stance phase, the LG muscle inhibited the ongoing EMG activity (see

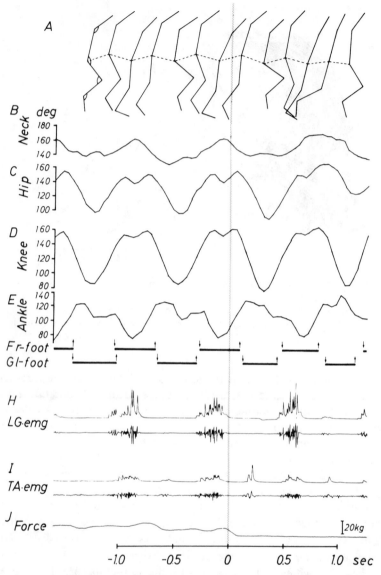

Figure 4—Stumbling corrective reaction as the rope was released within the STANCE phase. Movements analyzed are the same as those shown in Figure 5. A: Stick diagrams of 13 consecutive film frames are represented at 200 msec intervals for four steps, beginning with the knee flexion phase. The dotted line connecting each hip position indicated hip movement. B, C, D, and E: angular movements of the neck, hip, knee and ankle. F and G: the foot contact pattern of the right and left feet. A step cycle is subdivided into stance (thick line), and swing (between the thick lines) phases. The beginning and ending of the foot contact are shown by arrows. H and I: representations of the raw EMG, and the rectified and filtered LG and TA EMG are displayed in synchrony with movements shown in A, B, C, D, and E. J: force displacement against the forward running. Compare with Figure 2.

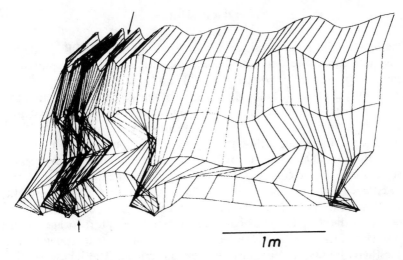

Figure 5 — Single limb movement analysis during stumbling corrective reaction as the rope was released within the stance phase. Stick diagrams are presented at 40 msec intervals. Trajectories of the head, shoulder, hip, knee, ankle, and toe are shown by horizontal lines connected at each plotted point, respectively. The timing of the rope release is indicated by arrows (↓ ↑).

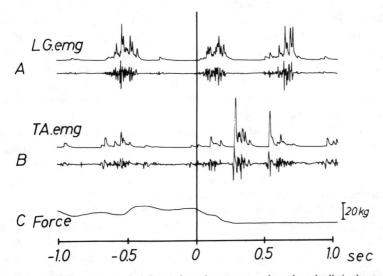

Figure 6 — Inhibitory response in LG muscle as the rope was released gradually in the stance phase. A and B: raw, and rectified and filter LG and TA activities, respectively, C: time course of rope tension during the forward running.

Figure 6). The activation period was shortened (240 to 190 msec), and the interval from the cessation to the next onset was also shortened (450 to 280 msec). The flexor TA muscle in the second burst, however, was extremely enhanced during the swing phase following the LG cessation.

Discussion

The results of the human study showed that during the SCR, EMG activity and kinematic data show some similarities to those reported in animal experiments by Forssberg (1979). Both findings suggest that tactile stimuli to the paw (animal) or the rope release (human) elicited the phasic influence of locomotor activity on the reflex pattern in extensor and flexor muscles, and induced compensatory movements. It is interesting to note that, although the triggering input was different between the animal experiments by Forssberg and the human experiments, the responses in the compensatory reaction adapt the locomotion to the varied circumstance in both cases. In the animal experiments, kinesthetic input for corrective reaction against mechanical touching might be elicited by the stimulation of hair and cutaneous receptors located in the limb (Forssberg, 1979). However, in these human experiments, the kinesthetic inputs are obviously less circumscribed. The kinesthetic inputs are clearly different from those involved in the animal experiments. In answer to the question, "What is the kinesthetic input in human stumbling corrective reaction?" we might speculate the following: The marked increase in the LG EMG activity during the stance phase was elicited as the rope was released within the swing phase (see Figure 2). When the subject was in the unloaded condition, the position of the upper extremity was observed to be forward with a slightly greater inclination. The neck was further extended and the head was exposed to a larger alternating vertical movement. The knee movement extended suddenly to a greater extent. These findings may be interpreted to be presumptive evidence for a vestibulo- and neck-spinal reflex contribution to the neuromuscular events. For example, Melvill-Jones and Watt (1971a, b) and Watt (1976) demonstrated the contribution of the vestibular system to the extensor hind limb by measuring muscular activity due to sudden free fall of a human subject from a suitable height above the ground. The vestibular nuclei are known to exert an excitatory influence on motoneurons innervating extensor muscles (Grillner et al., 1970). Evidence of a reflex pathway of the afferent input from neck muscles to hind limb extensor motoneurons was also reported by Kenins et al. (1978). In our data, the head and neck movement and lower limb extensor muscle activity showed a parallel relationship, i.e., the greater the range of neck angular movement, the greater the EMG activity (see Figure 2). Dietz et al. (1978, 1979) observed the gastrocnemius EMG activity during running, and pointed out that stretch reflex-induced muscle activity could mechanically contribute to the muscle force exerted by the extensor of the leg during ground contact.

When the rope was released rather gradually (see Figure 6C) the LG electrical activity during the stance phase was moderately inhibited, whereas the subsequent TA electrical activity during the succeeding swing phase was greatly enhanced (see Figure 6B). This tendency was augmented by the repeated experimental trials, indicating that neural circuits of SCR may be readily facilitated by training.

Additional TA electrical activity occurred during the early swing phase after cessation of LG muscle activity (see Figure 2I). This activity could be speculated to be effected by the adaptive nature of programming of the stumbling preventive reaction during sequential movement. In Forssberg's (1979) experiment when the paw was in contact with the ground during the support phase, a brief touch to the dorsum of the paw increased extensor activity and enhanced subsequent flexor burst. This flexor burst in our experiment was possibly caused by enhanced activity in the locomotor network during the activation of the flexors (Forssberg, 1979). Based on the latencies we observed one could presume a spinobulbospinal (Shimamura and Livingston, 1963) or even transcortical organization (Marsden et al., 1973; Everts and Tanji, 1974), a considerable delay being needed in both cases of the reflex.

A marked change appeared in the TA EMG during the swing phase when the rope was released during the end of the stance, or the beginning of the swing phases, and elicited fast and greater flexion movement (see Figure 4). It is assumed that this sudden change in temporal sequencing of muscle activity is a centrally programmed activation of extensors and flexors plus some additional involvement of fundamental physical laws of motion.

Acknowledgments

We thank Drs. R.S. Hutton and N. Hirai for their critique of this manuscript. We also thank Miss H. Ishida for typing the manuscript.

References

DIETZ, V., Schmidtbleicher, D., Ledig, T., and Noth, J. 1978. Timing of stance and swing phases and occurrence of phasic stretch reflex in running man. Eur. J. Physiol. 373: Suppl. R71.

DIETZ, V., Schmidtbleicher, D., and Noth, J. 1979. Neuronal mechanisms of human locomotion. J. Neurophysiol. 42:1212-1222.

DUYSENS, J., and Pearson, K.G. 1980. Inhibition of flexor burst generation by loading ankle extensor muscles in walking cats. Brain Res. 187:321-332.

EVERTS, E.V., and Tanji, J. 1974. Gating of motor cortex reflexes by prior instruction. Brain Res. 71:479-494.

FORSSBERG, H., Grillner, S., and Rossignol, S. 1975. Phase dependent reflex reversal during walking in chronic spinal cats. Brain Res. 85:103-107.

FORSSBERG, H., Grillner, S., and Rossignol, S. 1977. Phasic gain control of reflexes from the dorsum of the paw during spinal locomotion. Brain Res. 132:121-139.

FORSSBERG, H. 1979. Stumbling corrective reaction: a phase-dependent compensatory reaction during locomotion. J. Neurophysiol. 42:936-953.

GELLHORN, E. 1947. Patterns of muscular activity of man. Arch. Phys. Med. 28:568-574.

GRANIT, R. 1970. The Basis of Motor Control. Academic Press, New York.

GRILLNER, S., Hongo, T., and Lund, S. 1970. The vestibulo-spinal tract. Effects on alpha-motoneurones in the lumbosacral spinal cord in the cat. Exp. Brain Res. 10:94-120.

KENINS, P., Kikillus, H., and Schomburg, E.D. 1978. Short- and long-latency reflex pathways from neck afferents to hindlimb motoneurones in the cat. Brain Res. 149:235-238.

LEVINE, M.G., and Kabat, H. 1952. Cocontraction and reciprocal innervation in voluntary movement in man. Science 116:115-118.

MARSDEN, C.D., Merton, P.A., and Morton, H.B. 1973. Latency measurements compatible with a cortical pathway for the stretch reflex in man. J. Physiol. London 230:58-59.

MELVILL-JONES, G., and Watt, D.G.D. 1971a. Observations on the control of stepping and hopping movements in man. J. Physiol. London 219:709-727.

MELVILL-JONES, G., and Watt, D.G.D. 1971b. Muscular control of landing from unexpected falls in man. J. Physiol. London 219:729-737.

O'DONOVAN, M.J., Dum, R.P., and Burke, R.E. 1979. Force production, length changes and EMG activity in flexor digitorum longus (FDL) muscle during walking and jumping in unrestrained cats. 9th Soc. Neurosci., p. 380.

SEMANS, S. 1959. Physical therapy for motor disorders resulting from brain damage. Rehab. Literature 20:99-111.

SHIMAMURA, M., and Livingston, R.B. 1963. Longitudinal conduction systems serving spinal and brain-stem coordination. J. Neurophysiol. 26:258-272.

SMITH, J.L., Betts, B., Edgerton, V.R., and Zernicke, R.F. 1980. Rapid ankle extension during paw shakes: selective recruitment of fast ankle extensors. J. Neurophysiol. 43:612-620.

STOCKMEYER, S.A. 1967. An interpretation of the approach of Rood to the treatment of neuromuscular dysfunction. Am. J. Phys. Med. 46:900-956.

SUZUKI, S., Watanabe, S., and Homma, S. 1982. EMG activity and kinematics of human cycling movements at different constant velocities. Brain Res. 240:245-258.

WAND, P., Prochazka, A., and Sontag, K.H. 1980. Neuromuscular responses to gait perturbations in freely moving cats. Exp. Brain Res. 38:109-114.

WATT, D.G.D. 1976. Responses of cats to sudden falls: an otolith-originating reflex assisting landing. J. Neurophysiol. 39:257-265.

Electromyographic and Mechanical Aspects of the Coordination between Elbow Flexor Muscles in Monkeys

G. Peres and B. Maton
Laboratoire de Physiologie du Travail du C.N.R.S.,
Paris, France

B. Lanjerit and C. Philippe
École Nationale Supérieure des Arts et Métiers

Even in the case of very simple voluntary movements, such as elbow flexion, the external torque is the result of the torques exerted by several agonistic and antagonistic muscles. In humans it is only possible to record the external mechanical parameters of the movements and the myoelectric activities. From these data, one cannot calculate the torque produced by each muscle. Thus, the external torque or any other mechanical parameters of movement are often implicitly related to the electromyogram (EMG) of only one muscle of an agonistic group, the 'muscle equivalent' as termed by Bouisset (1973). Some experimental results have been put forward to be in agreement with this concept (Bouisset et al., 1977). Among them, the most significant are obviously 1) the existence of a linear relationship between work and integrated EMG of each of the three main elbow flexors (biceps brachii, brachialis, brachioradialis); 2) a well-determined chronology of muscle activities, i.e., muscle activities are nearly simultaneous despite a slight tendency for them to appear in the following order: biceps brachii (BB), brachioradialis (BR), brachialis (BA).

The purpose of the present work was to study the electromyographic and mechanical aspects of the coordination between elbow flexor muscles. From this analysis it was expected to define the characteristics of the motor command and to test the validity and the limits of the muscle equivalent concept. This is achieved by recording the tensile stress which is produced by each of the main elbow flexors in monkeys. Such an approach would be more interesting if it could be possible to extrapolate results from monkeys to humans. The experiment was composed of two parts. The first part dealt with the comparison of the relationships between EMG activities and mechanical parameters found in humans and those found in monkeys. The second part dealt with the relations between the force produced by each muscle and the external mechanical parameters.

Methods

The monkeys (Macaca fascicularis) used in this study were trained to perform flexion movements from a start position (elbow flexed at 90°) to a light spot placed in front of the animal. The onset of the spot indicated to the monkey that it had to make a movement. A correct response consisted of a phasic and smooth movement, voluntarily stopped in front of the spot or striking the stop switch. After each flexion, the forearm was reset in the start position by the experimenter. The experimental set up used was similar to the one used and previously described for humans (Bouisset and Maton, 1973). During the recordings, the monkey sat on a primate chair with its forearm strapped in a molded cast. This cast was part of a movable mechanical system which could be rotated in a horizontal plane, around a vertical axis. The forearm was supinated so that the supination component of the force exerted by the biceps could be neglected. The elbow was made approximately coincident with the rotation axis of the mechanical device. This device was equipped with a linear potentiometer and an accelerometer which allowed recording of displacement and tangential acceleration. The angular velocity was obtained by continuous differentiation of the angular displacement. The surface EMGs of BB and BR were recorded by means of small silver electrodes stuck to the skin with adhesive tape (bipolar derivation). The global intramuscular EMG of BA was simultaneously recorded by means of wire-electrodes. To quantify the excitation level of each muscle, EMG was integrated. Integrated EMG (Q) was given in the form of impulses whose number was proportional to the area of the full rectified EMG. 'Control recording' was carried out for 2 to 3 hours/day over a period of 1 week. After this period, the animal was deeply anesthetized with Nembutal and miniature force transducers were surgically implanted on BR and BB tendons, under aseptic conditions. The transducer used is a modification of the 'belt buckle' transducer suggested some years ago by Salmons (1969) and was somewhat similar to those used by Walmsley et al. (1978). A strain gauge (FLE 05 Wishay Co) was bonded along each side of a stainless steel frame (10 mm × 7 mm × 0.6 mm) and used in a simple DC Wheatstone bridge arrangement. The entire assembly was coated with M coat A (Vishay Co.) and then with Silastic rubber. The frame was placed

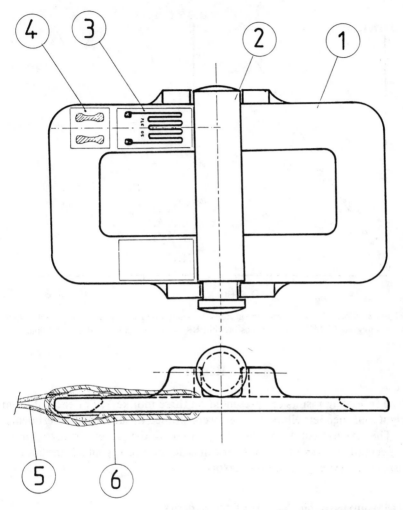

Figure 1 — Schematic diagram of the force transducer. 1 = frame, 2 = bridge axis, 3 = strain gauge, 4 = miniature plug, 5 = teflon insulated lead wires, 6 = coat layer.

on the tendon which was then pulled up through the frame, and held up by a stainless steel bridge axis across the frame, under the tendon. This axis was then coated with Silastic and the lead wires were passed subcutaneously to a connector on the monkey's skull. The transducer operated by measuring the strain induced in the frame by tension in the tendon. A static calibration was made before and after the experimentation. All the data were recorded onto an FM magnetic tape and transcribed on paper using an ink-jet recorder (band-width limited to 1200 Hz). Recording with implanted force transducers was carried out for 3 to 4 hours/day over a period of 4 to 5 days.

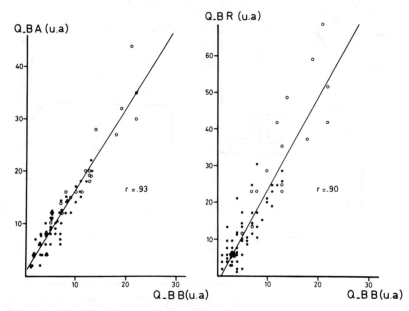

Figure 2—Correlations between integrated EMG activities of the three main elbow flexors. Q = integrated EMG, BB = biceps brachii, BR = brachioradialis, BA = brachialis.

Results

The following results are presented on three monkeys. For each of them right and then left elbow flexions were studied in successive experiments.

The experimental results are of two kinds: 1) electromyographic aspects of synergy of elbow flexor muscles in monkey and 2) mechanical aspects of synergy of elbow flexors.

Electromyographic Aspects of the Synergy
of the Elbow Flexor Muscles in Monkeys

As in human experiments, two main characteristics of synergy of elbow flexor muscles have been investigated: 1) the inter-relations between integrated EMGs and mechanical parameters, 2) the timing of muscle activities.

Inter-relations between EMGs and Mechanical Parameters. The excitation level of each elbow flexor was quantified by integrating the EMG between the onset of muscle activity and the peak of acceleration (θ''). The integrated EMG (Q) of each muscle was related to the corresponding values of work (W) or of acceleration peak (θ''). For each test, linear relations were found between Q and W ór between Q and θ''. These relations

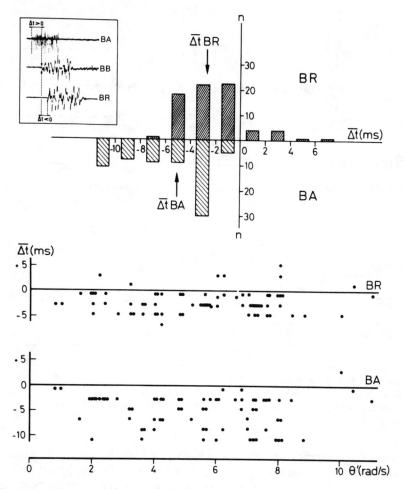

Figure 3 — Time intervals (Δt, see insert) between onset BB activity and that of the other flexors. At the top: time interval histograms of BR and BA onsets of activity. At the bottom: time intervals as a function of peak velocity (θ').

showed very little scatter as attested by the highly significant values of the correlation coefficients ($.89 \leq r \leq .98$; $p < .001$). Thus, linear correlations were found between the EMGs of the different flexors, as shown in Figure 2. Correlation coefficients were very significant at the level of $p < .001$ ($.87 \leq r \leq .92$ between Q_{BB} and Q_{BR} and $.90 \leq r \leq .93$ between Q_{BB} and Q_{BA}).

Timing of the Activities of the Muscles. The delays (Δt) between the onsets of elbow flexor activities have been studied as a function of the peak of angular velocity (θ'), taking as reference the onset of BB activity.

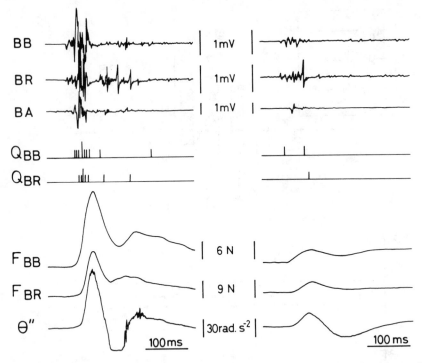

Figure 4—Records of a fast (left) and a slow (right) elbow flexion. From top to bottom: Raw EMG of BB, BR, BA. Integrated EMG (Q) of BB and BR. F_{BB} and F_{BR} force records from independent transducers in the same arm. θ'' = angular acceleration of the movement.

When the beginning of a muscle activity preceded that of BB, the delay was considered to be positive, as shown by the insert upper left of Figure 3. As it can be seen in the latter figure (bottom), the intervals between onsets were independent of the peak velocity of the movement. The values of these intervals were very short: the mean value of Δt BR was 2.8 msec and that of Δt BA was 5.1 msec (the interval histograms are shown at the top of the figure). Thus, it can be said that the muscle activities occur almost simultaneously despite a slight tendency toward the following chronology of the onsets of activities: BB—BR—BA.

All these EMG characteristics were found in intact monkeys as well as in animals with implanted transducers. So, it can be assumed with accuracy that the implanted transducers do not disturb the motor command.

Mechanical Aspects of Synergy of Elbow Flexor Muscles

Figure 4 shows two records of EMG and mechanical parameters during a

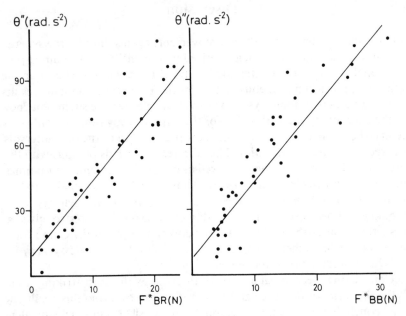

Figure 5 — Relations between movement peak accelerations and peak forces of BR (left) and BB (right). θ'' = peak accelerations, F* = peak forces.

fast (left) and a slow (right) elbow flexion performed by the same monkey as in the previous figures. The time course of BB and BR forces was similar to that of acceleration, at least up to the peak of acceleration. No measurable delay has been found between the onsets of the force curves nor between their peaks. The onsets and peaks of forces preceded those of acceleration by 6 ± 4 msec. After the peak acceleration, the forces decreased often more slowly than acceleration or even increased again if the monkey produced an isometric contraction against the stop target.

The peak amplitude of forces was observed to be related to the peak acceleration (see Figure 5). For each test, linear relations were found between these parameters with highly significant correlation coefficients ($.90 \leq r \leq .93$; $p < .001$).

The BA force was not recorded in any of the monkeys used in this study, because the tendon was not separable from the bone. So in the first experiment the force transducer which would have been implanted on the BA tendon was put solely between BA and BB to test the influence of eventual compression forces on the transducer. No signal was recorded from the latter. This means that compression forces from muscles or skin did not introduce significant bias in these records.

Discussion

In humans, during simple flexion movements against inertia, the synergy of elbow flexor muscles is characterized on the one hand by a direct proportionality between the integrated EMG of the main flexors and on the other hand by a simultaneous activity of these muscles. The same results were found here in monkeys. This was even valid for the slight tendency to the chronology BB — BR — BA for the onset of activities. The only difference between a human and a monkey is the length of the time intervals between the onset of activities. This difference depends obviously on the precision of the measurements as well as on the shorter conduction times in a monkey's nerves and muscles. Thus, it will be reasonable to extrapolate to a human the mechanical aspects of the synergy of a monkey's elbow flexors. The generality of these results also strongly suggests that relationships such as those between integrated EMG and force or work are fundamental characteristics of simple agonistic synergy involved in unidirectional movement.

Considering mechanical aspects of the synergy the most striking result is the similarity of the force and acceleration curves. From this result one may conclude that it is fairly reasonable to consider a muscle equivalent as representative of the whole of the group of flexor muscles; the force which is produced by the muscle equivalent being actually proportional to the external torque. Moreover, if we assume that the direction of the muscle force is given by the line connecting the origin and the insertion of the muscle, it appears from anatomical measurements that the lever arm of the force only varies by less than 10% within the angular sector which is comprised between the onset and the peak of acceleration. Thus, in the case of the present movements of short amplitude, the rotational torque exerted by the muscle may be considered as a constant fraction of the external torque. From the point of view of motor control it appears that the motor program does not take into account the anatomical and histological differences between the muscles of the group: the agonistic muscles are simultaneously activated, their excitation levels and force productions are proportionally graduated. Thus, in a simple voluntary movement of short amplitude the motor command seems to be a simple one.

References

BOUISSET, S. 1973. EMG and muscle force in normal motor activities. In: J.E. Desmedt (ed.), "New Developments in EMG and Clinical Neurophysiology." Vol. 1, pp. 557-583. Karger, Basel.

BOUISSET, S., Lestienne, F., and Maton, B. 1977. The stability of synergy in agonists during the execution of a simple voluntary movement. EEG Clin. Neurophysiol. 42:543-551.

BOUISSET, S., and Maton, B. 1972. The quantitative relation between surface and intramuscular electromyographic activities for voluntary movements. Amer. J. Phys. Med. 51:285-295.

SALMONS, S. 1969. Report on the 8th international conference in medical and biological engineering. Biomed. Engng. 4:467-474.

WALMSLEY, S., Hodgson, J.A., and Burke, R.E. 1978. Forces produced by medial gastrocnemius and soleus muscles during locomotion in freely moving cats. J. Neurophysiol. 41(5):1203-1216.

V.
GAIT
ANALYSIS

Motion and Role of the MP Joints in Walking

M. Fujita, N. Matsusaka, T. Norimatsu, G. Chiba,
T. Hayashi, M. Miyasaki, K. Yamaguchi, and R. Suzuki

Nagasaki University School of Medicine, Nagasaki, Japan

T. Itani

Nagasaki Rehabilitation College, Nagasaki, Japan

Recently gait studies were performed to analyze the motion and role of the metatarsophalangeal (MP) joints of the foot in level walking. The motion of the MP joints is considered to play an important role in walking.

The purpose of this paper was to analyze the motion of these joints in level walking by measuring angular changes of these joints, floor reaction force, and EMG, simultaneously.

Materials and Methods

Examinees were normal men who ranged in age from 20 to 26 years.

On a flat table for the walkway, in which Matake's force plate was fixed, each subject was asked to walk with bare feet at his regular pace and step length. By stepping on this force plate, vertical, forward-backward, and lateral components of the floor reaction force were recorded by a pen-writing oscillograph and a tape recorder. Simultaneously, EMGs of the extensor hallucis longus (EHL), extensor digitorum longus (EDL), flexor hallucis longus (FHL) and flexor digitorum longus (FDL) muscles were recorded by intermuscular wire-electrodes.

Angular changes of the first MP joint during level walking were measured by a new electrogoniometer consisting of a piece of discarded X-ray film, 10 cm long and 0.6 cm wide with a strain gauge attached to its middle part, and bridged across the MP joint by binding both ends onto the dorsum of foot and phalanx. Reliability of this electro-

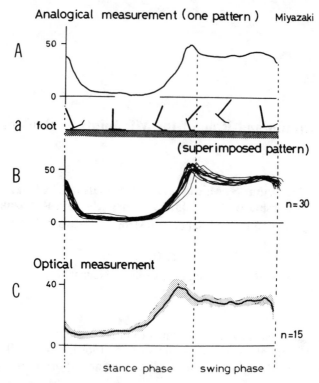

Figure 1—Angular changes of the MP joint. A and B were measured by a new elec-trogoniometer. C was measured by serial stick-pictures. a = the schema of the foot.

goniometer was confirmed before and after the experiment by calibra-tion.

Results

The motion of the MP joints in regular walking of a 26-year-old male is shown in Figure 1, in which A demonstrates successive angular changes of the first MP joint in one walking cycle measured by the elec-trogoniometer. The patterns of 30 walking cycles are superimposed in B, revealing sufficient reproducibility.

The first MP joint begins to flex rapidly as heel strikes and continues to keep about a 10° extended position during the foot flat phase. From the start of heel-off, it begins to extend and shows a peak just before the toe-off.

During the swing phase, the first MP joint maintains about 30 to 40° of extension continuously.

C in Figure 1 shows the average pattern of the angular changes of the

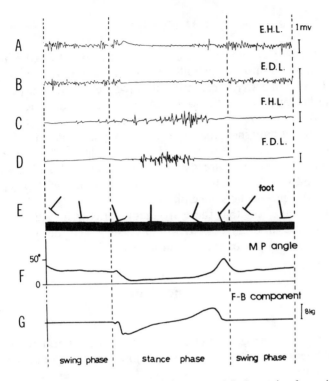

Figure 2 — Simultaneous records of EMGs, MP angle and floor reaction force. A to D = EMGs of the extrinsic toe muscles; E = the schema of the foot; F = angular changes of the first MP joint; G = forward-backward component of the floor reaction force.

fifth MP joint which is measured using serial pictures taken by a stick-picture camera. This pattern is somewhat different from those measured in the pattern A and B. It should be kept in mind that the peak of pattern C exists just prior to the peaks of patterns A and B.

Simultaneous records of the angular changes of the first MP joint, the forward-backward component of the floor reaction force, and the EMG of the extrinsic toe muscles are shown in Figure 2.

It is an important finding that the peak of the backward component of the floor reaction force exists somewhat prior to the peak of the extension of the first MP joint.

EHL and EDL are active from the last 10% of stance phase to the first 10 to 15% of the next stance phase. It would be related to the toe clearance at the point of the toe-off and during the swing phase. FDL and FHL discharge action potentials occur only during the midstance phase.

Discussion

In spite of many authors' studies, it is not yet resolved whether toe flexors contribute to push-off or roll-off in level walking. In our study FDL and FHL are not active at the moment of the toe-off, and the peak of the backward component of the floor reaction force appears prior to that of the peak of extension of the first MP joint.

The previous data show that the toes do not participate in the action of push-off in regular level walking. During the last 10% of the stance phase the other foot has already contacted the ground, and toes are not required for propulsion. That is to say, the MP joints move only passively during the end of the stance phase.

These results show that there are many varieties in active phases of EHL and EDL in every case. It would be suggested that these muscles contribute to stability of the foot in level walking.

Conclusion

In this study:

1. The new electrogoniometer was shown to be useful for measurement of angular changes of the first MP joint and a reproducible pattern was obtained.

2. Angular changes of the first MP joint, floor reaction forces, and EMG were recorded to analyze the role of this joint simultaneously.

3. During the stance phase, toe flexors do not contribute to push-off but only roll-off.

4. The role of the MP joints of the foot in level walking is only accessory and they act in roll-off to maintain floor contact during the stance phase.

References

BOJSEN-MØLLER, F. 1979a. Anatomy of the forefoot, normal and pathologic. Clin. Orthop. 142:10-18.

BOJSEN-MØLLER, F. 1979b. Significance of free dorsiflexion of the toes in walking. Acta Orthop. Scand. 50:471-478.

MANN, R.A., and Hagy, J.L. 1979. The function of the toes in walking, jogging and running. Clin. Orthop. 142:24-48.

SUTHERLAND, D.H. 1980. The role of the ankle plantar flexors in normal walking. J. Bone & Joint Surg. 62-A:354-363.

Human Walking Interpreted as Changes in Rotational Stiffness at Joints

Yoichi Tatara

Shizuoka University, Hamamatsu, Japan

Several models and analyses (Bresler and Frankel, 1950; Frank, 1970; Yamashita and Taniguchi, 1979) have previously been presented for the motion of the lower extremities for humans walking on a level plane. However, there still remains the question of how a human utilizes the muscles of the legs in walking. Further, the dominant variables in walking have been obscured by the numerous variables in the models cited.

The purpose of this study was to present one simple model for the motion of the normal leg during walking, using fewer variables and including the muscle tension and joint torques. As a result, active switches are presented for the joint torques during the stance and free swing periods, which were determined through a comparison of the analytical and measured results of the walking pattern.

Model of the Walking Leg

Suppose a human walks on a moving floor in a backward direction with the same speed as in walking forward. In this case, the upper body is almost still and the two legs move periodically. Hence, the body may be essentially regarded as a fixed end, and the legs regarded as vibrating links of an open chain. Walking motion in the plane containing the links can thus be reduced to the model of a kinematic chain with one fixed end and the other end free, with two (thigh and knee) joints having rotational stiffnesses as denoted by k_1 and k_2. To determine the dominant variables, heel and toe joints may be neglected because the upper joints exert much greater torques.

Our daily experience suggests that the stepping action with two legs in a steady walk on a level plane using a common pace is carried out almost involuntarily by the muscles of the legs which are controlled by some programs stored beforehand in the brain. The present model with a fixed end and joint torques will be available for analysis of the vibration of the legs excited by active joint torques.

Balance of Forces Inclusive of Muscle Tension

Our experience in walking also suggests that the joint torques in one cycle of each leg are not as great as when standing up or sitting down. Nevertheless, a floor reaction force, that is, the reaction of the body weight, has been considered as acting as the external torque on the stance leg (for instance, see Yamashita and Taniguchi, 1979). The introduction of the external torque to the present linkage model makes it impossible to perform an ordinary walking pattern without a strong hip joint torque of one stance leg almost equal to that when standing up, which is expressed by WL/2 (where W is the body weight and L is the length of extremity), as will be noted later. This is a contradiction.

This contradiction is solved here by considering the muscle tension on the stance leg. Normal stance on two legs is maintained mainly by active muscular tension exerted within the legs. The floor reaction force (R) is balanced by the vertical component of the muscle tension, and as a result the vertical movement of the upper body does not occur in the level walking. Hence, the floor reaction is considered to be independent of the external force affecting the walking movement of the stance leg on the level. (In the case of walking on a slope with an angle α, the component of the floor reaction force along the slope, $R\sin\alpha$, will act as an external force.)

Next consider the balance with respect to the horizontal direction in level walking. In the present model in which the moving body is fixed, the forward force (F), which is given to the upper body by stepping with the stance leg in consideration, is transformed so that it acts in the backward direction on the free end of the linkage. Although it balances the friction (F′) between the foot and the floor and the X-component of the muscle tension, it does not vanish but remains as an external force due to the translational movement of the walk.

Further, by considering the two-legged stance, another forward force (as noted by K) generated by the other stance leg will be given to the stance leg and it is considered as a force with forward (K_x) and upward (K_y) components. The total forward force is thus equal to the sum of F and K_x ($f = F + K_x$).

All of the relationships among the forces are shown in Figure 1 and are expressed by the following equations:

$$R = T_y = W - K_y \tag{1}$$

$$f = F + K_x = F' + T_x \tag{2}$$

where T_x and T_y are the respective components of the muscle tension T, W is the body weight supported by the stance leg (in the case of two legged stance, $W' \cong W/2$ in place of W), and f is the total forward force acting on the foot as the reaction of the forward movement.

Thus the model presented in Figure 2a for the stance leg and in Figure 2b for the swinging leg of free period is obtained. Note that the k_1 and k_2 at the thigh joint (1) and knee joint (2), respectively, are defined as active restoring torques which are in the vertical direction in terms of $k_1\theta_1$ and $k_2\theta_2$, and that in Figure 2b the weights of the leg act as external forces.

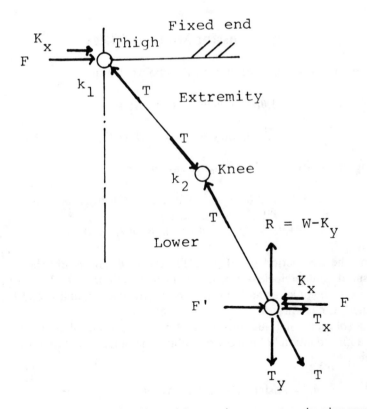

Figure 1—Balance of external and internal forces acting on one stance leg when we regard the moving upper body as the fixed end.

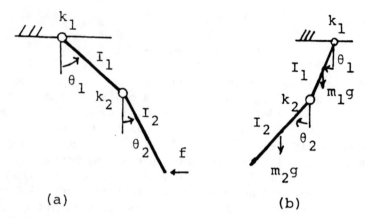

Figure 2 — Model of walking leg. a = Stance period, b = Free period.

Equations for Motion of Legs

The equations for the motion may thus be written as

$$I_1\ddot{\theta}_1 + k_1\theta_1 + k_2(\theta_1 - \theta_2) = 0 \tag{3}$$

$$I_2\ddot{\theta}_2 + k_2(\theta_2 - \theta_1) = -fL_2\cos\theta_2 \tag{4}$$

during the stance period (see Figure 2a), and

$$I_1\ddot{\theta}_1 + k_1\theta_1 + k_2(\theta_1 - \theta_2) + m_1gL_1\theta_1/2 + m_2gL_2\theta_1 = 0 \tag{5}$$

$$I_2\ddot{\theta}_2 + k_2(\theta_2 - \theta_1) + m_2gL_2\theta_2/2 = 0 \tag{6}$$

during the free period (see Figure 2b), where θ_1 and θ_2 are the angles measured counterclockwise in the vertical direction, L_1 and L_2 are the lengths of the links 1 and 2, m_1 and m_2 are the masses, and I_1 and I_2 are the inertia moments.

The solutions of Equations 3 and 4 are as follows: taking $\cos\theta_2 = 1$ (this approximation holds a considerably wide range of walking pattern), we have

$$\theta_1 = A_1\sin(\omega_1 t + \psi_1) + A_2\sin(\omega_2 t + \psi_2) - \tau \tag{7}$$

$$\theta_2 = B_1\sin(\omega_1 t + \psi_1) + B_2\sin(\omega_2 t + \psi_2) - (1 + 1/s)\tau \tag{8}$$

For simplicity, taking $m = m_1 = m_2$, $L = L_1 = L_2$, and $I_1 = I_2 = mL^2/3$, then

$$(\omega_i)^2 = \lambda\varkappa(2s + 1 \mp \sqrt{4s^2 + 1}) \tag{9}$$

$$B_i/A_i = (1 \pm \sqrt{4s^2 + 1})/2s, (i = 1, 2) \tag{10}$$

where

$$s = k_2/k_1, \varkappa = k_1/mgL, \tau = fL/k_1, \lambda = 3g/2L \tag{11}$$

Similarly, the solutions of Equations 5 and 6 may be obtained by

$$\theta_1 = A_1'\sin(\omega_1't + \psi_1') + A_2'\sin(\omega_2't + \psi_2') \tag{12}$$

$$\theta_2 = B_1'\sin(\omega_1't + \psi_1') + B_2'\sin(\omega_2't + \psi_2') \tag{13}$$

where

$$(\omega_i')^2 = \lambda[(2s + 1)\varkappa + 2 \mp \sqrt{(\varkappa + 1)^2 + 4s^2\varkappa^2}] \tag{14}$$

$$B_i'/A_i' = [\varkappa + 1 \pm \sqrt{(\varkappa + 1)^2 + 4s^2\varkappa^2}]/2s, (i = 1, 2) \tag{15}$$

The frequencies, amplitudes and phases are determined by the variables s, $\varkappa$ and τ which are assumed, incorporated with the initial conditions of θ_1, $\dot{\theta}_1$, and θ_2 and $\dot{\theta}_2$, which can be obtained from the measured data.

Measurement of Walking Pattern

The walking pattern of one cycle of one leg was measured with a stroboscope and streak camera for normal walking on a fixed level floor by two male subjects, A and B. The subjects wore sport trunks to which were attached strips on their right lower legs and feet, and painted with luminous paint like those used for cycling at night. The one cycle of walking with a slow pace was observed with a stroboscope flashing about 18 times/sec and by three or four successive photographs from a film obtained with a camera. Angles of the luminous stripes on the extremity (link 1) and the lower leg (link 2) were measured from the vertical direction with a universal projector on the negative photos, and plotted as θ_1 and θ_2 with time, as shown in Figure 3 for the subject A. Similar experimental results were also obtained for subject B, although not included here. The estimates of the stance and free periods were obtained from the photographs, where the movement of the swinging left leg could be read.

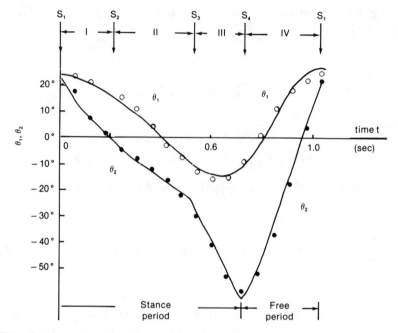

Figure 3—Measured results of $\theta_1(t)$ and $\theta_2(t)$ of angles of extremity and lower leg in the vertical direction, respectively, for subject A (white and black points) and mathematical results (solid curves).

The measured curves of $\theta_1(t)$ and $\theta_2(t)$ presented by white and black points in Figure 3 demonstrate several characteristics regarded as peculiar to human gait. They can be summarized as the slow decrease in θ_2 in the middle stage of the stance period, its rapid decrease in the last stage, and the rapid increase in both θ_1 and θ_2 in the free period (see Figure 3).

Also presented in Figure 3 is the fact that the stance period is clearly longer than the free period. This suggests the existence of a period of two-legged stance. Hence, the stance period is divided into three intervals, I, II, and III, where I and III correspond to the two-legged stance period and II to the one-leg stance period; in the periods I, II, III, and IV (where IV is the free period), the averaged body weight supported by one leg is thus regarded as W/2, W, W/2, and 0, respectively. Moreover, the periods I and III require equal duration, t_1, and the II and IV require equal duration, t_2. These relations are induced from the facts that the body weight, W, is supported alternately by one or two legs and the phase shift of two equal and symmetrical movements of the two walking legs is 180°. Thus, the stance period, t_s, and the free, t_f, are written as

$$t_s = 2t_1 + t_2, \quad t_f = t_2 \qquad (16)$$

Table 1

**Schematic of Two Tapes Describing
Individual Intervals and Combination Play of Two Walking Legs**

	Free period	Stance period		Free period		Stance period		Free period	
Left leg	I $W/2$ t_1	II W t_2	III $W/2$ t_1	IV 0 t_2	I $W/2$ t_1	II W t_2	III $W/2$ t_1	IV 0 t_2		Tape 1
Right leg	III $W/2$ t_1	IV 0 t_2	I $W/2$ t_1	II W t_2	III $W/2$ t_1	IV 0 t_2	I $W/2$ t_1	II W t_2		Tape 2
	Stance period	Free period		Stance period		Free period		Stance period	

These relationships are presented schematically by two tapes in Table 1, where the two program tapes expressing the intervals and the equations like Equations 7 to 15 for each leg motion are connected in parallel by combining period I with III and also II with IV. It is also interpreted that when walking a human attempts to decrease the one-leg stance period by the method of the combination because the one-leg stance requires more difficult body balance and the maximum muscle tension.

Switches of joint torques, k_1 and k_2, will thus be required at the beginning of each of the periods I to IV, because the average body weight supported by one stance leg is different among the periods; at the same time, the characteristics of the walking pattern of θ_1 and θ_2 as noted above appear in the periods II, III, and IV. These characteristics may be identified by the switches of joint torques.

Switches of Joint Stiffness

Using $t_1 = 0.21$ [s], $t_2 = 0.30$ [s], and $L = 0.40$ [m], the values of three parameters, s, $\varkappa$, and τ, were selected at the beginning of each of the periods I to IV. The initial conditions of $\theta_{1,2}$ and the derivatives with respect to t at each beginning were introduced from the measured data in Figure 3 into the calculation using Equations 7 to 15. The mathematical result was obtained after several selections of the three parameters according to trial and error, but it agrees very well with the measured result, as shown in Figure 3 together with the measured result. Table 2 shows the parameters, s, $\varkappa$, and τ, used in the last calculation and the

Table 2

The Change in the Active Joint Stiffnesses k_1 and k_2 and the Forward Force f

	I	II	III	IV
s	0.1	0.1	0.1	0.5
$\varkappa$	0.2	0.1	1.0	−0.4
τ	0.005	0.002	0.1	0
k_1 ($\times$ mgL)	0.2	0.1	1.0	−0.4
k_2 ($\times$ mgL)	0.02	0.01	0.1	−0.2
f ($\times$ mg)	0.001	0.0002	0.1	0
body weight	W/2	W	W/2	0

values of k_1, k_2 and f obtained from Equation 11. Note that without switches of k_1 and k_2 in every period results were impossible to obtain which were identical to the measured ones of θ_1 and θ_2, and an introduction of floor reaction W/2 into the period I, i.e., a solution of the equation with a term of (WL/2)sin θ_2 in the right side in Equation 4, led to k_1 = WL/2 so as to agree with the measured results of θ_1 and θ_2 in the period I; the case of k_1 = WL/2 may correspond to a torque such as that in standing up.

From Table 2, the changes in the joint torques, k_1 and k_2, are noted as follows:

1. Throughout the stance periods I, II, and III, the thigh stiffness, k_1, is much larger than the knee, k_2, i.e., k_1 = 10k_2.

2. In the one-leg stance period II, k_1 and k_2 take the minimum values. This interpretation requires a large amount of muscle tension of the legs to be employed in supporting the body weight and only slight muscle tension is utilized to act as the active joint torque.

3. In the period III, both k_1 and k_2 take the maximum values, and the forward force, f, also shows the maximum. The forward (driving) force for the walk is thus produced mainly in this period, perhaps by the push of the rear leg. The push action is considered to arise from the strong joint torques. The torques are directed counterclockwise due to the positive signs of k_1 and k_2 in a range of negative θ_1 and θ_2. Nevertheless, the push action is obtained by the relative movement of the links 1 and 2 such as the rapid decrease in θ_2 in Figure 3.

4. In the free swing period IV, both k_1 and k_2 are negative. The torques are directed clockwise in the range of negative θ_1 and θ_2 and counterclockwise in positive θ_1 and θ_2. This former behavior is understood by the

link 1 at the beginning of IV in Figure 3 which is almost at $\theta_1 = 0$, and the latter as the action to return to the stepping position at the beginning of the period I earlier than the time required by the free swing of the leg under gravity. In addition, the switches from positive to negative signs of k_1 and k_2 are considered to require alternation of contraction and relaxation of the muscles at the joints.

References

BRESLER, B., and Frankel, J.P. 1950. The forces and moments in the leg during level walking. Trans. ASME. 22:27-36.

FRANK, A.A. 1970. An approach to the dynamic analysis and synthesis of biped locomotion machines. Med. and Biol. Engg. 8:465-476.

YAMASHITA, T., and Taniguchi, T. 1979. Simulation of human walking characteristics by simple model. Proceedings of the Fifth World Congress on Theory of Machines and Mechanisms, pp. 851-854.

Relationships between Step Length and Selected Parameters in Human Gait

Hitoshi Ohmichi and Mitsumasa Miyashita
University of Tokyo, Tokyo, Japan

Walking velocity is a key measure for the analysis of human gait. For example, Margaria (1976) showed the energy consumption was related to the velocity, and Andriacchi et al. (1977) characterized normal and abnormal gait based on relationships between the velocity and the amplitude of ground reaction force. On the other hand, Ohmichi and Miyashita (1981) showed that the step length could identify exactly the vertical displacement and the change of momentary velocity of the trunk during walking.

The pelvic rotation is one of the major determinants of the complexity of human gait (Saunders et al., 1953). Therefore many investigators have measured the motion of the hip. Levens et al. (1948) recorded the pelvic rotation in the horizontal plane by means of pins which were inserted directly into the bones. Murray et al. (1964) presented the normal amplitude of the horizontal rotation for groups different in stature and age. Sutherland and Hagy (1972) developed convenient methods to photograph the angular displacement of the hip. Inman (1966) showed the typical patterns of both horizontal and frontal rotations of the pelvis. Zarrugh and Radcliffe (1979) adopted 'string transducers' to measure the angular and linear displacements of body segments while walking on a treadmill, and computed both horizontal and frontal rotations of the pelvis.

The present study was designed to examine the relationships between the step length and the pelvic rotations in both the horizontal and frontal planes in case of 'forced' walking, i.e., under various combinations of step length and step frequency. In addition, the resistance force exerted by the ground upon a foot was measured.

Methods

In the present study, the pelvic rotations in the frontal and horizontal planes are expressed as the vertical and horizontal differences between left and right hips in linear displacement, respectively. The linear displacement can be easily converted into the angular displacement using anthropometric data of the hip. The measurement procedures used in this study are described in detail in a previous report (Ohmichi and Miyashita, 1981). The vertical displacement of each trochanter major was detected by a string and potentiometer system contained in a motor driven cable car. The horizontal displacement of each trochanter major was recorded on a magnetic tape through 50 Hz square pulses from a pulse generator. The location of pulses fed were read by a tape recorder and a mini-computer system.

Healthy male adults A and B (similar body dimensions; 170 cm in height and 70 kg in weight) walked 15 m along the walkway at a constant velocity following the electric cable car driven at selected speeds (30 to 160 m/min). The subject adopted various step lengths or step frequencies. When the subject contacted completely a force plate (Kistler) with the right foot, the horizontal component of ground reaction force was integrated for calculation of the mean resistance force. In each trial, pelvic rotations during eight steps were averaged for the 15 m walk.

Results and Discussion

Figure 1 shows the time variation of the horizontal rotation (Rot_H) in linear displacement at 80 m/min with a step length of 72 cm for subject B. This curve agrees well with that reported by Levens et al. (1948). Rot_H increased from 3.5 to 11.5 cm as the step length (S) increased from 49 to 101 cm for the same subject ($Rot_H = 0.142 \times S - 4.26$, r = .947, n = 12). Several researchers have already reported the horizontal rotations during natural walking by means of cinematography. However, the 'normal values' within ± 5 degrees reported by these researchers (Levens et al., 1948; Murray et al., 1964; Inman, 1966; and Sutherland et al., 1972) were significantly lower than the values (5 to 19°) obtained in this study. Recently Zarrugh and Radcliffe (1979) reported that horizontal rotation increased from 5 to 15° with walking velocity (50 to 130 m/min) and that the horizontal rotation increased with step length in forced walking at 90 m/min. Their results are well supported by the present results. On the other hand, they stated that the frontal rotation was independent of step length during forced walking at a velocity of 90 m/min, though it increased with the velocity. To the contrary, the present study showed that frontal rotation (Rot_F) of the pelvis, i.e., maximum vertical difference between left and right hips increased linearly from 1.56 to 6.16 cm as the

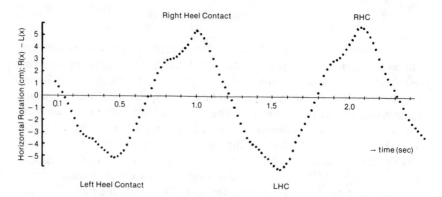

Figure 1—Time variation of horizontal rotation at 80 m/min with 72 cm step length (Subject B).

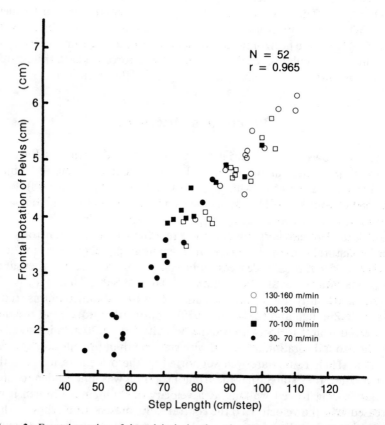

Figure 2—Frontal rotation of the pelvis during forced walking related to step length (Subject A).

Table 1

Correlation Coefficients Between Selected Parameters of Forced Walking

	Trials	Step length	Velocity	Step frequency
Frontal Rot$_F$	52	0.965	0.763	0.379
Horizontal Rot$_H$	12	0.947	0.575	−0.163
Resistance F	48	0.860	0.635	0.288

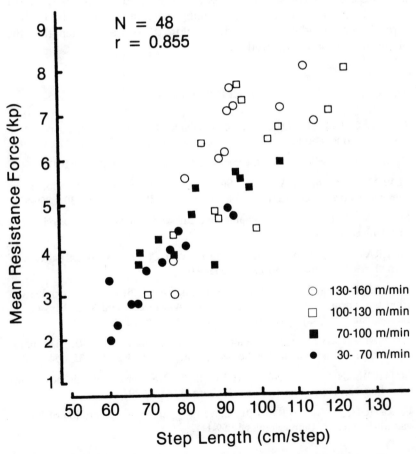

Figure 3—Mean resistance force acting on a foot related to step length (Subject A).

step length (S) increased from 45 to 112 cm for subject A (Rot_F = 0.070 × S − 1.62, r = .965, n = 52) (see Figure 2). Also the correlation coefficient between Rot_F and S was clearly larger than that between Rot_F and velocity (see Table 1).

The mean resistance (F) from the ground during one step period increased from 1.97 to 8.15 kp with the step length for subject A (r = .860 and n = 48) (see Figure 3). Since the duration of the resistance acting backward upon the foot only acted during 53 to 72% of one step period, the values of F were approximately 59% lower than the true mean amplitude of resistance.

In the present experiment, all the parameters such as Rot_F, Rot_H, and F were primarily determined by the step length rather than by the walking velocity or the step frequency, as is shown in Table 1. Furthermore, this was also the case in a previous report for the vertical displacement and the change of forward momentary velocity of the trunk. These findings suggest that the step length potentially could be one of the most essential determinants as well as walking speed in gait analysis. Further experiments should be conducted to explore the role of step length in the biomechanics of human walking.

References

ANDRIACCHI, T.P., Ogle, J.A., and Galante, J.O. 1977. Walking speed as a basis for normal and abnormal gait measurements. J. Biomechanics 10:261-268.

INMAN, V.T. 1966. Human locomotion. Canad. Med. Ass. J. 94:1047-1054.

LEVENS, A.S., Inman, V.T., and Blosser, J.A. 1948. Transverse rotation of the segments of the lower extremity in locomotion. J. Bone Jt. Surg. 30A:859-872.

MARGARIA, R. 1976. Biomechanics and Energetics of Muscular Exercise. Oxford University Press, London.

MURRAY, M.P., Drought, A.B., and Kory, R.C. 1964. Walking patterns of normal men. J. Bone Jt. Surg. 46A:335-360.

OHMICHI, H., and Miyashita, M. 1981. Analysis of the external work derived from the kinematics of human walking. In: A. Morecki and K. Fidelus (eds.), Biomechanics VIIB, pp. 184-189. University Park Press, Baltimore.

SAUNDERS, J.B., Dec, M., Inman, V.T., and Eberhart, H.D. 1953. The major determinants in normal and pathological gait. J. Bone Jt. Surg. 35A:543-558.

SUTHERLAND, D.H., and Hagy, J.L. 1972. Measurement of gait movements from motion picture film. J. Bone Jt. Surg. 54A:787-797.

ZARRUGH, M.Y., and Radcliffe, C.W. 1979. Computer generation of human gait kinematics. J. Biomechanics 12:99-111.

Reproducibility of Gait Analysis

T. Norimatsu, H. Okumura, M. Fujita, N. Matsuzaka,
G. Chiba, K. Yamaguchi, T. Hayashi, H. Kira,
R. Suzuki, and T. Okajima

Nagasaki University School of Medicine, Nagasaki, Japan

Kinesiological gait studies have progressed remarkably, but recent advances in the treatment of patients with locomotor disturbances have emphasized the need for better and readily available methods to evaluate the progress of therapy. In practice, performing gait analyses with patients is not practical because patients often cannot walk because of pain and other physical problems. On the other hand, if many parameters are recorded, it will require considerable effort and much time to analyze the data.

The purpose of this study was to develop a simple technique, especially in relation to the reproducibility of walking. To realize this purpose there were several points which had to be taken into consideration.

1. How many deviations are there in the walking cycles of the same subject?

2. How many frames are needed to analyze the walking gait exactly?

3. How many walking cycles are needed for the purposes of clinical treatment?

Materials and Methods

Fifteen walking cycles were recorded for the same subject, using the Nagasaki University Gait Analyzing and Computer System. Angular changes of body rotation, shoulder, elbow, knee, ankle, and metacarpophalangeal joint of the foot were measured by serial photographs (see Figure 1).

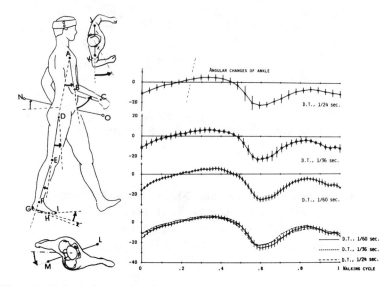

Figure 1—Method of measuring the angle (arrows indicate "plus").

To calculate the standard deviation of the different walking cycles, Cramer's rule was applied and a PC-8000 computer was used. The time durations of the stick picture camera used were 1/60, 1/36, and 1/24 sec. The mean pattern of the angular changes of 15 walking cycles are shown by solid or dotted lines, with the standard deviations indicated by the solid line.

Results

Angular changes of the shoulder girdle, shoulder, and elbow joints are illustrated in Figures 2a, b and c. Angular changes of the shoulder girdle and its standard deviation were small. There were no differences in pattern among 1/60, 1/36, and 1/24 sec time durations.

Angular changes of the shoulder were larger than for the shoulder girdle. The line of 1/36 sec time duration and 1/24 sec time duration were outside the range of one standard deviation for the 1/60 sec, during 30% to 57% of the walking cycle, but the patterns were quite similar.

Angular changes of the hip joint, pelvis, and pelvic tilt are shown in Figures 2d, e and f. For the hip joint, the 1/36 and 1/24 sec lines were higher than the 1/60 sec line during 40 to 60% of the walking cycle but there were no differences in the pattern.

In the pelvis and pelvic tilt, the angular changes were very small and there were no differences in patterns.

Angular changes of the knee, ankle, and metacarpophalangeal joint of

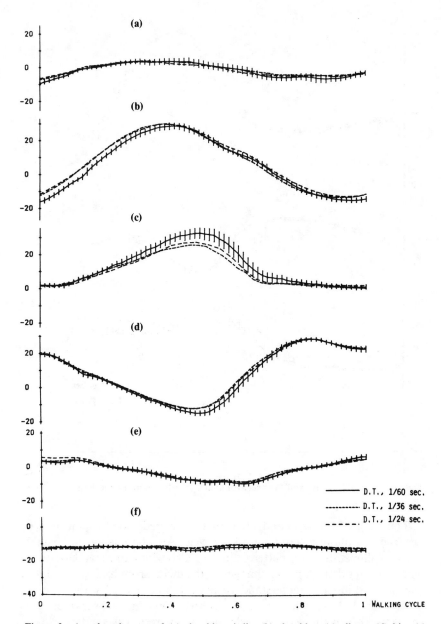

Figure 2—Angular changes of (a) shoulder girdle; (b) shoulder; (c) elbow; (d) hip; (e) pelvis; (f) pelvic tilt.

the foot are shown in Figures 3a, b and c. In the knee joint the traces showed almost identical patterns and the two lines were within one standard deviation of each other. The characteristics of the peak of the curve of the knee joint were different, however.

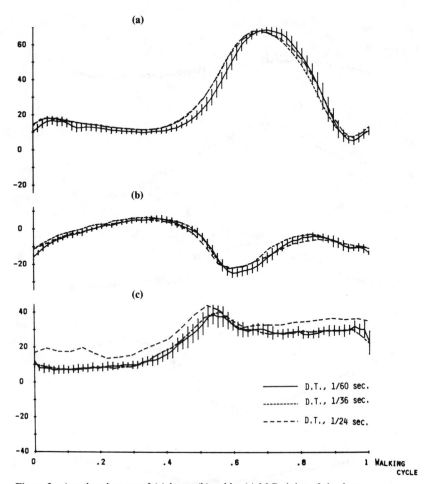

Figure 3—Angular changes of (a) knee; (b) ankle; (c) M.P. joint of the foot.

The peak of the 1/36 and 1/24 traces occurred sooner than the 1/60 sec trace for about 5% of the walking cycle. Angular change of the ankle showed almost the same pattern and were also within one standard deviation of each other. Angular changes of the metacarpophalangeal joint of the foot exhibited a large standard deviation compared with the other joints. The 1/24 sec trace was less than the 1/36 and 1/60 sec traces during 0 to 50% of the walking cycle, but the pattern was similar.

Discussion

The principal purpose of this study was to determine how many time durations would be required to precisely analyze human gait.

There were almost no noticeably large deviations in shoulder girdle, shoulder joint, hip joint, pelvis, pelvic tilt, and ankle joint with respect to their angular changes and patterns. In the elbow joint, longer time durations (1/36, 1/24) showed smaller angular changes compared with a shorter time duration (1/60 sec) but the pattern was exactly the same. Consequently, it is assumed that small angular changes make a great difference.

The knee joint showed the same pattern among the three durations but each peak was different. It was not possible to provide an explanation for this difference.

The metacarpophalangeal joint of the foot is a very important joint, but it is difficult to analyze its angular changes exactly. Therefore, attempts have previously been made to analyze it. The present results also indicate difficulty. The standard deviation was larger than for other joints and the long duration (especially 1/24 sec) trace was situated far outside one standard deviation. On the other hand, its pattern was very similar to the other traces.

Summarizing these results, there were almost no abnormal patterns in human walking, and therefore, 1/24 sec time duration is evidently sufficient to analyze human walking gait.

References

DUCROQUET, R.J., and Ducroquet, P. La Marche et les Boiteries.

MATSUZAKA, N. The pattern of the arm swing and the body rotation. Proceedings of JOBRS. 5:63-70.

MURRAY, M.P., et al. 1967. Gait as a total pattern of movement. Am. J. Phys. Med. 46(1):290-333.

SUZUKI, R. 1977. Gait analysis. Sogo Rehabil. 5(5):505-511.

Objective Evaluation of
Painful Heel Syndrome by Gait Analysis

Yoshihisa Katoh, Edmund Y.S. Chao, B.F. Morrey,
R. Keith Laughman, and Erich Schneider
Mayo Clinic/Mayo Foundation, Rochester, Minnesota, U.S.A.

Painful heel syndrome is a general term used to describe painful conditions affecting the inferior aspect of the heel. Although a number of possible etiologies have been proposed, the exact pathomechanics have not yet been established. Painful heel pad and plantar fasciitis are considered to be the two most common causes of this syndrome (Turek, 1977). Painful heel pad is due to the attenuation of the fat pad under the calcaneus (Kuhns, 1949). Plantar fasciitis results from inflammation at the insertion of the plantar fascia caused by excessive tension in the fascia (Brody, 1980).

Many treatment modalities have been tried. Some authors even recommend surgical treatment for severe, persistent cases. Generally, however, these diseases are believed to be self-limiting and symptoms usually disappear after several months. Therefore, conservative measures must be the first choice of treatment. Sponge rubber inserts (Eggers, 1957), high-heeled shoes (Stewart, 1980), UCBL shoe inserts (Campbell and Inman, 1974), and rubber or plastic heel cups (Snook, 1972) are examples of currently used orthotic devices. The mechanisms of action and effectiveness of these modes of treatment have not yet been established. Therefore, the objectives of this study were: 1) to establish an additional diagnostic method for the painful heel syndrome using gait analysis, and 2) to use this objective method in the evaluation of treatment efficacy.

Materials

To establish the normal values for subsequent comparison, 39 normal subjects with a mean age of 39 years (range from 21 to 63 years) were in-

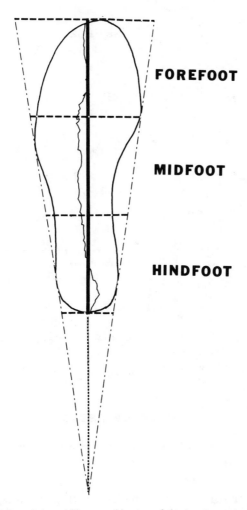

FOREFOOT

MIDFOOT

HINDFOOT

Figure 1 — Definition of the midline as a bisector of the two tangents drawn through the most prominent points at each side of the foot. Division of this midline into three parts defined the areas of forefoot, midfoot and hindfoot. The thin line represents an example of COP pattern of a normal subject.

vestigated. None of them had complaints referable to their lower extremities. Normal subjects wore their usual footwear during the evaluations providing normal data regarding the effect of different types of footwear. Because the rigidity of the sole and the height of the heels are considered to be important factors affecting the walking pattern, normal subjects were divided into three subgroups according to shoe types: rigid sole, soft sole, and high heels.

A total of 13 patients with a mean age of 55 years (range from 26 to 72 years) with significant heel pain were studied. These patients were

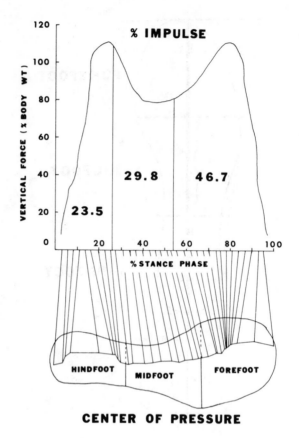

CENTER OF PRESSURE

Figure 2 — Vertical impulse of the hind-, mid-, and forefoot areas, expressed in percent of the total vertical impulse.

clinically categorized as follows: patients with painful heel pad (PHP) had local heel tenderness without other demonstrable cause of heel pain; patients with plantar fasciitis (PF) had increasing heel pain with passive toe dorsiflexion and additional tenderness along the plantar fascia. Nine feet with PHP and seven with PF were evaluated. The effect of treatment was then studied by observing the change in gait patterns with and without heel cups of two different types in nine feet with PHP and three feet with PF.

Method

All subjects walked at a normal pace over the walkway which contained a transparent piezoelectric force plate. Using a PDP 11/34 computer system, the three-dimensional dynamic ground reaction forces and the

Figure 3 — Walkway containing the force plate (FP) in the center and two sets of stainless steel tape mats (M) on both ends of the force plate.

center of pressure (COP) were determined. A video camera underneath the force plate captured the location of the foot or the shoe. The outline of the shoe was digitized from a videomonitor by means of a sonic pen digitizer and was combined with the COP pattern by the computer.

To normalize for different shoe sizes and shapes, tangents were drawn to the most prominent two points of the medial and lateral borders of the shoe (see Figure 1). The bisector of these two tangents was called the midline of the foot (or the shoe) and was used to divide the foot into hind-, mid-, and forefoot. The momentary COP under the foot is the point of application of the resultant ground reaction force vector. This point moves during the stance phase in a characteristic manner from heel to toe. An example of such a COP pattern in a normal individual is also shown in Figure 1. The COP pattern was used to determine the time interval of contact of each of the three areas of the foot. The vertical impulse (see Figure 2) was finally obtained from the area under the vertical force curve. The percentage of the total vertical impulse for a given period of time was determined for each of the three areas of the foot.

In addition, four foot switches were taped to each of the patients' shoes at the heel, fifth and first metatarsal heads, and great toe in order to provide the temporal and spatial gait factors. Two sections of the walkway, on either side of the force plate were instrumented with stainless steel tape mats to measure step length and width for each foot (see Figure 3).

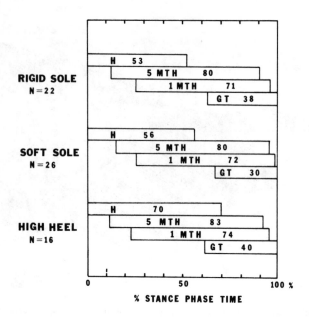

Figure 4—Percentages of stance phase time spent on four foot switches of 32 normal subjects with three different shoe types. Increased heel contact time for subjects with high heels was statistically significant (P < .001). H = Heel, MTH = Metatarsal Head, GT = Great Toe.

Results

The foot switch data of normal subjects (see Figure 4) revealed a consistent order of switch contacts, that is, heel, fifth metatarsal head, first metatarsal head and toe, with very similar contact durations among the various shoe types except for the high-heeled group. In this group, heel contact time was significantly prolonged (P < .001) compared to both other shoe types. Table 1 shows the impulse distribution for the three shoe types among normal subjects. No statistically significant differences were found between subjects with rigid sole shoes or soft sole shoes. High-heeled shoes, however, displayed a decreased forefoot impulse compared to the other two groups (P < .02).

As shown in Table 2, the average impulse of all normal individuals with shoes was 22% for the hindfoot, 37% for the midfoot, and 41% for the forefoot. In PHP patients, the pressure distribution was shifted from the heel toward the midfoot, resulting in an impulse pattern of 10%, 47%, and 43%, respectively, with both mid- and hindfoot changes being statistically significant (P < .05). Impulse distribution in the group of PF patients was shifted from the heel and forefoot to the midfoot with an impulse pattern of 14%, 54%, and 32%, respectively. In this group, the midfoot increase and forefoot decrease were statistically significant

Table 1

**Percentages of Impulse Distribution
in Various Shoe Types in Normal Subjects**

	Rigid Sole (N = 32)	Shoe Types Soft Sole (N = 42)	High Heel (N = 16)
Forefoot	41.8 ± 11.0	41.6 ± 9.1	34.5 ± 6.3*
Midfoot	37.9 ± 10.9	35.5 ± 12.5	41.6 ± 16.1
Hindfoot	20.3 ± 8.8	23.0 ± 9.1	23.9 ± 15.7

*Statistically significant compared to the forefoot impulse of rigid sole or soft sole shoes.

Table 2

Percentages of Impulse Distribution in Normals and Patients

	Normals (N = 78)	Without Heel Cup PHP (N = 9)	PF (N = 7)	With Heel Cup PHP (N = 9)	PF (N = 3)
Forefoot	40.7 ± 9.0	42.9 ± 9.7	32.0 ± 7.7* (P < .02)	44.3 ± 14.1	34.0 ± 4.6
Midfoot	37.1 ± 11.8	46.8 ± 12.4* (P < .05)	53.9 ± 9.4* (P < .001)	47.0 ± 10.7	56.3 ± 9.0
Hindfoot	22.2 ± 9.6	10.3 ± 6.8* (P < .001)	14.1 ± 12.4	8.7 ± 4.4	9.7 ± 5.0

*Statistically significant compared with normals.

(P < .02), but not the reduced hindfoot impulse when compared to normal.

Concerning the effect of treatment, no significant difference could be found. Similar effects for both types of heel cups were observed, however; i.e., reduced hindfoot and increased forefoot impulses. Subjectively, most patients walked at their own speed, the velocity was a good indicator of patient comfort during walking. The mean velocity of normal subjects was 83.0 m/min. In PHP patients, the mean velocity was 64.5 m/min and significantly slower (P < .01) than normal. However, with heel cups in place, they walked faster (71.2 m/min) than they did without cups. PF patients initially walked faster (71.4 m/min) than the PHP patients, although with heel cups inserted their velocity decreased to 68.8 m/min. These findings reflect the aggravation of foot pain caused by heel cups in PF patients, supported by the subjective opinion of all PF patients who complained of increased symptoms while using the inserts.

In patients with PHP, seven out of nine reported a decrease in symptoms with heel cups. No measurable difference between rubber and plastic heel cups was found.

Discussion

The value of measuring contact time of specific portions of the foot and the foot-floor impulse distribution during gait has been well recognized. High-heeled shoes are sometimes recommended as a conservative measure in the treatment of painful heel syndrome (Stewart, 1980). This is probably related to the impression that the plantar flexed attitude of the foot shifts the weight away from the heel to the anterior aspect of the foot. Our normal study, however, revealed a significantly decreased forefoot impulse with an increase in mid- or hindfoot impulse. Grundy (1975) has obtained similar findings. If we further consider the prolonged heel contact time from the foot switch pattern, it becomes obvious that high-heeled shoes actually increase the load on the posterior foot. This additional load could be detrimental for PHP patients. Therefore, the recommendation of wearing high-heeled shoes in the case of PHP has little scientific basis. However, the actual loading on the sole of the foot is expected to be different due to the inclined foot. The real therapeutic value of high-heeled shoes must be studied more carefully, based on detailed force analysis.

Compared to normal feet, obvious changes in the impulse distribution were found in pathological feet. These deviations reflected different underlying pathological mechanisms. PHP patients, whose pain is in the heel pad, try to avoid local pressure on the heel, reflected as a decreased hindfoot impulse. On the other hand, PF patients have reduced impulse in the forefoot, indicating that they are trying to avoid dorsiflexion of the forefoot, which would increase the tension in the plantar fascia at the terminal portion of the stance phase (Hicks, 1954). The significant difference of impulse distribution between these two patient groups has proven to be helpful for differential diagnosis.

Comparison of the impulse distribution of patients with and without heel cups revealed an anterior shift of the impulse using the heel cups. This change is desirable in patients with PHP, but increases the symptoms in patients with PF. This result was also verified by measuring the changes in velocity during walking.

Summary

To date, painful heel syndrome has received only minimal attention, and various forms of treatment are being empirically applied. Many patients afflicted with such pathology are quite disabled secondary to pain. The need for an objective evaluation technique for the disease itself and for the assessment of treatment effectiveness is obvious. The presented method of evaluation using the center of pressure pattern and the impulse distribution during the load-bearing period of gait clearly describes the characteristic changes between normal and abnormal feet which could not be observed through simple inspection. The technique has proven to be a diagnostic aid for distinguishing different etiologies of heel pain, and it appears promising as a means of determining treatment efficacy and rationale. It is hoped that this method of evaluation will be further tested to aid diagnosis and treatment evaluation for patients with other foot disorders.

References

BRODY, D.M. 1980. Running Injuries. CIBA Clinical Symposia. 32(4).

CAMPBELL, J.W., and Inman, V.T. 1974. Treatment of plantar fasciitis and calcaneal spurs with the UC-BL shoe insert. Clin. Orthop. 103:57-62.

CAVANAGH, P.R., and Lafortune, M.A. 1980. Ground reaction forces in distance running. J. Biomechanics 13:397-406.

EGGERS, G.W.N. 1957. Shoe pad for treatment of calcaneal spur. J. Bone Joint Surg. 39(A):219-220.

GRUNDY, M., et al. 1975. An investigation of the centers of pressure under the foot while walking. J. Bone Joint Surg. 57(B):98-103.

HICKS, J.H. 1954. The mechanics of the foot. The plantar aponeurosis and the arch. J. Anat. 88:25-30.

KUHNS, J.G. 1949. Changes in elastic adipose tissue. J. Bone Joint Surg. 31(A):541-547.

SNOOK, G.A., and Chrisman, O.D.O. 1972. The management of subcalcaneal pain. Clin. Orthop. 82:163-168.

STEWART, M. 1980. Campbell's Operative Orthopaedics. Vol. 2, 6th Ed., A.S. Edmonson and A.H. Crenshaw (eds.). St. Louis, The C.V. Mosby Co., pp. 1743-1744.

TUREK, S.L. 1977. Orthopaedics: Principles and Their Application. Third Edition, Philadelphia, PA, J.B. Lippincott Co., pp. 1312-1313.

Kinematic Analysis of Walking and Running in Twins

Tamotsu Hoshikawa
Aichi Prefectural University, Nagoya, Japan

Yoshinori Amano and Nobukazu Kito
Aichi University of Education, Aichi-Prefecture, Japan

Hideji Matsui
Nagoya University, Nagoya, Japan

The manifestation of the various characteristics in human beings is related to intrinsic factors depending on heredity and extrinsic factors, such as learning and training. Although there is ample evidence of the effect which both factors have on morphological, intellectual, and psychological aspects of character, there is little information on motor ability and movement patterns. It is necessary to determine the magnitude of influence of each factor in order to construct a curriculum for physical education and to select sport talent.

The purpose of this study was to estimate to what extent intrinsic or extrinsic components may account for individual differences in walking and running patterns employing monozygous and dizygous twins as subjects.

Methods

There were 18 pairs of monozygous and dizygous twins of both sexes and one set of triplets who served as subjects. Their ages ranged from 7 to 12 years.

The subjects were asked to perform a series of walks and runs with progressively increasing speeds on a motor-driven treadmill. After familiarizing them with the experimental procedures each began at a speed of 60 m/min for walking and 120 m/min for running, and pro-

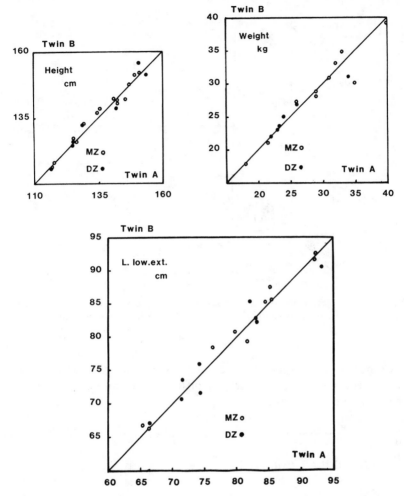

Figure 1 — Mean percentages of intrapair differences in monozygous and dizygous twins for body height, body weight and length of lower extremities.

ceeded to 80 and 100 m/min for walking and 150, 180, 210, 240, 270, and 300 m/min for running in accordance with their running abilities.

An electrogoniometer was used to obtain continuous recordings of the action of the knee joint. Foot-timing information was determined by a transducer sensitive to foot pressure secured to the ball and heel of the shoe.

The variables studied included durations of the swing and support phases, step frequency, step length, movement of the knee joint and its angular velocity, in addition to morphological parameters. Temporal variables and action of the knee joint were averaged for 10 to 15 successive steps.

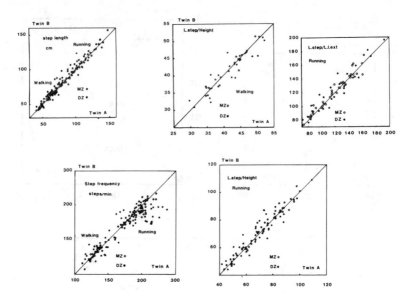

Figure 2 — Mean percentages of intrapair differences in monozygous and dizygous twins for step frequency, step length, ratios of step length to body height and step length to length of lower extremities.

Comparisons of the difference between monozygous and dizygous twins were made through intrapair differences computed by following the equation:

$$\frac{2(A - B)}{A + B} \times 100 = \text{Intrapair difference (\%)} \tag{1}$$

where A and B are twin brothers/sisters, respectively.

Results

Anthropometric Measurements

In body height, body weight, and length of lower extremities the intrapair differences between the two types of twins were not significantly different (see Figure 1).

Temporal Measurements

There was a definite difference, which was statistically significant, in the durations of the swing and support phases and in one cycle of running, but in the case of walking the differences were not significant except the duration of one walking cycle.

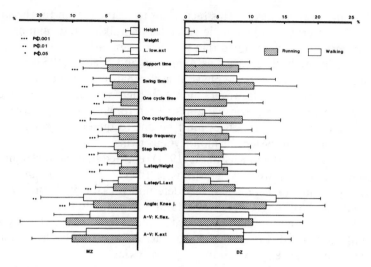

Figure 3 – Mean percentages of intrapair differences related to kinematic variables of walking and running in monozygous and dizygous twins.

Step Frequency, Step Length and the Ratios of Step Length to Body Height and Length of Lower Extremities

The intrapair differences for step frequency, step length, and the ratios of step length to body height and length of lower extremities yielded significant differences in monozygous twins as compared to dizygous twins except for the step length during walking (see Figure 2).

Action of the Knee Joint

Angular velocities of both phases of extension and flexion of the knee did not show any significant differences between the intrapair differences of monozygous and dizygous twins. There was a statistically significant difference, however, in the angle at the knee joint at maximum flexion during the swing phase of walking and running.

Discussion

Intrapair differences of the various variables obtained in this study were smaller, not only in monozygous, but also in dizygous, than that for maximal muscle power and histochemical property of muscle fiber for twins reported by Komi et al. (1976). Locomotion is a phylogenetic movement which depends quite closely upon integrated action of the reflexes and leg action during walking and running, and operates unconsciously in accordance with an already fixed program of sequence and timing of the reflexes. This fact seems to be the reason for the small intrapair difference in kinematic variables of walking and running.

Furthermore, it is assumed that geometric similarities play an important part in small differences, as Asmussen and Nielsen (1955) have indicated with reference to the physical performances of children. Results obtained in this study, however, show that there were no significant differences in anthropometric variables. Therefore, it could be assumed that slight intrapair differences must be due to qualitative similarity in the neuromuscular system of monozygous twins.

Komi et al. (1976) and Klissouras (1971) reported that the variability of maximal muscle power, skeletal muscle fiber composition, and maximal oxygen intake were genetically determined (93.4, 97.8 and 99.5%, respectively). Although there was a failure to estimate the degree of inheritability, it could be assumed that kinematic variables in walking and running are related closely with genetic factors based on the discussion of intrapair differences between two types of twins.

References

ASMUSSEN, E., and Nielsen, Kr.H. 1955. A dimensional analysis of physical performance and growth in boys. J. Appl. Physiol. 7:593-603.

KLISSOURAS, V. 1971. Heritability of adaptive variation. J. Appl. Physiol. 31:338-344.

KOMI, P.V., Viitasalo, J.T., Havu, M., Thorstensen, A., and Karlsson, J. 1976. Physical and structural performance capacity: effect of heredity. In: P.V. Komi (ed.), Biomechanics V-A, pp. 118-123, University Park Press, Baltimore, MD.

KOMI, P.V., and Karlsson, J. 1979. Physical performance, skeletal muscle, enzyme activities and fiber types in monozygous and dizygous twins of both sexes. Acta Physiol. Scand. Suppl. 462.

Gait Analysis after
Total Hip Replacement for Osteoarthritis

**Tomakazu Hattori, Setsumasa Yamazaki, Hiroshi Honjo,
Shunichiro Miki, Masami Hori, Kazuhiko Sawai,
Ryozoh Furukawa, and Shigeo Niwa**
Aichi Medical University, Aichi-ken, Japan

Total hip replacement is widely accepted for the treatment of osteo-arthritis of the hip joint (Charnley, 1979). We have performed this procedure for the treatment of osteoarthritis of the hip joint with careful selection of cases in patients over 55 years of age.

We have investigated gait analysis and motion analysis of the hip, knee, ankle and elbow joints in three dimensions with electrogoniometers (Mitsui, 1977a, 1977b; Furukawa, 1979).

This report concerns the clinical assessment of cases and the results which have been investigated objectively by gait analysis after total hip replacement for osteoarthritis of the hip joint.

Methods

An electrogoniometer as modified by Chao (1970) was utilized with some additional features added. This electrogoniometer had three potentiometers, which were oriented in each of three primary axial planes in a special linkage system. The electrogoniometer was attached to a pelvic belt and thigh belt, so that there was no limitation of hip movement in any direction. The movement was converted into changes in curent flowing through each of the three poteniometers and then connected to a microcomputer (see Figure 1).

Foot switches were attached to the soles of the shoes on the heel and toe to provide information on gait kinematics. The microcomputer processed these data, and the results were printed out in degrees.

The subjects, with the electrogoniometer and foot switches attached,

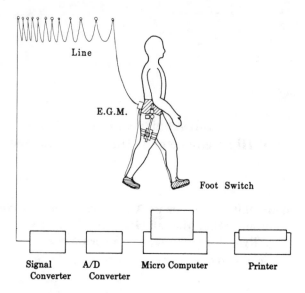

Figure 1 – Block diagram of data collection.

walked on a 5-m walkway for practice. After making any necessary adjustments, the subjects stood erect and were measured in this natural standing position as the starting position. Then they walked in free walk three times. In the analysis of these data, the first and the last strides were ignored and the middle three strides were measured and averaged.

Materials

This investigation included five subjects (two males and three females, with an average age of 60 years) and 52 patients (58 joints) with osteoarthritis of the hip joint who had received total hip replacements and rehabilitation services between 1974 and 1980 at the Aichi Medical University.

These patients were classified into three groups, monoarthritic (20 cases), bilateral arthritis of the hip joint (16 cases) and other joint afflictions (16 cases).

Total hip prostheses of the Charnley type were used in 48 cases (52 joints) and the Lindenhof type in four cases (4 joints).

The duration of follow-up was from 6 to 78 months (mean of 28.6 months).

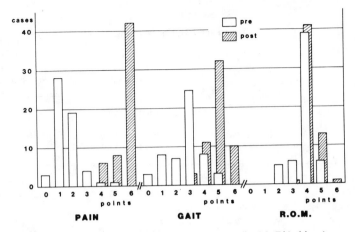

Figure 2—Clinical assessment of pre- and post-operation (by M. D'Aubigne).

Table 1

Range of Motion During Walking

		Mean S.D. (Degrees)		
		Sagittal plane	Frontal plane	Horizontal plane
Normal	5 cases	36.3° ± 8.8°	14.1° ± 4.1°	8.3° ± 1.8°
T.H.R.	30 cases	31.6° ± 9.0°	7.7° ± 3.6°	7.8° ± 3.6°
Mono O.A.	12 cases	30.4° ± 8.5°	7.6° ± 3.7°	7.8° ± 3.7°
Bilat. O.A.				
Mono ope.	8 cases	32.1° ± 6.4°	8.1° ± 2.2°	7.3° ± 4.1°
Bilat. ope.	3 cases	35.1° ± 8.7°	7.8° ± 4.2°	6.6° ± 2.9°
Others	7 cases	34.1° ± 9.3°	8.4° ± 4.3°	8.0° ± 3.4°

Results

The function of the hip joints before and after operation were evaluated by M. D'Aubigne's assessment method (D'Aubigne, 1954) (see Figure 2).

Remarkable improvements were observed in both pain and gait; there was no significant change in the range of motion of the hip joint, however.

Observing the range of motion of the hip joints during walking using the electrogoniometer (see Table 1), the angle in the sagittal and frontal planes decreased in patients after total hip replacement, but in the horizontal plane it was quite similar to normal walking (Lamoreux, 1971).

Table 2

Time Factor During Walking

		Cadence		Double support	
		Mean	S.D.	Mean	S.D.
Normal	5 cases	103.6 ± 18.6		16.8 ± 4.9	
T.H.R.	30 cases	114.0 ± 13.3		25.4 ± 8.0	
Mono O.A.	12 cases	106.1 ± 12.5		23.6 ± 5.0	
Bilat. O.A.		115.8 ± 13.5		28.3 ± 10.4	
Mono ope.	8 cases				
Bilat. ope	3 cases				
Others	7 cases	113.6 ± 11.1		23.9 ± 6.7	

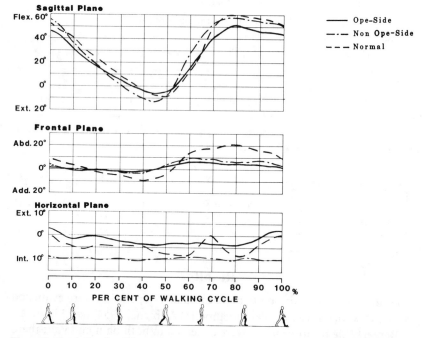

Figure 3 — Analysis of walking pattern.

Analyzing the time factor during walking using the foot switch and electrogoniometer (see Table 2), the cadence of a patient with mono-arthritis after total hip replacement revealed an almost normal value, but cases of bilateral and other joint afflictions showed larger values than normal.

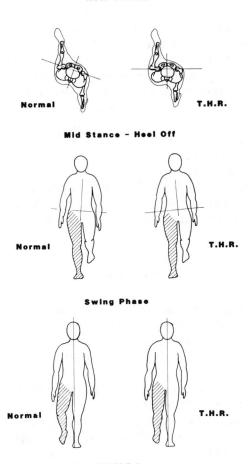

Figure 4 — Walking pattern (normal and T.H.R.).

The time of double support of the patient in a cycle of walking was prolonged in all cases after total hip replacement.

Further investigation of the monoarthritis and normals was carried out in order to clarify the pattern of walking after total hip replacement (see Figure 3). Range of motion of the hip joint during walking showed a similar pattern on the operation side, the non-operation side, and in normals in the sagittal plane.

In the frontal plane there were differences in the range of motion between the normal subjects and operation side, but no differences between the operation side and non-operation side.

In the horizontal plane, the operation side showed a noticeable decrease in the range of motion in comparison to normals and there was a tendency for external rotation to occur in comparison to the non-operation side.

Observations of the patterns in the frontal and horizontal planes during each phase of walking (see Figure 4) revealed that normal walking began with abduction and external rotation of the leg on heel contact, but the patient after total hip replacement did not show such motion on the operation side and non-operation side. Especially in the hip joint on the non-operation side, internal rotation on heel contact was present due to the decrease in the range of motion on the operation side.

During the mid-stance phase of normal walking, the pelvis was in an adducted position in the frontal plane, but on the operation side such a position was not evident.

In the swing phase of normal walking, the hip joint was in a position of abduction and was rotated externally during 70% of the swing phase, while the operation side showed slight abduction and no external rotation.

Discussion

The function of hip joints after total replacement is excellent and the patient can achieve almost normal capabilities for activities of daily living. It is clear, however, that the hip joint after operation differs from the normal hip joint anatomically, even though the artificial hip joint functions well.

It is suggested that the artificial joint after total hip replacement receives an abnormal torque which causes wear and a loosening of the prosthesis. From results previously mentioned, it is apparent that the hip joint after total replacement exhibits a decrease in range of motion in the frontal and horizontal planes.

Further studies are suggested in which analyses of the stresses in the hip joint during walking may be undertaken.

References

CHAO, E.S. 1970. The application of 4 × 4 matrix method to the correction of the measurements of hip joint rotations. J. Biomechanics (Great Britain) 3:459-471.

CHARNLEY, J. Low Friction Arthroplasty of the Hip. New York: Springer-Verlag, 1979.

D'AUBIGNE, M.R., and Postel, M. 1954. Functional results of hip arthroplasty with acrylic prostheses. J. Bone and Joint Surg. 39A:961.

FURUKAWA, R. 1979. The hip range of motion normally necessary for daily activities in Japanese way of living. Jap. Phy. The. and Ocup. The. 13(3):177-185.

HATTORI, T. 1980. The geometric analysis of hip joint motion in three dimensions using the electrogoniometer. Annual Meeting of Jap. Orth. Biomechanics Research Society.

LAMOREUX, L.W. 1971. Kinematic measurements in walking. Bulletin of Prosthetics Research.

MITSUI, T. 1977a. Knee motion analysis of children during walking by electrogoniometer. Annual Meeting of Jap. Ortho. Biomechanics Research Society.

MITSUI, T. 1977b. Biomechanics of the knee joint in relation between hip and ankle. J. Central Japan Ortho. and Trauma. 20:107-111.

VI.
ERGONOMICS
AND
POSTURE

Keynote Lectures

Human Engineering Aspects in System Design

Günter Rau

Helmholtz-Institut for Biomedical Engineering,
Aachen, West Germany

Ergonomics is an interdisciplinary science concerned with the interaction between humans and the technical environment they produce. According to the by-laws of the International Ergonomics Association a definition of ergonomics is given by the task an ergonomist is expected to do as indicated by Figure 1 (Bernotat and Rau, 1981). Ergonomics is often used as a synonym for human engineering.

In the beginning, ergonomics was mainly based on the classical human sciences, especially psychology and physiology. With the development of

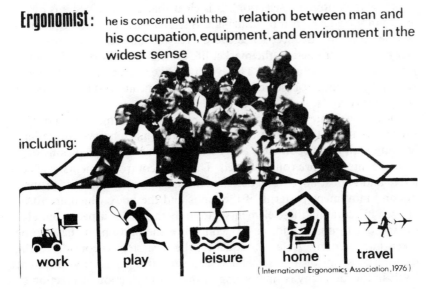

Figure 1—Definition of ergonomics (Bernotat and Rau, 1981).

514   Rau

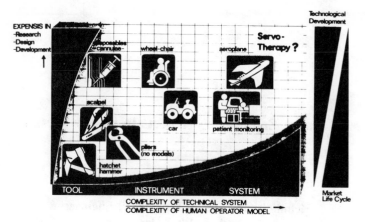

Figure 2 — Increase in complexity of a human operated system causes increase of expense in research, design, and development as indicated for ergonomics in medicine.

more sophisticated technical devices and instrumentation, the engineering sciences have made a signifcant contribution to the improvement and knowledge in ergonomics. A proposal for a university curriculum in ergonomics has been developed (Bernotat and Hunt, 1977). Applications in the design, use, and maintenance of new machines and large scale technical systems have become increasingly important.

As indicated by Figure 2, the complexity of modern man-machine systems has increased rapidly. Tools formerly were improved by a strategy known from the biological evolution characterized by 'trial and error' or 'mutation and selection'. It is clear that complex systems of to-day cannot be developed by utilizing this strategy but have to be planned and designed in advance, since the degree of complexity requires too many system parameters, the market life cycle is very short, and technological development is rapid. The evolution-like strategy is used only occasionally in special cases of simulation as a means within the system design process.

The biomechanics of human movement is one of the main components which has to be taken into account in a man-machine system design. Starting with anthropometric analysis, the geometrical dimensions of a machine interface are roughly fixed. Then, the control elements, such as levers and knobs, have to be designed according to the dynamic properties and movement patterns of the limbs and the body which are to be utilized in actuating these elements. Not only the motor aspects, but also the sensory feedback and sensory control of the movements have to be taken into account. As a consequence, the next step to consider is the decision behavior of the user of a system since this behavior at different levels in the hierarchy is finally responsible for any action or reaction of the user.

However, the ultimate aim of an ergonomic design is directed to the overall performance of the total system comprising all man-machine components. To optimize the system's performance, different areas of ergonomics contribute to the solution by using a systematic approach.

Within the classical field of biomechanics, ergonomics has been concentrated on occupational biomechanics in the sense used by Tichauer (1978). This approach is supported by psychological data, electrophysiological signals (ECG, EMG, etc.), cardiovascular (blood pressure, etc.) as well as metabolic (oxygen, catecholamid, etc.) indicators. During the last century, human occupational activities have shifted more and more from heavy physical work to supervisory control of very complex systems (Sheridan and Johannsen, 1976). Consequently, a shift in emphasis from the physical workload to the mental workload is involved (Bernotat, 1979).

Man-machine systems which are still increasing in complexity now demand an ergonomic design in advance. The main areas of ergonomics taking part in the design process as well as typical design steps will be characterized in subsequent sections of this article (Bernotat, 1979; Bernotat and Rau, 1981).

Areas of Ergonomics

In the design of a new technical system of high complexity, different tasks have to be performed concurrently. Simultaneously, two criteria have to be fulfilled, which are the high over-all performance of the man-machine system and the high work satisfaction for the use which must be within the physical and mental work tolerances during the operation of the system. The design can be conceived within four main problem areas as illustrated in Figure 3. These have been defined and developed from practical experience. Special problems occur at the 'interface' which have to be considered in the design of system aspects already existing within the areas.

The four main areas are as follows (Bernotat and Rau, 1981).

1. Personnel Ergonomics

The aim of personnel ergonomics is the adaptation of a human to the technical system by selection and training according to the desired physical, mental, and skill characteristics of the human user. Also, the formation of a crew and the organization and planning of team work must be a part of the task which is characterized as human-human interaction.

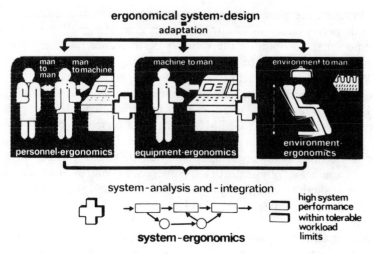

Figure 3—Areas of ergonomics (Bernotat and Rau, 1981).

2. Equipment Ergonomics (Anthropotechnics)

Anthropotechnics involves the adaptation of the machine or technical parts of the system as well as the complete operational process to the capabilities and limitations of the human user through an appropriate design.

This includes the anthropometric layout of the machine, the suitable design of visual and acoustical displays, the selection, arrangement, and combination of the control elements as input devices.

3. Environmental Ergonomics

The physical environment has to be adapted to the complete system and especially to the human user, including such aspects as temperature, humidity, illumination, vibration, noise, etc., according to the requirements and limitations of the human user. These aspects are collectively known as environmental ergonomics.

4. System Ergonomics

System ergonomics consists of a system-oriented approach in describing, analyzing, and integrating all the elements and their interrelationships into the overall system by applying the methods of 'systems engineering' to the design procedure of the man-machine system.

It can be expected that system ergonomics will be more important as the systems become more complex (Doering, 1976). Computer languages such as SAINT are being developed for simulation of the man-

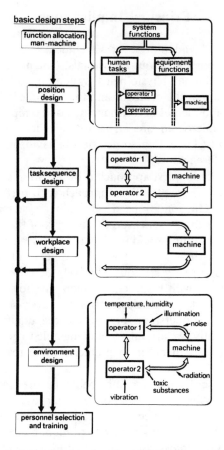

Figure 4—System-ergonomics in research and application (Bernotat and Rau, 1981).

machine system behavior (Kraiss, 1981), which may turn out to be, among others, especially appropriate to include aspects for the design of decision making and control movement components.

In the system design process, a special design methodology has to be followed as indicated by Figure 4 (Bernotat and Rau, 1981). First of all, the function allocation between a human and a machine has to be defined. This requires an extended analysis of the mission and the possibility of the existing similar systems. Then the job or operator position design follows, which involves defining who is doing what and when. The next step is to design the tasks and task sequences for each operator position. After these definitions, the design and layout of the individual work places (which includes the geometry, design, and layout of displays and controls, seat design, etc.) as well as environmental design have to be performed. Finally, by selection and training, the appropriate personnel are provided to make the total man-machine system operational.

Principal Implications

A prerequisite for the adaptation of a technical system to humans is to know as much as possible about individuals. One must have knowledge not only about the human physiological system, including motor performances, but also about the behavior and the multiple influences that can modify physical and mental performances. For example, if it is planned to utilize modern microcomputer structures as a decision aid in complex situations, one must know human decision behavior in detail as well as the individual differences.

In life sciences the simplest approach is observation and description of human behavior. A more effective approach is to perform experiments whereby one or several conditions are varied in a systematic manner and the reaction of the subject is measured. The crux of the matter is that the human subject is so complex and is influenced by such a large number of factors at the same time, some of which are not controllable, that it is very difficult to accurately and quantitatively describe a human's inner behavior. Therefore, with respect to human psychological behavior, it may be possible to describe selected human processes in a black-box manner but not to understand or analyze the underlying functional mechanism. Presently, we are far from being able to go from analysis to higher order synthesis in order to predict human reaction in a quantitative manner.

One tool to describe the dynamic behavior of complex systems is mathematics, including statistics. For this purpose, however, existing mathematics are still insufficient to model the human being even for the present because of the limited knowledge of human behavior. To circumvent this difficulty, in most cases simplified experiments are performed in which only a few conditions are varied and results are interpreted in the form of lists, graphs, or models in order to apply them to the design of a new technical system. The better the analyses, syntheses, and models become in the future, the more effectively technical instrumentation can be adapted to humans (Bernotat, 1979).

The scientific approach described needs an interdisciplinary team and has to follow the methodological line indicated above. In most cases this goal cannot be achieved in a short time. One of the first steps is to design checklists which are valuable tools for a systematic design, evaluation, and supervision of a system.

It is obvious that a mission-oriented system analysis and design can be performed only by an interdisciplinary team as already mentioned. The cooperation of experienced users will be indispensable because of their knowledge of the mission and of the existing systems. A crucial point is the evaluation and validation of a design which results in a solution based on laboratory experiments. In any case, it has to be proven that the resulting solution is also optimal or near an optimum in the performance of the specified mission in the actual situation.

Examples of Control Movements in Computer-aided Man-machine Interaction

After having discussed the concepts and research areas of ergonomics, some examples of interaction design should indicate the possibilities offered by modern technology which is currently available. These examples are tasks within medical applications, and the user of the systems interacts more or less as a supervisor with the system. The design is focused on the aspects of equipment ergonomics (anthropotechnics).

Interactive Evaluation of ECG-traces

In using multi-electrode leads of the thoracic electrocardiograms such as the method of 'precordial mapping' from a single electrophysiological heart cycle, about 50 to 200 leads are recorded simultaneously. These 200 curves have to be evaluated by the physician in a short period of time. After an extended analysis, a rather simple solution turned out to be most appropriate. This is shown in Figure 5 (Rau, 1979).

The curves are displayed on an electronic graphic display. The user moves a cursor by moving a small wheel with the fingers of his right

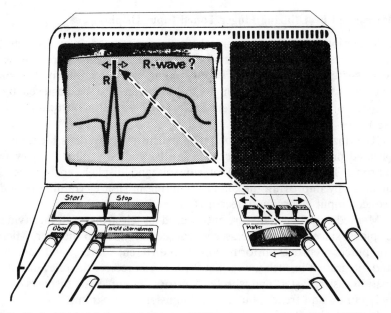

Figure 5 — Interactive evaluation of curves, ECG trace as an example (Rau, 1979).

hand. A rolling ball is much more difficult to operate, but it would allow a movement of the cursor in both x and y directions. This is not necessary in this particular task, since after having attached the cursor to the ECG trace, only the information reduced to the x-axis has to be put in. The moving wheel has been adapted to the task as well as to the sensory and motor characteristics of the hand movements by varying the moment of inertia, diameter, and damping empirically in order to find optimum performance. On the other hand, the system is programmed according to the task and in an interactive mode asks the specific points of the trace to be evaluated. After positioning the cursor on the point in question, the user actuates a push button with his left hand which probably already rests very near to or even on the button. For the technically innocent user, only two buttons and one wheel need to be utilized after starting the procedure. By this procedure and arrangement the 50 curves with three to four values each are evaluated within about 5 min.

The selection and design of the visual display, control elements, and the complete procedure have been performed as a result of an extended analysis by utilizing ergonomic design principles. Meanwhile, a semi-automatic procedure has been worked out but the first concept had to be modified slightly as far as the interaction was concerned. Now a pattern recognition algorithm evaluates most of the curves automatically and the remaining 10 to 15% of the curves, which are not recognized at a sufficient level of reliability, are processed by the human user as described above.

Interaction by Utilizing Finger Touch Input Displays

Information input into a technical system during a complex task can be normally performed by means of a number of different control elements (e.g., switches, push buttons, knobs, rolling ball, etc.), information output from the machine system to the user can also be achieved by a number of displays of different kinds (e.g., visual, acoustical, tactile, etc.).

By utilizing digital microcomputer structures, nowadays highly complicated program sequences can be started and controlled by very few buttons. Therefore, the area of controls and displays are reduced in quantity by improving the quality resulting in shorter training periods, reduced input errors, and decreased mental work load.

Marked improvements are introduced by electronic graphic (video) displays, since very complex scenes can be presented to the user. Microcomputers preprocess information from different sources, and compress and integrate it in such a way that the user can receive the relevant information easier, faster, and with fewer errors. Examples are displays of not only curves and trends, but also columns and pictograms.

Particularly effective is a combiantion of input and output on an elec-

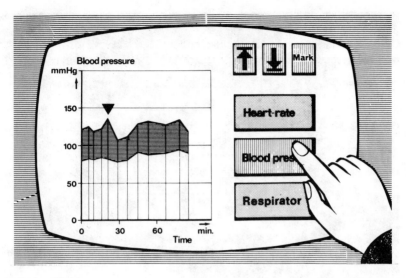

Figure 6 – Principal operation of a finger touch input display in an anesthesia supervision and recording system (Rau, 1979).'Heart rate', 'Blood pressure', 'Respirator' are defined as 'virtual push buttons'; 'pressing' it by touching the screen with the finger, a corresponding graph is displayed on the screen.

tronic display while in addition the user can be guided or supported. One technical solution is the electronic picture display with touch input (finger touch input display). This approach has already been suggested for application in air traffic control (Gaertner and Holzhausen, 1977) and is now being introduced to a new generation of measuring instrumentation. The medical application described in the following example is concerned with anesthesia supervision and record (Rau, 1979; Trispel et al., 1979) (see Figure 6).

In addition to the picture on the screen as information output, a measuring device can detect the position of a finger touching the screen. If within the picture a 'push button field' is synthesized, displayed, and defined as a button, it can be activated by touching it. In the given example, the physician can 'switch on' different curves by selecting the desired one. This combination of display and control is in many cases superior to conventional elements (Gaertner and Holzhausen, 1979).

In a study concerning anesthesia recording (Trispel et al., 1980), an experimental system based on this concept has been developed and extended to the possibility of analog input functions. As already mentioned, the first step was an exact analysis of the mission 'anesthesia'. With knowledge of the process including all complications, an information structure can be designed which enables the anesthetist to communicate with the technical system adequately during operation. The result of this analysis was to know, e.g., which kind of information is

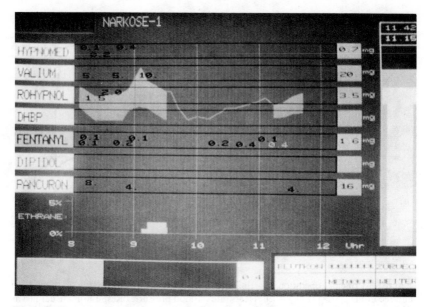

Figure 7—Example of a page in the anesthesia recording system, where the display is combined with a manual data entry of medication dosages (Trispel et al., 1980).

needed or desired by the anesthetist during which phase of the mission.

Part of the example is a record of a given amount of drug delivered to the patient. The anesthetist selects by touching the push button, 'drugs', a specified 'page' of the 'information system' which is thereafter displayed. An example is shown in Figure 7.

Touching the sensitive field defined as 'FENTANYL' activates this part of the system, and the system reaction is indicated to the user by means of a color change. Simultaneously, the horizontal bar at the bottom left is activated for adjustment and quantitative input; the exact value is displayed digitally, the range of values (sensitivity of input) is adjusted automatically according to normal drug delivery as found during the previous analysis. This 'analog potentiometer' can then be adjusted by touching and moving the finger continuously along the bar. In addition, a total amount of drug delivered as well as the time and single doses are displayed. For corrections or recording of previously delivered dosages the 'time setting potentiometer' (vertical bar on the right edge) can be used.

In a similar manner, other pages of the anesthesia information system can be selected which can report on the status of respiration, the condition of the cardiovascular system, etc. Another important advantage in this combination of control elements with the electronic screen is the possibility of displaying complicated curves of vital parameters as functions of time in the background which would never be possible in conventional display and control technology. It has turned out that color as a code dimension is indispensable in many complex tasks.

This example indicates future possibilities of interaction between a human and a machine system. The interaction has become very simple by using interactive displays, by structuring the information in adequate 'display files and pages' according to the results of the preceding analysis, and by integrating the display and control. Of course, on the one hand, the biomechanics of control movements has become simpler, but on the other hand, the optimization has to be carried out experimentally concerning, for example, size and interdistances of the 'virtual push buttons and sliders', etc.

Adaptive, Automatized Procedures for Sensory Tests

Diagnosis of disturbances in perception or in the sensorimotor system of the human being can be supported by automatized adaptive systems. Here also, the interaction between a human (patient or subject) and the technical system is guided. The guidance is performed according to a strategy which the physician would use during an investigation. Interesting applications in visual perception began years ago (Bouma et al., 1971), but it turned out that the strategy of the automatic vision test could be essentially improved (Trispel et al., 1979). One difficulty is the design of the input device which has to be actuated by the subject as a response to a stimulus. Compared to the vision test, a more complicated test is the detection of depth perception.

When using a TV monitor for presenting a stimulus stereoscopically, one half picture is presented to the left, the other half in sequence to the right eye. The picture is synthesized in such a way that the subject sees the objects spatially. Figure 8 shows the two cursors which are displayed to the eyes alternatively as described, the subject with normal sight sees one cross (object) stereoscopically.

The subject's task is to position the cross in a specified location in space by movement of the object in the orthogonal plane (x, y) by actuating the 'ball element' with the right hand, and moving the object along the depth axis (z) by actuating the finger stick with the left hand.

The control devices indicated in Figure 8 schematically are the results of long development of the procedure, even though the solution now looks very simple. The advantage of the arrangement is its compatibility, for even children can perform the task quickly with high accuracy after a very short instruction and training period.

The design can most probably be applied to other difficult tasks of 3-dimensional positioning. Biomechanics of movement and movement control have to be considered when the geometrical dimensions, mechanical and dynamic characteristics and location of such a combination of

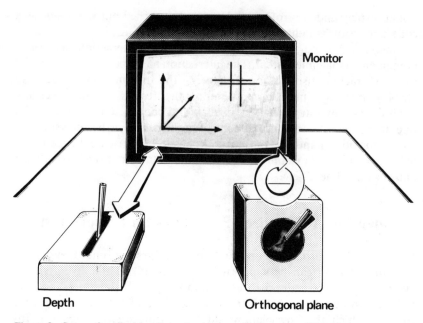

Monitor

Depth   **Orthogonal plane**

Figure 8—Set up for display of simple stereoscopically perceptible objects and control elements for adjustment of the objects in 'space'.

control devices are selected. Range of operation, sensitivity of reaction related to the movement of the controlled object (e.g., the cross in Figure 8 in the example presented, position of a manipulator), and accuracy have to be chosen based on the analysis of the task involved.

Some Future Goals and Developments

The examples indicate that very complex functions are shifted from the human user to the machine system while the system becomes more 'intelligent', especially by the use of microcomputers. As a consequence of complex interactivity, the user is freed from unnecessary tasks of handling the machine system, and is guided or supported reliably and quickly. The first goal is high performance of the system combined with reliability of operation and safety with respect to mistakes made by the user. A higher degree of complexity of a machine system demands a higher degree of automation if simple control has to be guaranteed. In this context, the biomechanics of fine control movements when operating control devices in its broadest sense is only one but is an important section in human engineering design.

Finally, the design task is to optimize the total man-machine system according to its purpose, task or mission without over- or underloading the user.

Therefore, the design of the displays and controls can be accomplished only after analysis of the specific mission and a first preliminary design approach. Universal solutions are not of much value, only some principles and guidelines have general application to the design process. In the future, new technologies will broaden the design of unconventional interfaces between humans and machine systems including communication by voice and speech. The flexibility of the interface design can still be improved by the availability of more flexible microcomputer structures. However, in ergonomics design the main problems to be solved do not seem to be technological in nature. A better understanding of the human user regarding sensory and motor performance as well as decision behavior are of greater importance to systems design which, as previously mentioned, can be effectively performed only by interdisciplinary groups matching the specifications of technological systems with the abilities and limitations of the human being.

References

BERNOTAT, R. 1979. Die ergonomische Gestaltung der Kommunikation Mensch-Maschine (Ergonomic Design of Man-Machine Communication). NTG-Fachberichte, Bd. 67, S. 3-14, VDE-Verlag, Berlin.

BERNOTAT, R., and Hunt, D. (eds.). 1977. University Curricula in Ergonomics.Forschungsinstitut fuer Anthropotechnik, Meckenheim (West Germany).

BERNOTAT, R., and Rau, G. 1981. Ergonomics in Medicine. In: Ghista, Reul, Rau (eds.), Perspectives in Biomechanics, pp. 381-898. M.B. Gordon Publ., NY.

BOUMA, H., Cardozo, B.L., van Dun, J.J.A.M., and Valbracht, J.C. 1971. An automatized routine test for the determination of visual acuity. IPO Annual Progress Report No. 6, 131-136.

DOERING, B. 1976. Analytical Methods in Man-Machine System Development. In: Kraiss and Moraal (eds.), Introduction to Human Engineering, TUV Rheinland, Koeln.

GAERTNER, K.-P., and Holzhausen, K.P. 1977. Human Engineering Evaluation of a Cockpit Display/Input Device using a Touch Sensitive Screen. Proceedings of the 25th Meeting AGARD Guidance and Control Panel.

KRAISS, K.-F. 1981. A Display Design and Evaluation Study using Task Network Models. IEEE Transactions on Systems, Man, and Cybernetics, SMC-11: 339-351.

RAU, G. 1979. Ergonomische Überlegungen bei der Gestaltung komplexer medizinischer Instrumentierung unter Einsatz von Mikroprozessoren (Ergonomic considerations in the design of complex medical instrumentation under control of microprocessors). Biomedizinische Technik Bd. 24, pp. 10-15.

SHERIDAN, T.B., and Johannsen, G. (eds.). 1976. Monitoring Behavior and Supervisory Control. NATO Conference Series. Plenum Press, NY.

TICHAUER, E.R. 1978. The Biomechanical Basis of Ergonomics. John Wiley & Sons, NY.

TRISPEL, S., Rau, G., and Guenther, K. 1979. Direkter Zugriff auf Bildschirminformation über Beruehreingabe (touch input) und Anwendung in komplexen klinischen Systemen (Direct access to screen display information by touch output and applications in complex clinical systems). Biomedizinische Technik Bd. 24, pp. 64-65.

TRISPEL, S., van Ackeren, G., Rau, G., and Reim, M. 1979. Eine adaptive Messtrategie zur automatischen Bestimmung des Fernvisus (An adapted measurement strategy for automatic visual acuity of detection). Berichte d. Dtsch. Ophthalmol. Ges. 76:359-362.

TRISPEL, S., Rau, G., Schlimgen, R., and Kalff, G. 1980. Human factors in medical real-time dialogue systems: A case study. Proceedings of the Human Factors Society, 24th Annual Meeting, pp. 590-594.

TRISPEL, S., and Rau, G. 1981. Die Mensch-Maschine-Schnittstelle in der Medizin (The man-machine-interface in medicine). Helmholtz-Institut Forschungsbericht 1979/80, pp. 93-98.

The Static Analysis of Heavy Manual Exertion

D.W. Grieve
Royal Free Hospital School of Medicine, London, England

A method of task-analysis, using the postural stability diagram (PSD) and some anthropometric data on strength necessary for its implementation now exists (Grieve, 1979a, b; Pheasant & Grieve, 1981). The purpose of PSD analysis is to simultaneously assess interactions of strength, task demands, frictional limitations, and other environmental constraints. A model of the body which predicts both the static forces in any direction that can be exerted with given placements of the hands and feet, and the space requirements of the standing operator while doing so, would have considerable value. The collection of empirical data that encompasses all possible conditions is impractical and not economical. A predictive model which incorporates rules governing the musculoskeletal system and anthropometric data is to be preferred, both for convenience and interpolation in novel situations, and for the spin-off, e.g., in clinical studies, that can be expected from its validation. The success of computerized biomechanical models by Chaffin and his collaborators (Chaffin, 1969; Martin & Chaffin, 1972) in predicting the forces developed in static lifts, pushes, and pulls suggests that several critical features were correctly identified. In the present studies, postures adopted during maximal exertions were recorded, in order to create a link between strength and space requirements of the standing operator. Their analysis, which is the subject of this article, emphasizes features which, it is hoped, can provide the basis for detailed validations of multipurpose models of exertion.

The factors which limit a maximum manual exertion in an unconstrained stance have not been positively identified in the living. It is clear that a standing person, however muscular, cannot press steadily downwards with more than body weight; neither can a person push or pull horizontally if the center of gravity is vertically above the center of pressure at the feet. These are trivial cases. In general, a resultant static manual force (sagittal plane, zero torque) can be discussed in terms of two orthogonal components. The first (let us call it the gravity component) is due to the weight of the body exerting leverage about the center of foot pressure, and it acts perpendicular to the foot-hand axis. The second, or radial component, acts along the foot-hand axis and does not affect the balance of the lever. A chosen direction of exertion implies that the components are in a fixed ratio.

Either component could limit the exertion in a chosen direction. The gravity component reaches a limit when the center of gravity reaches a limit of its horizontal range relative to the center of foot pressure. The radial component is limited by the capacity of the muscles and ligaments to withstand a thrust or tension along the foot-hand axis. The muscles and ligaments have a dual role in maintaining the configuration of the body lever, which creates the gravity component, and maintaining the radial component which is a fixed multiple of the other. Maximal static exertion requires the adoption of a posture in which the dual roles are performed in an optimum manner.

Procedure

The subjects (10 men and nine women) were lightly clad and shod, and stood on a non-slip floor. They exerted steady maximal forces on a bar-handle force transducer, using symmetrical postures of their own choosing. Exertions were made in each of eight directions in the sagittal plane, approximately 45° apart, when the handle was either 1 m or 0.25 m high and the toes were either under the bar or 0.5 m to the rear. The total manual force and its direction were measured from an X-Y recording of the vertical vs. horizontal components. The subject was photographed at the instant when a maximal, steady force was achieved.

Body markers were initially attached to the standing subject. Tabs on the skin overlying the fibular malleolus, the lateral epicondyles of the femur and humerus, and a point 5 mm anterior to the tempero-mandibular joint indicated the centers of the ankle, knee, and elbow joints, and the center of gravity of the head and neck. Encircling bands around the arm and thigh identified the levels of the centers of gravity of those segments. To minimize problems due to skin movement, the hip and shoulder centers were determined by extrapolation along the axes of the limb-segments from their centers of gravity. Straws, perpendicular to their metal mounting plates, indicated the directions in which the centers of the C7-T1 disc, the T11 body, and the L5-S1 disc lay, at assumed distances below the skin surface. Landmarks on the projected photographs were measured with a digital trace reader; scaling and parallax corrections were applied to obtain the vertical and horizontal coordinates of all points in the sagittal projection, from an origin on the floor below the bar. Limbs were treated as pairs. The locations of the joints and weight centroids of the eight segments were determined. Segment and joint angles were computed, together with the net horizontal and vertical forces and net torques at each joint.

Most of the data in this article relates to the males with toes placed under a handle of 1 m height. To emphasize the effects of the direction of exertion upon selected variables, a curve-fitting procedure was adopted in preference to presenting scatter plots. The average direction of exertion used by the subjects in each 45° sector in the sagittal plane was determined, together with the eight corresponding means and standard errors of a selected variable. The means and standard errors were separately fitted by periodic cubic spline functions of the direction of exertion. Interpolations of the standard error were not permitted to fall below the given values at adjacent knots on the spline. The smooth curves and the shaded surrounds (e.g., see Figure 2) based on the cubic splines, represent a predicted mean ± 1.96 standard errors (or 90% confidence limits). A scatter plot of original points is superimposed on the spline functions in Figure 2 to illustrate the effectiveness of the procedure.

Description of Posture and the Use of Body Weight

The stick figures, which appear as a key to some of the illustrations, depict the *average* locations of joint centers adopted by the male group. For some directions of exertion, such as pushing (see Figure 1), lifting, and pressing, all subjects adopted similar postures. In contrast, individuals varied considerably when pulling or pulling/pressing. The variability is reflected in the scatter of locations of the centers of gravity. The graphs in Figure 1 show how the standard error (x 1.96) of the mean vertical and horizontal locations of the centroids change with the direction of exertion. Differences between subjects are more apparent in the vertical (Y) direction than the horizontal (X) direction. The horizontal locations are functionally important when making use of body weight; subjects appear to find similar solutions to this problem. Any other postural variations affecting the vertical location presumably relate to making optimum use of the available musculature. Common solutions to this problem are apparent in lifts, pushes, and presses, whereas a range of individual solutions were observed for pulling and pulling/pressing.

Figure 2 (top left) shows that manual strength is strongly influenced by the direction of exertion. By taking moments about the center of foot pressure, a relationship (the 'equation of static exertion') can be shown between the vertical and horizontal components of manual force. Since manual torque is negligible in these exeriments, we have:

$$LIFT/W = (h/x_p). \quad PUSH/W = (x_p - x_g)/x_p \qquad (1)$$

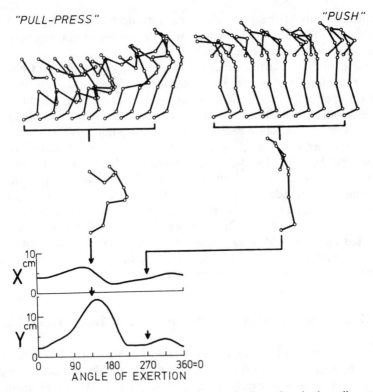

Figure 1—Postures adopted by nine men during maximal exertions in the pull-press and push directions, giving rise to the mean postures shown. The plots (X and Y), of 1.96 × standard errors of the mean positions of the center of gravity, show that postural differences between subjects give rise to more vertical than horizontal scatter when they occur.

where: x_p and x_g are the distances of the centers of foot pressure and of gravity behind the hands, when the hands are at height (h) above the ground. The plot of LIFT/W versus PUSH/W in Figure 2 (top right) shows a manual force vector (OP). The equation of static exertion is represented by the line P'QP. The manual force has a component (OQ) which is solely due to the weight of the body leaning against the handle. It contributes force (OR) in the desired direction. There is also a component (QP) due to voluntary muscle action and not predictable from the posture, which contributes RP in the desired direction. The fraction (OR/OP) of the total manual force which can be accounted for by the use of body weight, varies from 0 to 100%, depending on the direction of exertion. Equation 1 was used to determine x_p. The shifts of weight and pressure centroids as the direction of exertion was altered are shown in Figure 2 (bottom left), from which the fraction (OR/OP) was calculated (bottom right). The high fractions when pushing or pulling show that the postures were then important for the deployment of body weight; al-

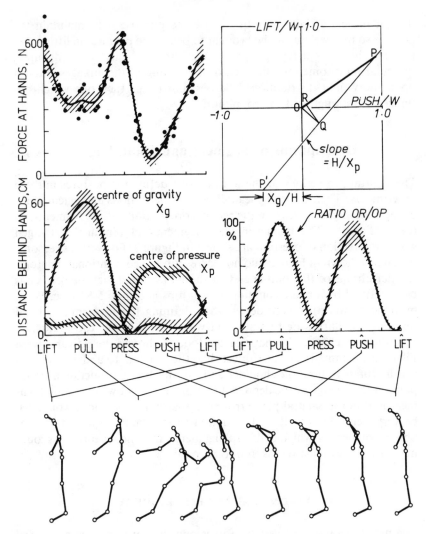

Figure 2 — Top right: Postural Stability Diagram, showing manual force vector (OP), whose head lies on the line P'QP which represents the equation of static exertion. The gravitational (OQ) and radial (QP) components of the force contribute OR and RP in the desired direction.

Top left: Maximal manual forces plotted as a function of the direction of exertion, including a scatter plot of the individual efforts.

Lower left: Horizontal shifts of the center of gravity and the center of foot pressure as the direction of exertion changes.

Lower right: The fraction of the manual force which is attributable to the gravity component.

All curves are cubic splines. Shading represents 90% confidence limits for the mean.

though muscle activity may be intense, its purpose is to maintain the posture so that weight can be brought to bear. The postures in lifting and pressing have a different significance: strength then depends upon muscular (and sometimes ligamentous) coupling of the linkages and not upon the weight distribution. The best posture is then one which optimizes the strengths of the couplings.

Descriptions of Segment and Joint Angles

The sagittal projections of the joint centers define the postures in these experiments. Their locations lead directly to definitions and calculations of segment angles and, from the relative orientations of the segments, to the joint angles. The manner in which segment and joint angles changed with the direction of exertion are shown in Figure 3. For various reasons, the angles shown in Figure 3 do not correspond to conventional, anatomical definitions of the joint angles. It should be noted that the pelvis has been defined by a plane containing both hips and the L5-S1 disc. Wrist movements were not considered when defining the forearm-and-hand segment. Similarly, the foot segment was defined by the toe-ankle line, regardless of changes in shape of the foot. Although the postures were bilaterally symmetrical, the subjects had the freedom to abduct and outwardly rotate at both the shoulder and the hip. From a functional standpoint, a knowledge of the degree of flexion at the elbow (for example) as measured in the sagittal plane from the anatomical position, would not be very helpful. The elbow is required to transmit torques in various oblique planes for which collateral ligaments and unconventional groupings of muscles must be taken into account.

Forces and Torques at Joint

The net horizontal and vertical forces and net torques at joints varied with the direction of exertion as shown in Figure 4. Each joint transmitted the same horizontal force while the vertical force increased downwards from the hands to the ankles according to the weights of the superincumbent parts.

The torques developed at the five pairs of adjacent joints varied in a coordinated manner, as can be seen in Figure 5. The smooth plots were obtained from the cubic spline functions of the means which are featured in Figure 4. No attempt was made to incorporate the confidence limits. It should be noted that these results were found for exertions on the 1 m handle with the toes below it; they will not necessarily apply at other handle heights and toe placements. Electromyography will be required to clarify how the muscle groups contribute to the net torques which were

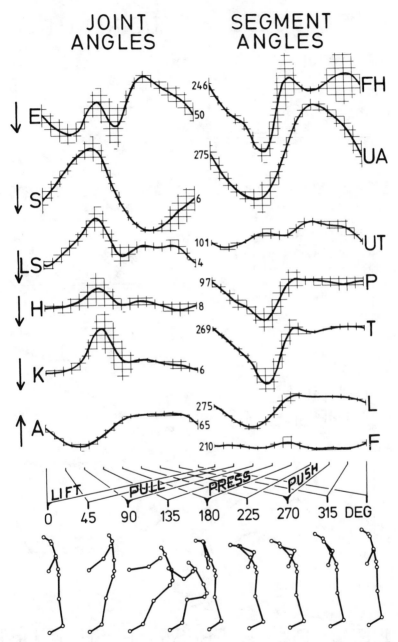

Figure 3 — Cubic splines showing the changes of mean joint and segment angles as functions of the direction of exertion. FH, UA, UT, P, T, L, and F refer to forearm and hand, upper arm, upper trunk, pelvis, thigh, leg, and foot, respectively; E, S, LS, H, K, and A refer to elbow, shoulder, lumbosacral, hip, knee and ankle angles, respectively. The stick figures show mean postures in the eight principal directions. Cross hatching, with 10° spacing, shows the 90% confidence limits of the means.

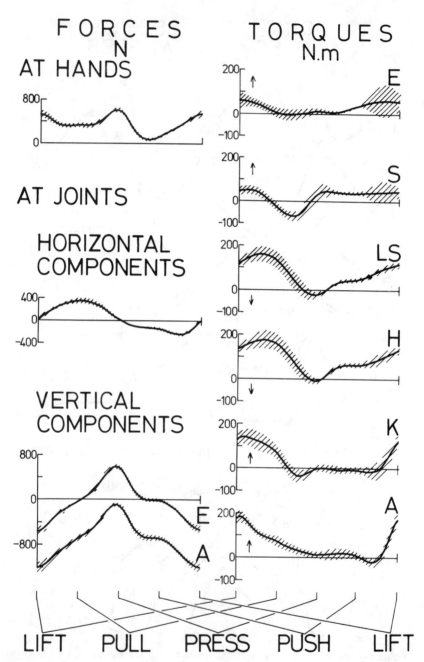

Figure 4 – Forces and torques as functions of the direction of exertion (cubic splines of the means, with 90% confidence limits shaded).

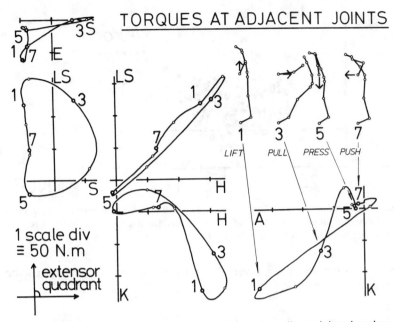

Figure 5—The relationships between net torques developed at adjacent joints, based on the cubic splines shown in Figure 4. Four of the mean postures are numbered and inset as a key to the changes of coordinated activity that occur.

measured. Unless co-contractions feature prominently for purposes of stability, the lower limb appears to make little demands upon the hip flexors, knee extensors, or anterior tibial muscles. The simultaneous demands for hip extensor, knee flexor, and ankle flexor torques, and the manner in which the foot-base is employed, hint strongly at the combined use of hamstrings, gastrocnemii, and the deep flexors of the leg. The torques at the hip and lumbosacral joints are very highly correlated, which is true of *all* the experiments, and is a simple consequence of the proximity of these joints. When the torque-relationships for the shoulder-lumbosacral and the lumbosacral-hip combinations are compared, the marked contrast between them suggests a complex shift of relationships between the musculatures in adjacent regions of the trunk as the vertebral column is traversed from shoulder to pelvis. One might expect such a result when the pectoral muscles, serratus anterior, trapezius, and latissimus dorsi are involved.

It has already been noted that the most rapid changes of posture with the direction of exertion, and the greatest individual variations, are associated with the transition from pulling to pressing. This transition is less likely to arise in everyday activities than shifts between the other quadrants. It can also be seen from the torque-relationships in Figure 5 that the pull (posture 3) and press (posture 5) conditions are in strong

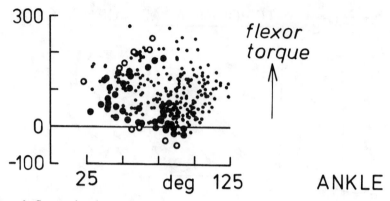

Figure 6 — Scatter plot, from *all* the observations on 10 men at two handle heights and two foot placements, of ankle torque as a function of ankle angle. The larger symbols refer to exertions on the 1 m-high handle with toes beneath the bar.

contrast, often requiring reversals of torque to pass from one to the other.

Torque-angle Relationships

The scatter plot in Figure 6 shows, for *all* exertions by male subjects, the torques at the ankle as a function of ankle angle. There is no evidence that the greater torques can be generated at each angle in these particular types of exertion. All the torques may be sub-maximal and ligaments may play an important role in the extreme postures. If subsidiary experiments were devised to search for true maxima, two complications could be anticipated. The existence of multi-joint muscles implies the influence of postures and demands upon adjacent joints. Torques at the ankle, for example, are influenced by posture at both the knee and ankle. A functional 'surface' of observed torques as a function of the angles of both joints (1 m handle, toes beneath hands) is shown in Figure 7, derived from the cubic spline functions which appear in Figures 3 and 4. To recognize whether maximal capacity of the muscles and ligaments at the ankle were utilized, three variables must be considered simultaneously. A similar exercise at the knee would introduce the hip posture as a fourth variable. The second complication, especially relevant to the upper limb, is that subsidiary tests of strength must be done with the same orientations of the limbs relative to the sagittal plane, not necessarily those normally studied in formal tests of strength at the joints.

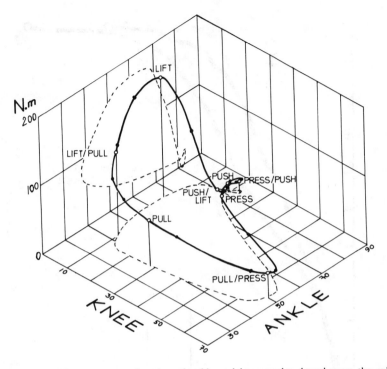

Figure 7—Ankle torque as a function of ankle and knee angles, based upon the cubic splines of the means which appeared in Figures 3 and 4.

Stresses at the Lumbosacral Joint

The lower lumbar region is commonly regarded as at risk during heavy exertions, especially during lifting. A compressive stress at the lumbosacral disc was computed for every direction of exertion and all hand and toe placements. No allowance was made for intra-abdominal pressure. The disc was assumed to have an area of 1300 mm² and to be subjected to the sum of the net compression in the L5-S1 to C7-T1 interdiscal direction, plus the compression, associated with extensor torques, due to erectores spinae muscles acting 50 mm posterior to the centroid of the disc. The frequency ogive (male) in Figure 8 indicates a 95th %ile stress of 4.3 N/mm², and a maximum observation of 7 N/mm² for voluntary exertions which were distributed evenly in all directions. The highest stresses were associated with lifting forces on a low handle. It must be assumed that the actual stresses were somewhat lower due to the presence of intra-abdominal pressure. The 95th %ile stress corresponds almost exactly to the compressive force limit (578 Kp) which was incorporated in Chaffin's earlier models (1969).

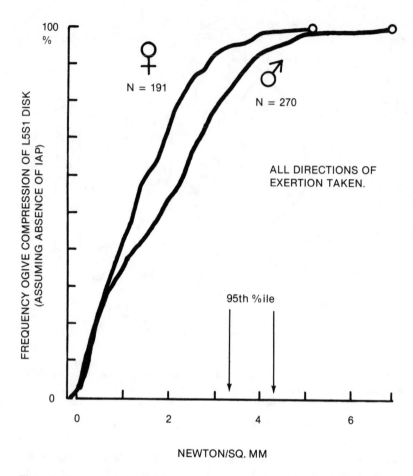

Figure 8 — Frequency ogives for the computed compressive stresses on the lumbosacral disc, in all exertions, i.e., all subjects, all directions of exertion, both handle heights, and both toe placements (see text for details).

Conclusion

Although only a few of the experimental results can be included here, a selection has been presented of the features which emerge when subjects are studied during the voluntary performance of maximum static exertions. The exertions and the postures were presumably regarded as optimal by the subjects in those circumstances. Although further studies are envisaged, they can never be comprehensive. Their value is seen in the further validation of computerized biomechanical models, which are versatile. If computer models can predict *both* the forces *and* the postures adopted in a set of human experiments, their application to work-place and task-design would be strengthened. Confidence in the underlying rules of the model should also lead to clinical applications, especially to improved discussion and functional assessment of patients with neuromuscular and articular impairments.

Acknowledgments

The author is indebted to Miss J. Randall and Mr. P. Livesley, for contributions during B.Sc projects, to Dr. S.T. Pheasant for discussion, and to Miss T. Rashdi and Mr. B. Pike for technical and photographic assistance during the research program.

References

CHAFFIN, D.B. 1969. A computerized biomechanical model—development of and use in studying gross body actions. Biomechanics 2:429-441.

GRIEVE, D.W. 1979a. The postural stability diagram (PSD): personal constraints on the static exertion of force. Ergonomics 22:1155-1164.

GRIEVE, D.W. 1979b. Environmental constraints on the static exertion of force: PSD analysis in task design. Ergonomics 22:1165-1174.

MARTIN, J.B., and Chaffin, D.B. 1972. Biomechanical computerized simulation of human strength in sagittal-plane activities. American Institute of Industrial Engineers Transactions 4:19-28.

PHEASANT, S.T., and Grieve, D.W. 1981. The principal features of maximal exertions in the sagittal plane. Ergonomics. (in press)

A.
Ergonomics

Biomechanical Analysis of Loads on the
Lumbar Spine in Sitting and Standing Postures

G.B.J. Andersson, R. Örtengren, and A.L. Nachemson
Sahlgren Hospital, Göteborg, Sweden

A.B. Schultz
University of Illinois at Chicago Circle,
Chicago, Illinois, U.S.A.

Low back pain (LBP) is responsible for more sickness resulting in absence days in industry than any other disease or injury. Although no definite proof exists, it is generally believed that mechanical factors are important in its development, and it is well known that mechanical stress in patients with LBP increases their pain (Andersson, 1981). Most preventive methods are based on a mechanical hypothesis and are aimed at reducing the load on the lumbar spine.

There are no direct methods presently available, however, to measure spinal loads. As a consequence the load-pain relationship is difficult to establish precisely, and exposure cannot be quantified in epidemiological research. Further, the preventive programs are general in nature, and when several options are available, priorities cannot be made based on quantitative criteria.

The loads imposed on the lumbar spine by different physical activities can be estimated from measurements of myoelectric back muscle activity, intradisc pressure, and intraabdominal pressure (Andersson, Örtengren, and Nachemson, 1977; Örtengren and Andersson, 1977; Davis, 1981; Nachemson, 1981; Schultz et al., 1981). Load measurements by these methods are not always practical, however. They are partly invasive, equipment and personnel costs are high, and the data analysis is often complex.

Biomechanical model analysis can also be used to assess back loads (Schultz and Andersson, 1981). While external loads can often be calculated with reasonable accuracy, major assumptions must be made in these models to estimate the internal forces. Validation experiments are therefore required before the model can be used with confidence.

The purpose of this article is to summarize experiments, reported in detail elsewhere, in which the internal loads required to equilibrate the body in various well-controlled and fully quantifiable static postures were analyzed using mechanical models (Andersson et al., 1980; Schultz et al., 1982a, b, c). Predictions of back muscle contraction forces were then compared to measurements of the myoelectric activity of some back and abdominal muscles, and predictions of the compression of the spine were validated through disc pressure measurements.

Material and Methods

Subjects and Positions Studied

The back muscle contraction forces were estimated in three studies, and validated using electromyography. The studies were made on three different occasions on a group of 10 male subjects with a mean age of 29 years, a mean height of 182 cm and a mean weight of 74 kg. All studies were made with the subjects placed in a reference frame, which permitted accurate control of body posture. In the first study, the subjects were sitting on well-adjusted office chairs at tables individually adjusted in height (Andersson, Örtengren, and Schultz, 1980). They were asked to hold the right hand in 10 different positions over the table (see Figure 1). In each position the subjects first held no weight, then a 20 N weight, and finally a 40 N weight.

In the second and third series, the subjects were standing in the reference frame with their pelvises securely strapped against a padded pelvic support. Forty-two symmetrical tasks were studied in the second series, six in which horizontal forces were applied to the trunk from a harness, and 36 in which weights were held close to or at a distance from the trunk and in upright and forward leaning postures. Twenty asymmetrical tasks with lateral bending and rotation were studied in the third series.

Selected postures from the series described were chosen to estimate spine compression and these estimates were then validated through measurements of intradisc pressure. This study was made on four volunteers, three female and one male. The ages were from 19 to 23 (mean-22 yrs.), heights from 165 to 187 (mean-174 cm), and weights from 57 to 78 (mean-64 kg).

Configuration Measurements

The distance between the center of the L3 disc and mass center locations

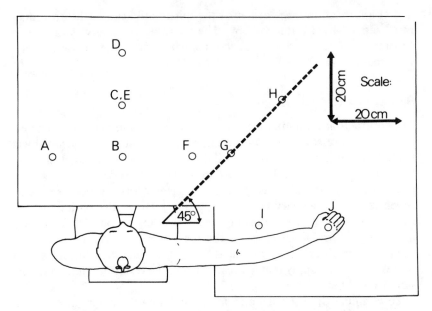

Figure 1—Positions at the table in which the hand was held with and without a weight.

of the major body segments above that level were measured by taping targets to the approximate mass center locations of the head, trunk, arms, and hands, and by estimating the lever arms for the trunk extensor muscles and the abdominal muscles from anthropometric measurements. Horizontal scales were fixed to the reference frame, and for each experimental position the location of the targets was estimated, in the sagittal and, if necessary, in the frontal plane, to within a few cm using the scales and plumb bobs.

Measurements of Myoelectric Activity

The myoelectric signals were picked up by 12 bipolar, recessed surface electrodes, placed 3 cm lateral to the midline on both sides of the spine at the L1, L3, and L5 levels, and over the rectus abdominus muscles, and the internal and external oblique abdominal muscles. The signals were amplified, filtered and stored on magnetic tape, and subsequently analyzed for rms amplitude.

Measurement of Intradisc Pressure

The intradisc pressure was measured by means of a subminiature pressure transducer built into the tip of a needle, as previously described by Nachemson and Elfström (1970). The transducer was inserted via a lateral approach into the center of the third lumbar disc, the procedure being guided by TV-fluoroscopy. The transducer signal was calibrated and recorded on a strip chart recorder from which the pressure values were read.

Biomechanical Analysis

The contraction forces in the trunk muscles needed to perform each task and the compression loads on the spine were computed for each subject from a biomechanical analysis. Full details concerning the analysis have been outlined in an earlier report (Schultz and Andersson, 1980). In brief, the analysis was based on the requirement that all body segments superior to an imaginary transverse cutting plane at the L3 level of the spine must remain in equilibrium during task execution. There are two main steps in the analysis. First the net lower-trunk support reaction needed for equilibrium is computed. Then the trunk muscle contraction forces needed to supply that net reaction and the resulting spine compression are estimated.

To compute the net reaction, one must know the magnitudes of the externally-applied loads and the weights of all body segments superior to the cutting plane. The externally applied loads are the measured horizontal forces or the weights held. The weights of the head, each arm, and the trunk above the L3 level were assumed as 5, 4.45, and 36.1% of the subject's body weight. The points at which these forces and weights are applied must also be known. Here, those data were available for each task and each subject from the measurements taken. Once all of these data are entered into the six equations expressing the requirements for equilibrium, the six components of the net reaction can be computed.

The trunk muscle contraction forces and the spine compression loads were then estimated from the net reaction by using a biomechanical model of the L3 cross-section, and solving the model equations by a linear-programming technique. The major muscle groups spanning the lumbar region were represented in the model by five bilateral pairs of single muscle equivalents. The L3 motion segment was assumed to resist compression, lateral shear, and antero-posterior shear, but to have no significant moment resistance. In solving these 13 unknown internal forces so as to produce the net reaction required for equilibrium, two objective functions were used. One led to the minimization of the compressive load on the L3 motion segment. When that criterion was used the maximum contraction intensity was limited to 100 N/cm^2 in any muscle. The second led to an approximate minimization of the largest muscle contraction intensity. The maximum allowed contraction intensity was then first set to 5 N/cm^2, and if no solution was found at that intensity, the allowed intensity was then increased by 5 N/cm^2 and another solution trial was made.

Data Evaluation

For each of the tasks, the mean values and standard deviations over the subjects of the intradisc pressure and the myoelectric signal activity at each electrode location were calculated. For the principal test of the validity of the assumptions made in the biomechanical analysis, the mean measured intradisc pressure was correlated with the mean predicted compression on L3 by a linear regression analysis. In addition, the validity of the predictions of the muscle contraction forces was tested by comparing the mean contraction forces calculated with the mean measured myoelectric signal amplitudes using linear correlation techniques.

Results

Intradisc Pressures

The disc pressure ranged from 270 kPa in relaxed upright standing to six times that value (1620 kPa) in the most strenuous task examined; standing, holding 80 N in partially-outstretched arms with the trunk flexed to 30° at the hips. Generally, when the trunk was upright and required to support only the weights of the superior body segments, the intradisc pressure was approximately 300 kPa whether sitting or standing. Holding additional weights or resisting horizontal forces resulted in pressures in the range of 400 to 800 kPa. Merely flexing the trunk by 30° and holding the arms partially outstretched, on the other hand, created an intradisc pressure in excess of 1000 kPa.

Predicted L3 Compressions and Muscle Contraction Forces

The mean predicted compression on the L3 disc ranged from 340 N to 2350 N. Complete tabulations of all data are available upon request.

Most of the activities studied tended to flex the trunk. As a result, the biomechanical analysis predicted that the anterior trunk muscles would seldom contract. Contractions of the anterior muscles were needed chiefly in the experiments in which lateral bending, extension, and twisting were resisted. The maximum predicted contraction forces were 130 N in each rectus abdominus muscle, 420 N in the left-side oblique abdominal muscles, and 120 N in the right-side oblique abdominal muscles. In contrast, the latissimus dorsi and erector spinae muscles contracted regularly. The lumbar parts of the latissmus were often predicted to contract with their maximum allowed intensity, corresponding to a force of about 240 N on each side. The erector spinae muscles were predicted to contract with forces up to 890 N on each side of the back.

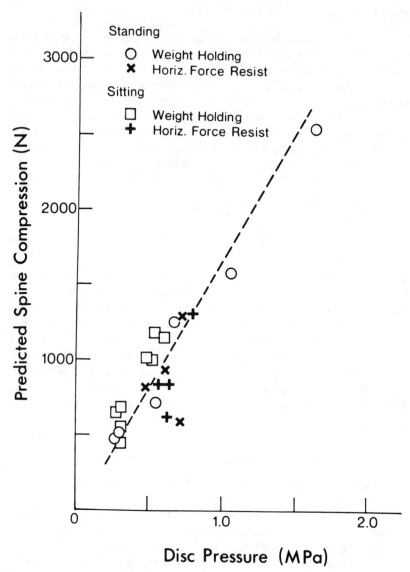

Figure 2 — Predicted spine compression and measured disc pressures.

Correlation between Intradisc Pressure and Predicted Spine Compression

The measured intradisc pressures and the predicted compressions at the L3 level were in good agreement (correlation coefficient 0.94; see Figure 2).

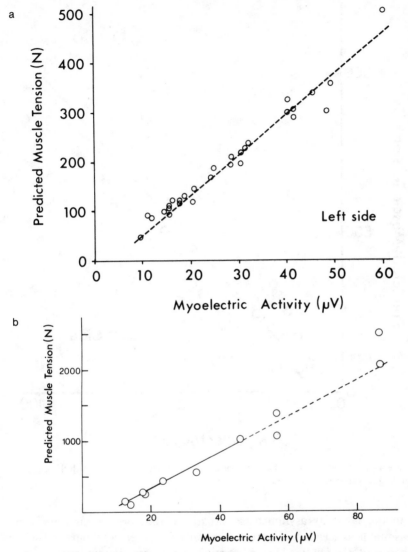

Figure 3—Predicted erector spinae tensions and myoelectric signal levels with the subjects, symmetrical postures. a = sitting, b = standing.

Correlations between Myoelectric Signal Levels and Predicted Contraction Forces

The measured myoelectric signal levels of the erector spinae muscles correlated well with the predictions. This was particularly so in symmetrical postures. Figures 3a and b show sample data from experiments with subjects sitting and standing.

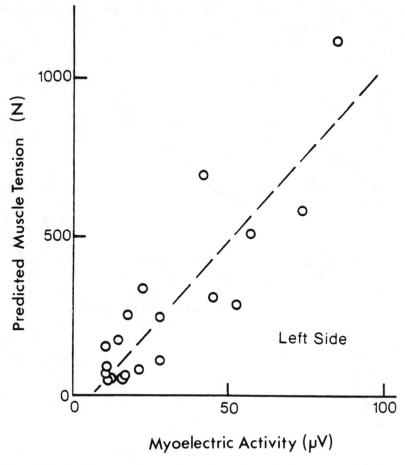

Figure 4—Predicted erector spinae tensions and myoelectric signal levels with the subjects standing, asymmetrical postures.

In the sagittally asymmetrical situations, the correlations were also reasonable as seen in in Figure 4. Generally for the erector spinae muscles a better fit was obtained using the minimum contraction intensity criterion. In the anterior trunk muscles, the predicted muscle forces did not correlate quite as well with the measured myoelectric signal levels (correlation coefficients 0.67 to 0.77).

Discussion

The studies show that estimates of trunk muscle tension can be made using relatively simple input data. The predictions of muscle tension are better in sagittally symmetrical situations than in sagittally asymmetrical ones. They are still quite reasonable in the asymmetrical ones, however. The study also shows that spine compression can be estimated with reasonable accuracy.

Biomechanical analysis may be a reliable, quick, safe, and inexpensive method long needed to assess spine loads at work and in research. The study also shows that myoelectric activity measurements can be used as direct indicators of muscle contraction force. Such measurements can thus be used in less well-controlled configurations or configurations difficult to quantify mechanically where tension prediction would be impractical.

It is obvious from the studies that to load the spine as little as possible the external loads should be kept low, and the work should be carried out close to the body.

Acknowledgments

Supported by grants from the Swedish Work Environment Fund, the Swedish Medical Research Council, and US Public Health Services Grants NS 15100 and OH 00514 and Development Award AM 00029.

References

ANDERSSON, G.B.J. 1981. Epidemiological aspects on low back pain in industry. Spine 6:53-60.

ANDERSSON, G.B.J., Örtengren, R., and Nachemson, A. 1977. Intradiscal pressure, intra-abdominal pressure and myoelectric back muscle activity related to posture and loading. Clin. Orthop. 129:156-164.

ANDERSSON, G.B.J., Örtengren, R., and Schultz, A. 1980. Analysis and measurements of the loads on the lumbar spine during work at a table. J. Biomech. 13:513-520.

DAVIS, P. 1981. The use of intraabdominal pressure in evaluating stresses on the lumbar spine. Spine 6:90-92.

NACHEMSON, A., and Elfström, G. 1970. Intravital dynamic pressure measurements in lumbar discs. Scand. J. Rehab. Med. Suppl. 1.

NACHEMSON, A.L. 1981. Disc pressure measurements. Spine 6:93-97.

ÖRTENGREN, R., and Andersson, G.B.J. 1977. Electromyographic studies of trunk muscles, with special reference to the lumbar spine. Spine 2:44-52.

SCHULTZ, A.B., Andersson, G.B.J., Örtengren, R., Björk, R., and Nordin, M. 1982a. Analysis and quantitative myoelectric measurements of loads on the lumbar spine when holding weights in standing postures. Spine 7:390-397.

SCHULTZ, A.B., Andersson, G.B.J., Örtengren, R., Haderspeck, K., and Nachemson, A.L. 1982b. Loads on the lumbar spine: validation of a biomechanical analysis by measurements of intradiscal pressures and myoelectric signals. J. Bone Joint Surg. 64(A):713-720.

SCHULTZ, A.B., Andersson, G.B.J., Haderspeck, K., Örtengren, R., Nordin, M., and Björk, R. 1982c. Analysis and measurements of lumbar trunk loads in tasks involving bends and twists. J. Biomech. 15:669-672.

An Ergonomic Investigation of Window-use Capacities of Physically Handicapped Adults

D. Dainty, S. McGill, M. Mason, C. Cotton, and W. Morrison

University of Ottawa, Ottawa, Canada

The opening and closing of windows in the home is a task most of us take for granted. To the handicapped individual, however, even the simplest of window hardware may present an impossible obstacle. The purpose of this study was to evaluate, in part, the abilities of handicapped people to open, close, lock, unlock, remove, and clean various types of windows. To achieve this end, an attempt was made to isolate and measure those abilities which were required to perform the activities mentioned. The relative abilities of the handicapped for these tasks were compared with those of non-handicapped persons so that estimations of operating capacities could be realistically determined.

Methodology

Subjects

There were two groups of subjects, non-handicapped and handicapped. The non-handicapped subjects were physically healthy males and females (N = 22) with a mean age of 27.8 years and a mean height and weight of 172.2 cm and 70.4 kg, respectively. The handicapped subjects were chosen for their various disabilities such as quadriplegia, hemiplegia, stroke, cerebral palsy, diabetic neuropathy, etc. This group (N = 22) had a mean age of 47.6 years and mean height and weight of 169.2 cm and 64.3 kg, respectively.

The differences in the ages of the two groups were statistically significant, however, the heights and weights were not. These last two factors were more important in this study, for the ability to reach and use the body weight effectively was essential in many cases to unlock, open and clean the windows, especially for the handicapped subjects.

Tests

Test Rig. The test rig consisted of four windows mounted in separate frames to form a rectangular box open on one side and at the top (see Figure 1). Window A, a crank-open (casement) type of window was mounted on one end of the rectangle. Window B, a horizontal slider, and window C, a guillotine type of window, were mounted on the closed side of the rectangular box. This last window opened in two ways, either from the side or from the top.

Strength Tests. To evaluate the grip strength of each of the handicapped subjects, two devices were used. One was a standard dynamometer for measuring grip strength, the other a sphygmomanometer preset to 40 mm Hg pressure. The two instruments were used, as a number of the subjects had difficulty in gripping the standard dynamometer. The sphygmomanometer is a tool used to measure grip strength by physiotherapists in clinical evaluations. Correlations were obtained to investigate the similarities between right and left grips and also between dynamometer and sphygmomanometer results.

Force Evaluations. The torque and force evaluations were determined using transducers attached directly to the windows. The torque required to turn the crank handle in window A was determined by having each subject attempt to turn a torque measuring device attached to the crank handle while the window was locked.

The force requirements were measured using a Lebow load cell with a capacity of 2,225 N (500 lb). Calibration curves were obtained by statically loading the load cell from 0 to 1335 N (300 lb).

The load cells were placed in series with the opening mechanism of each of windows B and C such that the window was slightly opened and a static force simulating the opening maneuver could be recorded. For window D the load cell was attached to the lever handle for locking the window and the force measured statically at the point where the handle began to unlock the window. The mean of three trials was used as the experimental value in all force and torque measures.

Opening Times. To determine this value each subject was timed by a standard manual stopwatch while unlocking and fully opening each window. Three trials were given and the mean of the three was used in analysis. Similar procedures were followed to obtain closing times, including latching the window. All subjects were given instruction and practice runs on opening, closing and locking of all windows before they were timed.

Figure 1 — Windows tested: A) Crank open (casement); B) Horizontal slider; C) Guillotine; D) Swing open to side and from top.

Cleaning. The final test was the ability to clean the windows, both inside and outside. To assist in quantifying this procedure, a grid of 5 cm squares was placed on each window. The subjects were then asked to reach and touch, in any manner possible, all of the surfaces of the windows. The total number of squares touched was recorded as a percentage of the total number of squares on the window.

Results and Discussion

Strength Tests

Pearson product moment correlations were calculated for pairs of data across hands for each of the devices used and across devices for both hands. The main influence causing the low correlations was the test of the right hand using the sphygmomanometer where the scores ranged from 29.3 mm Hg to 300 mm Hg. A large variance occurred in the right hand, not seen in the other test score results.

Although the hand grip dynamometer is a standard testing device used with physically normal people it may not be suitable for people with a handicap who generally have lower strength than normal individuals (Cotton and Dainty, 1978). Further investigations on large numbers of handicapped people using other strength-measuring devices need to be undertaken before valid and reliable testing protocols for grip strength in handicapped individuals can be established.

Force Evaluations

In order to compare the force characteristics across windows and between groups, a univariate analysis was performed (Carlson and Timm, 1980). The mean values for each of the windows for both groups were highly significant across all four windows ($p < .0001$) showing that the handicapped group had much more difficulty opening every window than did the non-handicapped subjects. The major difference occurred in window C (see Figure 2).

The results also showed that within each group there was a significant difference among windows (Rao's $F = 60.9$, $p < .0001$). The greatest force was developed on the last window (window D) where a lever was held and turned. If this type of locking mechanism was combined with a modified crank handle type of opener, we feel the results would be quite beneficial to the handicapped. The lever provides a large moment to be created for opening with minimal force, whereas the crank opener provides for minimal body movement and consequently minimal muscular activity as compared to the push, pull, or lift type of windows where considerable body movement is required.

Opening Times

A multivariate analysis (Timm and Carlson, 1973) was performed on the timing data to discover what differences existed and where they existed. The mean times for opening and closing windows are shown in Figure 3. The windows listed as D_1 and D_2 are for the side opening and top opening functions of window D. Both group differences (Rao's $F = 14.5$, $p <$

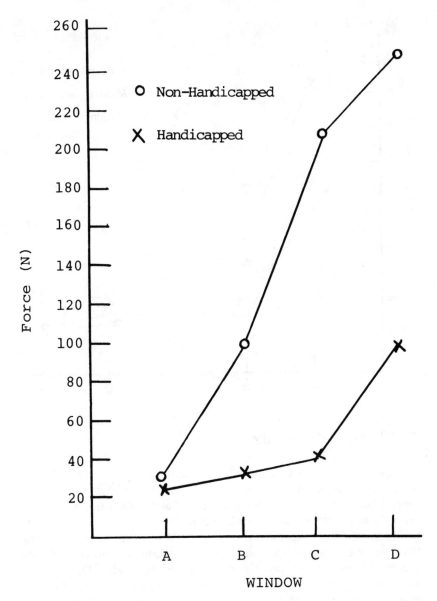

Figure 2 — Mean values of force generation.

.0001) and window differences (Rao's F = 34.9, p < .0001) were statistically significant.

The first window (window A) was the slowest opening and closing for both groups as well as being significantly slower for the handicapped group. This slowness may present problems in an emergency. The slowest of the three manual (as opposed to mechanical, i.e., crank opening)

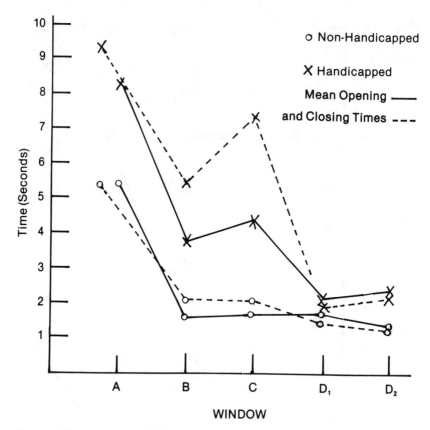

Figure 3—Mean opening and closing times.

windows was window C, especially for the handicapped group. One of the problems associated with the slowness was the inability of the handicapped subjects, especially those in wheelchairs, to position themselves properly to lift the window to full opening.

Finally, the data suggested that window D in either of its configurations was the fastest to open and close. Depending on how critical is the factor of speed of opening, the choice would be window D. However, subjectively evaluating the windows, most members of the handicapped group preferred the crank-open type window found in window A.

Cleaning

No statistical evaluation was made of the groups' abilities to clean the windows. Except for a few isolated cases most subjects were able to clean all of windows A and D with no difficulty. Window B presented difficulty only to those who were unable to open it.

Window C presented a different problem, common to both groups. As the bottom portion was lifted, the upper portion descended so that the windows were aligned with one another when half-open. Approximately 90% of the window was cleaned by members of both groups.

General Discussion

The accessibility and ease of manipulation of all windows was observed and subjective evaluations recorded. Although window A performed the poorest on the objective tests, it was preferred by the handicapped subjects because of the need for limited movement by the operator. Perhaps modifications in the handles for turning the winding mechanism and unlocking would make this a better window than in its present configuration.

Both windows B and C required the operator to slide a portion of the window, thus requiring gross body movements to complete the task. Add to the awkwardness of opening, the difficulty in cleaning, either in terms of having to remove the window (window B) or inaccessibility to a portion of it (window C) and the suitability of either of these windows in a handicapped person's environment becomes questionable.

Window D was the superior window in this testing protocol. An advantage to window D in terms of maneuverability and cleaning ease was the fact that it opened inwardly.

As a final note to the general discussion, two major problems were identified with the operation of the windows by the handicapped subjects. First, the angle of the force applications were hindered a great deal especially by those using walking aids or restricted to a wheelchair. Second, the size and mechanical operation of some of the hardware were large obstacles because of the lack of sensation and/or fine motor control in the hands of some subjects. Consequently, several considerations need to be made concerning the use of various windows and hardware in homes occupied by handicapped people.

Acknowledgment

This research was supported by a grant from the National Research Council of Canada.

References

CARLSON, J.E., and Timm, N.H. 1980. Full Rank Multivariate Linear Model. Presented at the American Education Research Association Meetings, Boston, April, 1980.

COTTON, C.E., and Dainty, D.A. 1978. Analysis of the Physical Exertion of Door Use. Technical Report Division of Building Research, N.R.C.

TIMM, N.H., and Carlson, J.E. 1973. Full Rank, Multivariate Linear Model Program. Presented at American Statistical Association Meetings, New York, December.

Electromyography, Biomechanics, or Inquiries in Ergonomic Investigations?

Bengt Jonsson
National Board of Occupational Safety and Health,
Umeå, Sweden

Disorders of the locomotor system such as tendinitis, peritendinitis, epicondylitis, low-back pain and muscular tenderness are among the most common vocational disorders. Although it is commonly assumed that frequent incidents of muscular over-loading or of muscular fatigue may be an etiological factor in the development of some musculoskeletal disorders, our knowledge about the pathogenesis and prevention of such vocational disorders is still very limited. As a consequence, ergonomic investigations designed to ascertain the relation between the physical load during work and disorders of the locomotor system ought to be given high priority in occupational medicine.

This article deals with some of the methodological problems which are often present in ergonomic investigations. It should be remembered that the majority of workers have a variety of different work tasks, thus necessitating long-term studies in the actual work situation. It should also be remembered that the ergonomist usually is a multi-disciplinary specialist, often with a background in behavioral sciences, and with little or no training in advanced physiological or technical methods of research.

Epidemiology of Vocational Disorders of the Locomotor System

In the epidemiological evaluation of the etiology of vocational disorders one has to analyze the relation between *exposure* and *effect*. In the evaluation of disorders of the locomotor system which are assumed to be caused by over-loading during work, the physical strain during work is the exposure and the disorders are the effect. Unfortunately, every person is exposed to some degree of physical strain and the effect of this physical strain may vary from no discomfort to a severe disorder. Therefore, the epidemiologist may have to evaluate the relation between *dose* and *response*. The degree of physical strain on individual parts of the locomotor system is the dose and the disorder or the severity of the disorder is the response.

The ergonomist who wants to measure the degree of physical strain on individual parts of the locomotor system may use among other methods subjective estimation of the work load, electromyography, or biomechanical evaluation. The effect of physical strain on the locomotor system may be studied either by inquiries or medical examination.

The methods used by the ergonomist in order to evaluate or measure the exposure of the subject to physical load during work ought to be such that: 1) objective quantitative measurements of high reliability and high validity are obtained, 2) they are simple enough to be used without special training or expensive equipment, 3) the analysis of the results is not time consuming and does not require the use of highly sophisticated equipment, and 4) the methods allow long-lasting recordings with the subjects free to move around at their places of work, and do not require continuous visual contact between the subjects and the recorder.

Inquiries

An inquiry regarding the occurrence of musculoskeletal complaints during a specific period of time, e.g., during the last year, may result in a general idea of the incidence of such complaints in different parts of the locomotor system. Such a picture will be specific for the group of subjects or for the occupation under study. The incidence of discomfort (during the last year) and the prevalence of discomfort usually are related to each other. The incidence of sick-leave due to disorders in different parts of the locomotor system usually coincide with the incidence and the prevalence of discomfort in the corresponding parts of the locomotor system.

Inquiries may also be used to outline what the subjects believe to be the background of their complaints. Such an analysis is far from objective and should not be used as the only method to study the exposure in an epidemiologic investigation. On the other hand, such an inquiry may be an acceptable instrument to study how the workers experience their work environment.

Ergonomic Evaluation

The simplest way to record the exposure to physical strain on different

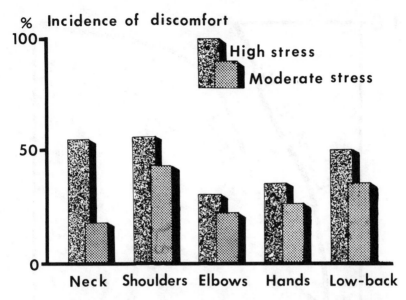

% Incidence of discomfort

High stress

Moderate stress

Neck Shoulders Elbows Hands Low-back

Figure 1—Incidence of discomfort (during the last year) in different parts of the locomotor system in groups where the physical strain on the corresponding parts of the locomotor system have been evaluated as high or moderate, respectively.

parts of the locomotor system is to make a subjective ergonomic evaluation of the physical load during work. Such an evaluation may be based on a graduation of the physical load into light, moderate, or heavy load. This evaluation is fast and easy to perform. On the other hand, it is neither an exact nor an objective method. Different investigators may get quite different results. Despite the lack of objectivity, there is a fairly good correlation between this subjectively determined load on individual parts of the locomotor system and the incidence of pain or discomfort in the corresponding body regions (see Figure 1) indicating that there may be a correlation between physical strain during work and disorders of the locomotor system.

Electromyography

Electromyography offers an excellent way to obtain an objective recording of the muscular strain in the actual work situation (Jonsson, 1978). The use of telemetry, furthermore, offers an opportunity to record the muscular strain in a subject who is moving freely over a fairly large area. Unfortunately, the myoelectric recordings have to be restricted to a few muscles which means that the results of the myoelectric study are highly dependent on the choice of adequate muscles.

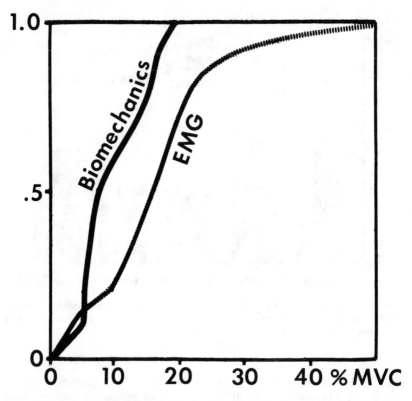

Figure 2 – Amplitude probability distribution function of the myoelectric signal from the right trapezius muscle (EMG) and the cumulative distribution of the biomechanically evaluated torque on the right shoulder (Biomechanics). Both the myoelectric results and the biomechanical evaluation have been expressed in a percentage of the maximal voluntary force of contraction in the right shoulder (% MVC). The analysis was performed on a work task in an electronics plant.

The myoelectric signal may be evaluated both with respect to quantitative and qualitative determinations of the relative load on individual muscles during work and with respect to the incidence of muscular fatigue during work. Determination of the amplitude probability distribution function of the myoelectric signal (Jonsson, 1978; Hagberg, 1979) makes it possible to obtain a myoelectric profile of the muscular work performed (see Figure 2). The amplitude probability distribution curve will indicate the level of the static component of the muscular work, the mean level of muscular load and the level of the peak loads. Analysis of the changes in the myoelectric signal which are characteristic of muscular fatigue may reveal incidents of muscular fatigue during work (Lindström et al., 1970; Hagberg, 1981).

Vocational electromyography has the advantage of allowing objective quantitative as well as qualitative measurements of high reliability and validity. Furthermore, electromyography will allow long-lasting record-ings with the subjects free to move around at their places of work, and there is no need for continuous visual contact with the subjects. On the other hand, electromyography requires a highly trained investigator and expensive equipment, and the analysis of the results is usually fairly time-consuming.

Biomechanical Evaluation

Another objective way to record the physical strain on individual parts of the locomotor system is by biomechanical evaluation. Several different methods are available but most of them require expensive equipment or are impossible to use with subjects who are moving around over a large area at their places of work. One method which might be useful for the ergonomist is the recording of the movements of different body segments during work followed by a calculation of the load on individual parts of the locomotor system, e.g., calculation of the torques generated about individual joints or the force exerted by individual muscle groups. The major problem with this type of biomechanical analysis is the kinematic recording. Analysis of the movements of the body segments from kine-matographic recordings or from video recordings are extremely time-consuming and require continuous visual contact with the subject. The use of light emitting diodes such as those used in the Selspot system re-quires both expensive equipment and continuous visual contact with the subject. Kinematic recordings by means of goniometers are at present usually impossible to perform because there are no small, simple goniometers allowing multijoint analysis with many degrees of freedom in each joint. It should be remembered that the subjects in most cases have to wear their normal protective clothes.

In those situations where a biomechanical analysis is possible to per-form, it may also be possible to calculate a curve of the cumulative distribution of load levels (see Figure 2), a curve which might closely resemble that of the corresponding amplitude probability distribution curve of the myoelectric signal.

Those biomechanical methods which are currently available for an ergonomist usually allow objective quantitative measurements of high reliability and validity. On the other hand, these methods are usually not simple to use and they will often require expensive equipment. Further-more, many of these methods cannot be used in situations where the sub-ject must be free to move around or where visual contact with the subject cannot be maintained continuously.

Concluding Remarks

Presently, there does not seem to exist any single method which combines exactness and simplicity, while providing valuable information in the study of people at their places of work. The methods available for those ergonomists who want to study the physical strain on individual parts of the body during work will in most cases still be a subjective ergonomic evaluation in combination with inquiries. Electromyography and biomechanical evaluation should mainly be reserved for the specialist. Among these two latter methods, electromyography seems to be the one which at present is best suited for vocational studies.

References

HAGBERG, M. 1979. The amplitude distribution of surface EMG in static and intermittent static muscular performance. Eur. J. Appl. Physiol. 40:265-272.

HAGBERG, M. 1981. Muscular endurance and surface EMG in isometric and dynamic exercise. J. Appl. Physiol.: Resp., Environ., Exercise Physiol. 51:1-7.

JONSSON, B. 1978. Kinesiology—with special reference to electromyographic kinesiology. Contemp. Clin. Neurophysiol. 34:417-428.

LINDSTRÖM, L., Magnusson, R., and Petersén, I. 1970. Muscular fatigue and action potential conduction velocity changes studied with frequency analysis of EMG signals. Electromyogr. 4:341-356.

Ergonomic and Medical Factors in Shoulder/Arm Pain among Cabin Attendants as a Basis for Job Redesign

Jörgen Winkel, Berit Ekblom, and Bengt Tillberg
University of Luleå, Luleå, Sweden

Pain in the musculo-skeletal system is a frequent problem in many occupations (Andersson, 1971; Maeda, 1977). It might be caused by physical strain in work and/or during leisure time. Concealed, specific medical factors might also contribute to or cause the disorders. Therefore, all these factors have to be under control in order to assess the significance of work stress.

Cabin attendants (C/A) and their work with an airline (SAS-Sweden) have been studied. In a questionnaire answered by 799 C/A, 21.5% reported having suffered from pain in the neck and 12.8% in the right wrist during the last 12 months (Orring and Östberg, 1980). There were 15 persons, according to The Work Injuries Insurance Law, who reported to the Swedish National Social Insurance Board. They expressed doubts about whether the complaints should be accepted as work injuries. The patients, however, associated their disorders with serving coffee in the cabin. Thus, an ergonomic evaluation of this task and a medical examination of patients were carried out to assess the significance of work stress as a basis for job redesign.

Material and Methods

Ergonomic Evaluation

Ordinary work in the cabin was filmed continuously (1 f/sec) during nine runs representing eight working hours for C/A. From these films, time and motion studies were carried out.

In the laboratory, six healthy, female C/A aged 24 to 30 years were studied. The subjects were selected for ordinary body size and weight and

Figure 1 — a = The ordinary pot; b = The experimental pot with "better" ergonomic design of the handle.

normal muscular strength in the investigated muscle groups. Each subject served coffee in a mock-up for 15 min. The task was randomly repeated three times using: 1) the ordinary pot (see Figure 1a), 2) a pot with a 'better' ergonomic design of the handle (see Figure 1b), and 3) another work method excluding the pot, distributing the cups one-by-one from a service cart at which they were filled from a tap by the subjects. The pot/cup was held in the right hand and the subjects maintained a constant work rate equal to an average assessed from the above mentioned work studies in the cabin.

These experiments were followed by a test in which five of the subjects held a full, ordinary pot until exhaustion. The arm-position was the same as during transport of the pot in ordinary work. Between the experiments, 15 min of rest were given.

The circulatory strain was assessed from the heart rate monitored continuously during the experiments. Every 5th min the overall level of rated perceived exertion (RPE) was assessed by the subjects on a numerical scale (Borg, 1970). On a diagrammatic representation of the human body, divided into 27 parts, the subjects indicated discomfort areas before and after the experiments according to Corlett and Bishop (1976). The degree of discomfort in the selected body parts was indicated on a seven-point scale.

The muscular load was evaluated by vocational electromyography (see Figure 2). After a pilot study of 10 different muscles in the low back, the shoulder girdle, and the arm, four of these were selected for the main study — the descending part of the trapezius, the infraspinatus, the biceps, and the extensor carpi radialis longus muscles, all on the right side. The myoelectric activity was recorded on tape by bipolar surface electrodes. Test contractions gave the EMG-force relationship for the investigated muscles by a simultaneous recording of myoelectric activity and force during a slowly increased submaximal contraction for shoulder elevation, outward rotation of shoulder, elbow flexion and radial flexion

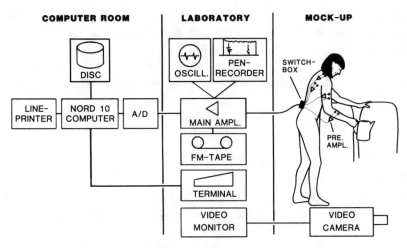

Figure 2—Set up for studying muscular load and fatigue.

of wrist. The muscular load during work was evaluated by estimating the amplitude probability distribution function (APDF), thus exposing the static, the median and the maximum contraction levels for the time studied (Hagberg, 1979). The myoelectric and force signals were determined as root-mean square (RMS) values by computer-aided analysis from the tape recorder (Ericson and Hagberg, 1978). By power function regression analysis of the EMG levels versus force levels during the test contractions, the EMG-force relationship was established. Thus the APDF of the EMG-signals during work could be transformed to an APDF of contraction levels for the different muscles. The distribution of load levels was evaluated for the first five min of each experiment to avoid influence of fatigue on myoelectric signal amplitudes.

Development of muscular fatigue during the exhaustion test was evaluated by regression analysis of the EMG-amplitude (RMS) and mean power frequency (MPF) versus time. Differences in test parameters between the three experimental tasks and changes in RMS and MPF during the exhaustion test were tested for significance by the student's t-test ($P < 0.05$).

Medical Examination

Ten of the 15 patients were studied—one man and nine women (mean age: 34 years, mean employment time: 8 years). The remaining were stationed abroad or had finished their employment with the company. The medical history included activities during leisure time. A general clinical examination was conducted as well as specific orthopedic assessments of back, neck and extremities, and anthropometric data including grip strength. Laboratory examination included hemoglobin, sedimentation rate, glutamic-oxaloacetic (GOT) and glutamic-puryvic transaminases (GPT), rheumatoid factor (Waaler-Rose), antinuclear factor (ANF) and creatine kinase. These laboratory tests should be able to screen any medical disorder giving complaints in the musculoskeletal apparatus.

Results

Ergonomic Evaluation

Analysis of the films indicated that, on the average, C/A serve coffee for about 15 minutes during short runs. A working day may comprise four to six runs. On longer runs, the task is repeated several times. Furthermore, the dominant hand was loaded during several other tasks, e.g., serving juice, and distributing meal trays.

The laboratory study showed a significantly higher overall rating of perceived exertion during work with the pots compared to work with the alternative method. No significant differences in heart rate could be shown between any of the experimental tasks (mean heart rate ≃ 90 beats/min). After work with the pots an increase of discomfort in one or several body parts was reported by all the subjects. The complaints were all limited to the right arm and shoulder and the perceptions corresponded to areas in which the muscular load was high. No such correlation was shown for work with the alternative method.

High static load level was found for all the investigated muscles especially during work with the pots (see Figure 3). Serving coffee without using a pot, however, reduced the static load level significantly for the infraspinatus and the extensor carpi radialis muscles. Furthermore, both the median and the maximum load levels were significantly reduced for all four muscles. An improved ergonomic design of the pot handle did not significantly reduce any of the load levels for the investigated muscles.

During the exhaustion test all the subjects showed significant electromyographic indications of muscular fatigue within about one min in the infraspinatus and the extensor carpi radialis muscles—earlier in the latter, in agreement with a high initial relative load (31.5% MVC).

Medical Examination

The medical history revealed a trauma in one patient (diagnosed as lesion of discus triangularis in the dominant hand). Exceptional sport activities stressing the shoulder girdle and arm were carried out by two patients. The results of the clinical examinations are shown in Table I. Many pa-

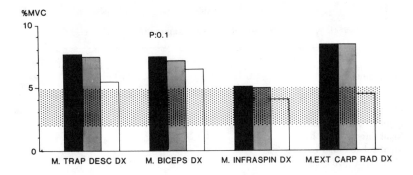

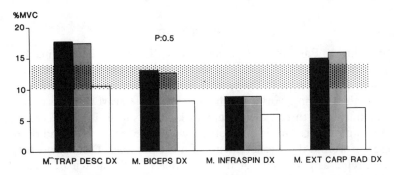

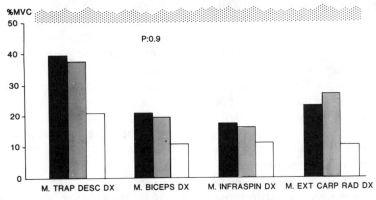

Figure 3 — Muscular load levels during the three experimental tasks. MVC = maximal voluntary force of contraction; black staples = ordinary pot; hatched staples = experimental pot; blank staples = alternative work method; P: 0.1 = static load level; P: 0.5 = median load level; P: 0.9 = maximum load level; dotted areas = suggested threshold limit values according to Jonsson (1978).

tients had more than one complaint, and therefore the total number exceeds 10. The laboratory tests showed nothing indicating specific medical disorders.

Table 1

Results of Clinical Examinations

Complaints	No	Dominant hand	
		Right	Bilateral
Right trapezius desc.	4	3	1
Right tendinitis of biceps	4	2	2
Right lateral epicondylitis	4	3	1
Right subacromial bursitis	4	2	2
Right wrist (Mb Quervain)	2	2	0

Discussion

According to Pandolf (1978), an overall RPE rating is determined by two prominent factors: a local muscular exertion and a central cardio-pulmonary exertion. In the present study the overall level of RPE rating was highest during work with the pots while the heart rate was about the same during all three experimental tasks, thus suggesting local strain as a dominant factor. This is in accordance with the localization and rating of discomfort and the measurements of muscular load. Discomfort is often interpreted as a warning of the inadequacy of the match between man and his work (van Wely, 1970). Hence, development of discomfort in the right arm and shoulder during the experiments with the ordinary pot may be an indication of a potential risk factor in the job. This statement is in agreement with the results of the electromyographic portion of this study. Jonsson (1978) has suggested threshold limit values for 'long-lasting' work. These are illustrated in Figure 3 and propose an overload first of all on the descending part of the trapezius, the biceps, and the extensor carpi radialis muscles. However, the investigated task was performed for relatively short periods of time. Therefore, somewhat higher load levels might be acceptable if the dominant hand was not loaded by other tasks during the working time.

Apart from these speculations the patients suffered from pain in tissues mainly related to those three muscles showing the highest load levels (see Table 1). The fast development of EMG-indications of muscular fatigue during the exhaustion test suggests the same changes even during transport of a full pot from the galley to the cabin in the aircraft.

The company and the safety representatives have suggested a better handle for the pot in order to solve the problem. None of the results of this study suggest that this would bring about an ergonomic improvement. The measurements of perceived exertion, discomfort, and muscular load suggest a change to a work method that reduces the load in the hand. This can be done without reducing performance.

Conclusion

The study suggests that local physical strain during cabin work was a predominant cause of the reported disorders. The problem would primarily be solved by a change of work methods.

References

ANDERSON, J.A.D. 1971. Rheumatism in industry: A review. Br. J. Ind. Med. 28:103-121.

BORG, G. 1970. Perceived exertion as an indicator of somatic stress. Scand. J. Rehab. Med. 2-3:92-98.

CORLETT, E.N., and Bishop, R.P. 1976. A technique for assessing postural discomfort. Ergonomics 19(2):175-182.

ERICSON, B.-E., and Hagberg, M. 1978. EMG signal level versus external force: a methodological study on computer aided analysis. In: E. Asmussen and K. Jørgensen (eds.), Biomechanics VI-A, pp. 251-255. University Park Press, Baltimore, MD.

HAGBERG, M. 1979. The amplitude distribution of surface EMG in static and intermittent static muscular performance. Eur. J. Appl. Physiol. 40:265-272.

JONSSON, B. 1978. Kinesiology. With special reference to electromyographic kinesiology. In: W.A. Cobb and H. Van Duijn (eds.), Contemporary Clinical Neurophysiology, 34:417-428. Elsevier Scientific Publishing Company, Amsterdam.

MAEDA, K. 1977. Occupational cervicobrachial disorders and its causative factors. J. Human Ergol. 6:193-202.

ORRING, R.M., and Östberg, O. 1980. Cabin attendants' working environment. Technical Report. 1980:74T. Department of Human Work Sciences, University of Luleå, Sweden.

PANDOLF, K.B. 1978. Influence of local and central factors in dominating rated perceived exertion during physical work. Perceptual and Motor Skills 46:683-698.

VAN WELY, P. 1970. Design and disease. Appl. Ergonomics 1(5):262-269.

Effects of Two Different Load Carrying Systems on Ground Reaction Forces During Walking

H. Kinoshita and B.T. Bates
University of Oregon, Eugene, Oregon, U.S.A.

Throughout recorded history people have devised ways for carrying materials from place to place. Even in modern, technologically advanced societies, human load-bearing is still an essential part of some occupational tasks and a number of recreational activities. Although a number of physicians associated with the military and industry have reported that habitual carrying of excessive loads can lead to injuries of the lower extremities (Renborn, 1953; Noro, 1967; Stanitski et al., 1978), the mechanics of lower extremity function during load carrying while walking have not been well documented. It is believed that the study of ground reaction forces (GRF) during load carrying can provide relevant information about the mechanics of gait under load-carrying conditions, and can assist in the understanding, diagnosis, and prevention of lower extremity injuries associated with carrying tasks. One purpose of this study, therefore, was to examine the effects of carrying loads of varying magnitude on the characteristics of the three components of GRF.

Over the years, military and industrial personnel and recreational enthusiasts have attempted to design optimal load carrying systems. Recent research findings supported the idea that a front-and-back distributed load-carrying system reduced forward lean of the trunk and placed fewer physiological demands upon the carrier than conventional "backpack" carrying systems (Datta and Ramanathan, 1971; Ramanathan et al., 1972). Consequently, the second purpose of the study was to compare the effects of both a distributed load-carrying system and a conventional "backpack" carrying system on the GRF.

Procedure

Five healthy males ranging in age from 29 to 42 years served as subjects

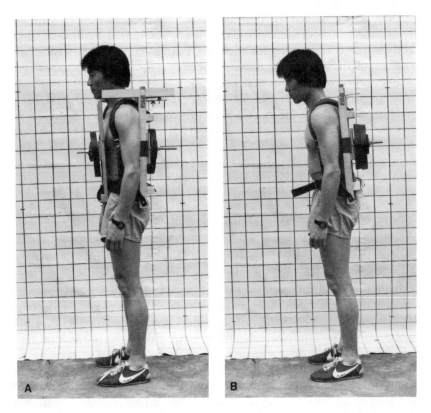

Figure 1—Subject standing while wearing both versions of the pack frame. A) back load only with 20% body weight; B) double-pack system with 20% body weight.

for the study. The mean height and weight of the subjects were 170.5 ± 4.44 cm and 66.98 ± 5.81 kg, respectively. The experimental setup consisted of a Kistler force platform (Model 9281 A) set into a walkway, interfaced via a TransEra A/D converter to a Tektronix 4051 Graphic Calculator. Data were sampled at 417 Hz. A photoelectric timing system was used to control walking speed over a 4 m interval.

A pack frame system consisting of a rigid wooden frame whose dimensions corresponded to those of a commercially available pack (Kelty, Sonora) was developed for the study. The frame was constructed in such a way that a front pack frame could be attached to the backpack frame to create a double-pack carrying system. A weight attaching mechanism for both parts of the frame was fixed in the geometrical center of the frames. The frame was adjustable in the anteroposterior plane so that the front section could be positioned the same distance from the chest of each subject. Figure 1b shows a subject wearing the pack for the back-only, and Figure 1a the front-and-back load-carrying conditions.

Data were obtained for 10 trials for each of the following conditions: 1) normal gait without any external load, 2) external loading of 20% body weight carried using the backpack system, 3) 20% body weight carried using the double-pack system with an equal distribution of the weight between the front and back, 4) external loading of 40% body weight carried using the backpack system only, and 5) 40% of body weight carried using the double-pack system. Acceptable trials required that the subject contact the force platform in a normal stride pattern with a walking speed between 1.17 and 1.33 ms^{-1} (4.5 ± 0.3 km/hr). The average speed was chosen based upon optimal physiological speed for men carrying loads as discussed in the literature (McDonald, 1961).

Results and Discussion

The means and standard deviations for each 10-trial subject-condition were computed. The data presented are the mean values for all subjects by condition and the mean of the subject standard deviations for each condition. Figure 2 contains a representative set of force curves identifying the variables. Force and impulse values were normalized by dividing by body mass and time parameters were expressed as % of the total support time.

Selected variables describing the vertical force components are given in Table 1. The vertical force curves were characteristically bimodal and similar for all five conditions. During the no-load carrying condition (C1) relative time to the two maximum peaks (Tz1 and Tz3) and the minimum point (Tz2) occurred at approximately 20, 75, and 45%, respectively, from heel contact. As the load was increased, an apparent increase in Tz2 was observed, while Tz1 and Tz3 remained nearly constant. Significant increases in the three force variables (Fz1, Fz2, and Fz3) and in the total vertical impulse (Iz) were found to be approximately proportional to the increase in the system weight.

Similar trends were found in the selected variables describing the average anteroposterior force curves (see Table 2). The increases in the maximum braking and propulsive forces (Fy1 and Fy2) and the impulses (Iy1 and Iy2) were all nearly proportional to the increases in system weight. The total impulse values were between 12.7 and 13.5% of the corresponding vertical components.

The mediolateral data are given in Table 3. The magnitude of the impulses for all conditions was less than 3% of the corresponding vertical impulses. The first impact force (x1) was in the lateral direction as the foot moved medially with an average maximum force of 0.51 N/kg body mass under the no-load carrying condition (C1). Increases in both the force and impulse values (Fx1, Fx2, Ix1, and Ix) were concomitant with the increases in load, but the increases were not as systematically propor-

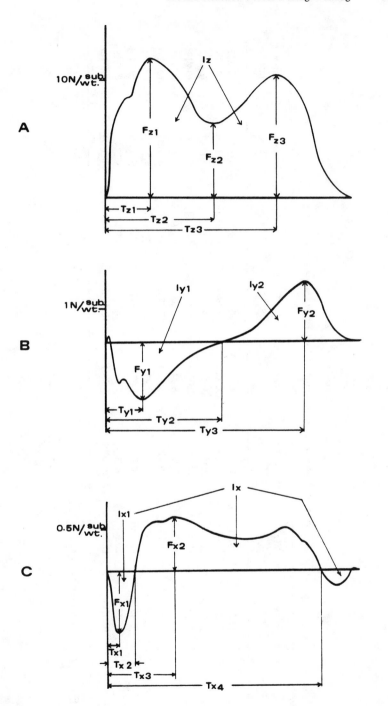

Figure 2—Graphic representation of ground reaction force variables. A) vertical, B) anteroposterior, C) mediolateral.

Table 1

Selected Variables Describing the Vertical Force Components[a]

Condition	Fz1 N/kg	Tz1 %	Fz2 N/kg	Tz2 %	Fz3 N/kg	Tz3 %	Iz N.S/kg	Support Time Sec.
1	11.74 (0.318)	19.71 (1.470)	7.22 (0.264)	44.70 (2.000)	10.78 (0.286)	75.22 (1.306)	6.003 (0.147)	0.7837 (0.013)
2	14.27 (0.440)	20.42 (1.200)	8.13 (0.342)	45.58 (2.020)	12.94 (0.362)	74.98 (0.820)	7.172 (0.154)	0.7832 (0.014)
3	14.08 (0.378)	19.77 (1.244)	8.30 (0.210)	46.14 (2.364)	12.87 (0.252)	74.75 (1.366)	7.097 (0.161)	0.7813 (0.014)
4	16.33 (0.452)	20.89 (1.340)	9.14 (0.278)	47.97 (1.674)	14.51 (0.411)	75.55 (1.274)	8.221 (0.177)	0.7854 (0.012)
5	16.53 (0.470)	20.66 (1.168)	9.51 (0.260)	47.05 (1.952)	14.54 (0.412)	75.30 (0.948)	8.274 (0.171)	0.7898 (0.012)

[a]See Figure 2 for symbol identification.

Table 2

Selected Variables Describing the Anteroposterior Plane Force Components[a]

Condition	Fy1 N/kg	Ty1 %	Iy1 N.S/kg	Ty2 %	Fy2 N/kg	Ty3 %	Iy2 N.S/kg
1	−2.24 (0.101)	15.20 (1.260)	0.39 (0.033)	49.97 (2.904)	2.25 (0.104)	83.76 (0.936)	0.37 (0.018)
2	−2.84 (0.222)	15.55 (1.086)	0.48 (0.034)	50.67 (2.358)	2.76 (0.148)	83.98 (0.700)	0.46 (0.023)
3	−2.72 (0.192)	14.84 (1.164)	0.46 (0.031)	50.76 (2.070)	2.75 (0.136)	83.78 (0.900)	0.45 (0.023)
4	−3.24 (0.248)	15.06 (1.370)	0.55 (0.050)	51.33 (2.388)	3.24 (0.156)	83.79 (1.044)	0.55 (0.028)
5	−3.15 (0.238)	15.46 (1.442)	0.56 (0.037)	50.78 (2.092)	3.24 (0.140)	84.05 (0.934)	0.55 (0.026)

[a]See Figure 2 for symbol identification.

tional to the increases in system weight as were the corresponding vertical and anteroposterior components. The temporal variables associated with the first lateral force (Tx1 and Tx2) were relatively constant for all load carrying conditions as compared to those describing the other two temporal variables (Tx3 and Tx4). The larger mean standard deviations associated with Tx3 and Tx4 for all five conditions showed greater instability of these temporal variables for all conditions.

The second purpose of the study was to compare the two types of carrying systems through an examination of the GRF variables. The statistical analysis using a two-factor repeated measures ANOVA design resulted in significant interactions for more than half of the force and impulse variables as well as half of the time variables. Consequently, no simple conclusions could be drawn regarding the effects of the two carrying systems on the GRF. Examination of the mean group data, however, revealed that the differences between the two carrying methods were rather small in an absolute sense and, therefore, probably have little bearing on lower extremity injuries.

A kinematic analysis of the film data is in process to investigate other potential differences that might exist. It must also be stressed that subjects in this study were required to walk for only short distances. Since physiological studies have shown that different carrying systems have produced long-term effects upon subjects, a subsequent investigation should be undertaken to examine the effects of fatigue induced by the two carrying systems on the GRF.

Table 3

Selected Variables Describing the Lateral Plane Force Components[a]

Condition	Tx1 %	Fx1 N/kg	Ix1 N.S/kg	Tx2 %	Fx2 N/kg	Tx3 %	Tx4 %	Ix N.S/kg
1	4.98	-0.506	0.023	9.77	0.468	19.68	86.98	0.170
	(0.598)	(0.082)	(0.004)	(0.692)	(0.074)	(3.702)	(2.476)	(0.023)
2	5.05	-0.608	0.029	9.61	0.504	22.73	86.43	0.196
	(0.678)	(0.114)	(0.007)	(2.484)	(0.115)	(5.120)	(2.460)	(0.032)
3	5.09	-0.646	0.031	9.77	0.483	20.88	86.84	0.184
	(0.616)	(0.096)	(0.006)	(1.076)	(0.127)	(6.604)	(3.192)	(0.031)
4	4.94	-0.650	0.030	9.33	0.563	22.60	86.54	0.204
	(0.560)	(0.121)	(0.006)	(1.016)	(0.135)	(5.056)	(3.410)	(0.042)
5	4.95	-0.673	0.034	9.95	0.533	21.45	85.34	0.212
	(0.656)	(0.103)	(0.006)	(0.075)	(0.111)	(3.426)	(3.938)	(0.038)

[a]See Figure 2 for symbol identification.

References

DATTA, S.R., and Ramanathan, N.L. 1971. Ergonomic comparison of seven modes of carrying loads on the horizontal plane. Ergonomics 14(2):269-278.

MCDONALD, I. 1961. Statistical studies of recorded energy expenditure of man. Nutrition Abstracts and Reviews 31(3):739-762.

NORO, L. 1967. Medical aspects of weight carrying. Industrial Medicine and Surgery 36(3):192-195.

RAMANATHAN, N.L., Datta, S.R., and Gupta, M.N. 1972. Biomechanics of various modes of load transport on level ground. Indian Journal of Medical Research 60:1702-1710.

RENBOURN, E.T. 1954. The knapsack and pack: An historical and physiological survey with particular reference to the British soldier. Journal of the Royal Army Medical Corps 100(1):77-88.

SAVOCA, C.J. 1971. Stress fractures: A classification of the earliest radiographic signs. Radiology 100:519-524.

A Force Platform for the Evaluation of Human Locomotion and Posture

P. Bourassa and R. Therrien

Université de Sherbrooke, Sherbrooke, Québec, Canada

The dynamic study of human locomotion and posture has benefited from recent developments in force platform design and construction (Matake, 1976) as well as from computerized data acquisition and treatment systems (Cavanagh and Lafortune, 1980). The force platform design reported in the present study takes advantage of a very high semiconductor gauge factor to increase the stiffness of the suspension elements and obtain a much higher natural frequency than would be possible with standard strain gauges.

A force platform was designed and instrumented in this laboratory in order to measure the force and moment vector components of Equation (1) generated by the foot-to-ground impact in the course of physical activity. The main objective was to obtain a low cost, sensitive, yet reliable and stiff platform.

Method

The structure of the developed platform is shown in Figure 1. It consists of a welded frame of square tubing and four stiff steel ring suspension elements in the horizontal plane bolted at each corner. Surfaces were ground to high precision in order to minimize the internal stresses resulting from the fixation of the top plate. The 16 semiconductor strain gauges used had a gain factor of 50 and a linearity behavior around 0.2% and were cemented, cured, and wired into eight Wheatstone bridges. Half of the gauges were in the horizontal plane and responded to the vertical force and to the moment vectors in the horizontal plane producing output voltages V_1 to V_4 in Equation (1); the other half produced output voltages V_5 to V_8, and were cemented on the side walls of the rings at a

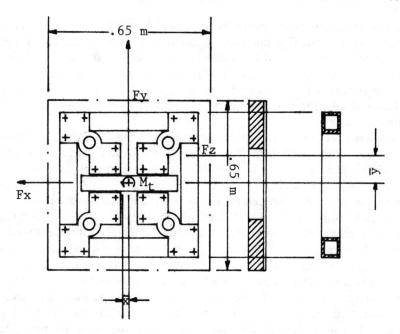

Figure 1 — Force platform structure and force diagram.

predetermined angle in order to respond to the horizontal force components and to the normal torsion vector. The best location of these strain gauges was found through a strain study of the structure by finite element analysis which yielded the strain and strain field gradient at each location along the rings. The model also gave the components of the correspondence matrix linking the mechanical variable vector to the measured strain and measured voltages vector as shown in Equation 1.

$$
\begin{bmatrix} Fz \\ \overline{x}Fz \\ \overline{y}Fz \\ Fx \\ Fy \\ Mx \end{bmatrix} = \begin{bmatrix} C_1 & C_1 & C_1 & C_1 & 0 & 0 & 0 & 0 \\ C_2 & -C_2 & -C_2 & C_2 & C_4 & C_4 & 0 & 0 \\ C_3 & C_3 & -C_3 & -C_3 & 0 & 0 & C_5 & C_5 \\ 0 & 0 & 0 & 0 & C_6 & C_6 & 0 & 0 \\ 0 & 0 & 0 & 0 & 0 & 0 & C_7 & C_7 \\ 0 & 0 & 0 & 0 & C_8 & -C_8 & -C_8 & C_8 \end{bmatrix} \begin{bmatrix} V_1 \\ V_2 \\ V_3 \\ V_4 \\ V_5 \\ V_6 \\ V_7 \\ V_8 \end{bmatrix} \quad (1)
$$

Figure 1 illustrates the structure and defines the variables. There were eight instrumental differential amplifiers of the type AD522 mounted with adjustment potentiometers on a common printed circuit board located directly in the platform, together with its required power supply. Thus, noise and unwanted signals were minimized.

The output voltages were fed into a 12-channel data acquisition system which transformed them into a numerical hexadecimal format. They were stored as such, to minimize space on floppy discs. Data could then be processed locally and immediately into a decimal format and multiplied by the calibration matrix giving the mechanical generalized force vector. A 6809 Southwest minicomputer was used to process the data and take the Fast Fourier Transform (FFT) of any of the output time variables.

A static calibration test was done with fixed heavy weights at various $\bar{x}$, $\bar{y}$ locations to yield the calibration constants, C_1 to C_3 of Equation (1). With a pulley system, an inclined force to the horizontal plane was applied to a center weight to produce values of C_4, C_5, C_6, and C_7. C_8 was obtained with an offset center weight and the inclined loading tension. The calibration was checked dynamically for the vertical component, C_1, C_2, C_3, through the evaluation of the impulse integral obtained through a pendulum impact. This procedure has the advantage of eliminating the static gravity weight effect from the output measured value.

Various problems arose during the development and testing, such as the thermal response of the strain gauge, the induced strains from temperature gradients, or the tightening of the elements and the top plate. The strain gauges were insulated and the top plate was provided with spherical joints in order to avoid the problem of induced strains. The initial top plate was light but not rigid enough. The platform exhibited a high natural frequency. A hardwood gymnasium-type top plate was then installed. It provided good damping but was not dimensionally stable. Then, the wood platform was fixed over the initial aluminum plate. The resulting top plate was then rigid, and provided good output exempt from surface flexural vibration. A new design is now underway to keep the stiffness high and lower the top plate mass and restore the desired natural frequency well above the measurement field.

Applications and Discussion

A first use of the platform was made in the evaluation of the influence of the shoe in the dynamics of jogging. An oval course including a straight access level runway of 10 m was installed for this experiment. The force platform was inserted close to the end of this runway, and was rigidly mounted over the concrete floor. The platform output voltages were fed into the data acquisition unit and to the computer for direct evaluation. Three subjects, 20 to 28 years old and weighing 55 to 68 kg, were selected. Various runs were measured for each subject, running barefoot, then with three different pairs of jogging shoes; the first pair was an inexpensive one with little recognized absorption; the second pair was a well-recognized training shoe, and the third pair was a racing flat of good reputation. The results are shown in Table 1, and typical force-time trac-

Table 1

Average Results for Each Shoe and the Barefoot Condition

	Maximum rearfoot force ratio	Maximum forefoot force ratio	Rate of loading	Rate of unloading	Pendulum test
	$(P_r/W)_{max}$	$(P_f/W)_{max}$	$k_r \ s^{-1}$	$k_f \ s^{-1}$	$(F_t)_{max} \ N$
Barefoot	4.40	3.05	840	−50.3	
Jogging shoe 1	3.39	3.39	420	−50.7	1640
Training shoe 2	3.97	3.24	334	−48.7	1005
Racing flat 3	3.35	3.01	242	−45.6	822

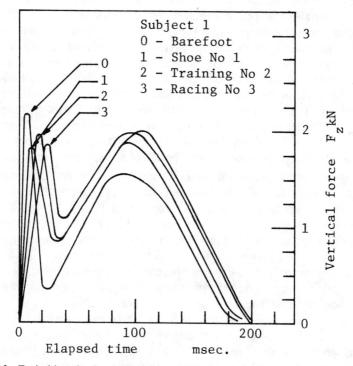

Subject 1
0 - Barefoot
1 - Shoe No 1
2 - Training No 2
3 - Racing No 3

Figure 2—Typical impulse for various shoes.

ings are presented in Figure 2.

For each of the four cases, the maximum recorded impact force is quoted in Columns 1 and 2, in a non-dimensional way. Column 1 gives the maximum rearfoot initial contact force divided by the subject weight.

The second column gives the maximum forefoot force-to-weight ratio. As can be seen, there is little to be concluded since there seems to be little evidence of a benefit from wearing a shoe. These results may be an indication of a possible adaptation of the subject's running pattern to the level of forces transmitted to the foot, and corroborate the results obtained by Therrien et al. (1981). Column 3 gives the rate of loading, k_r, of the nondimensional rearfoot contact force in the early rising recorded pulse, and Column 4, the final unloading rate of the k_f nondimensional forefoot propulsive force. This k_r variable takes a very high value in the barefoot condition and decreases appreciably as the subject wore a progressively better shoe. These very same shoes, which were tried alternatively by all subjects, had already been evaluated in the laboratory for their rearfoot and forefoot sole shock absorption. Under a constant momentum pendulum impact test, they had produced a uniformly varying force pulse and the maximum transmitted force during the impact is reported in the last column. This force is gradually lower from Shoe 1 to 3 indicating the proper sequence of better shock absorption. This provides a possible indication that the rearfoot loading rate for a runner is a better indication of the absorption performance of a jogging shoe, especially if we consider that the loading rate occurs over a fairly short time period of the order of 5 to 20 msec over which the runner has possibly little or no feedback. The fourth column shows nonsignificant variation. The unloading rate occurs over some 50 msec or more and could be better related to the propulsive pattern of motion. These constant k_f values could perhaps indicate here that the subjects try unconsciously to keep the same velocity barefoot as with any pair of shoes.

Fast Fourier Transformation (FFT) and Power Spectral Density (PSD) were obtained in each case through data processing with a FFT subroutine. A typical plot of a PSD appears in Figure 3. Here one can observe many interesting features among which is the energy around the natural frequency of the platform. Another important power distribution appears at a lower frequency and requires an analysis of the complete system as a two-degrees-of-freedom system, consisting of the masses of the runner and the platform, the rigidity of the platform, and the shoe stiffness. The lower part of the spectrum is believed to be more related to the runner's pattern, including the influence of the shoe. The remainder of the diagram is, therefore, more related to impedance reactions which of course are to be found on all kinds of running tracks, not only on a force platform. Therefore, it should be noted that covering a platform with a particular mat will not necessarily yield the same PSD as that found if the mat did cover its own foundation.

The new force platform was also used in the study of human posture. Posture analysis has been made using data from the $\bar{x}$ and $\bar{y}$ coordinates of the center of pressure as an individual is asked to remain for a fixed length of time, with open and closed eyes, on one or two feet, over the

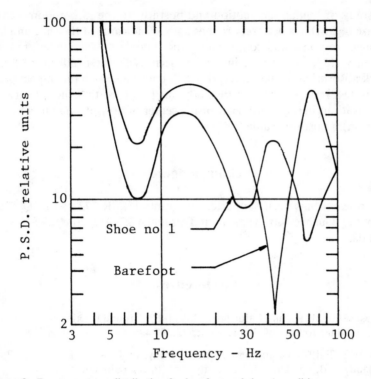

Figure 3—Power spectrum distribution for barefoot and shoe 1 condition.

force platform. In these cases, the motion is slow and the force platform natural frequency poses no problem. The spectra have also been obtained for these cases. The posture problem is a case of a stable or unstable control system. Each individual has a unique feedback response. Little variation was obtained in the x, y coordinate system. It is believed that the posture analysis in the $\bar{x}$, $\bar{y}$ plane is, as such, too limited. The problem of stable posture or sway is a problem of rapidity of response and is best seen in the velocity axes $\dot{\bar{x}}$ and $\dot{\bar{y}}$ as in a previous dynamic study (Perocheau et al., 1981). The data processing for the coordinate velocities have shown a better discrimination of the various parameter conditions and it is thought that, similar to other mechanical cases, the surface integral in these new generalized coordinates should be more relevant as a means of assessing stability characteristics of the subject in various posture conditions.

Conclusion

The main characteristics of a newly developed strain gauge instrumented force platform have been presented and applications have been made to the study of human locomotion and posture. In the locomotion study, the use of PSD has been reported as a good indicator of the shoe and instrument response and complements the time data. The loading rate in the early stage of the foot impulse is seen as the best indicator of the shock absorption of the shoe and is well correlated with the impact pendulum test results. The posture study has proposed that the velocity of the variations in the coordinates of the center of pressure may be a better criterion in posture evaluation.

Acknowledgments

This research has been supported by grants HCSR 7722 and EQ 1783 from the Quebec Government and grant A-2713 of the CRSNG of Canada.

References

CAVANAGH, P.R., and Lafortune, M.A. 1980. Ground reaction forces in distance running. Journal of Biomechanics 13:397-406.

MATAKE, T. 1976. On the new force plate study. In: Paavo V. Komi (ed.), Biomechanics V-B, pp. 426-431. University Park Press, Baltimore.

PÉROCHEAU, A., Bourassa, P., and Laneville, A. 1981. Sur les limites de stabilité et de dérapage des véhicules (Stability limits and skidding of road vehicles). Proceedings of the Eighth Canadian Congress of Applied Mechanics, Moncton, Vol. 1, pp. 379-380.

THERRIEN, R., Prince, P., and Bourassa, P. 1981. Influence du mode de réception et du port d'un soulier de jogging sur la dynamique du contact pied-sol: une étude préliminarie (Reception pattern and jogging shoe influence on the foot ground impact). Proceedings of the Eighth Canadian Congress of Applied Mechanics, Moncton, Vol. 2, pp. 771-772.

B.
Posture

Time Series Analysis of Postural Sway and Respiration Using an Autoregressive Model

Kazuyuki Takata, Hidetatsu Kakeno, and Yosaku Watanabe
Toyota Technical College, Toyota-shi, Japan

Shinya Takeuchi
Aichi University of Education, Kariya-shi, Japan

Investigations concerned with postural sway of human beings have been performed in several fields, but there are very few time series analyses (Bräver and Seidel, 1979).

This study involved stochastic modeling using an autoregressive model. This method was applied to mental patients in order to attempt to determine the differences between normal and abnormal individuals. Furthermore, the researchers estimated the power spectrum in order to study the frequency component included in the postural sway. Using the power spectrum it was also possible to estimate the influence exerted by respiration on postural sway.

Methods

Condition 1

Female mental patients (n = 39) from 19 to 54 years of age and 11 normal males from 19 to 48 years of age were chosen as subjects.

These subjects stood on a force plate with heels together with the feet at an angle of about 45°, under two conditions, eyes open and eyes closed.

Condition 2

This exercise was performed to study the influence of respiration on postural sway. Normal male college students (n = 16) from 19 to 20 years of age were chosen as subjects. They performed the exercise as follows: a) inhale without conscious control, b) inhale following the motion of a spot moving vertically with determined frequency on the CRT. The CRT was placed 3 m in front of the subject and at the level of the eyes. Frequencies of motion of the spot were 10, 15, and 30 cycles/min.

Recording

In Condition 1, components of sway in the sagittal and frontal planes detected by the force plate were recorded with an A/D converter with a sampling rate of 2/sec, over an interval of about 4 min. In addition to the sway components, the respiration curve was recorded with a thermistor in the nostril during Condition 2.

Time Series Analysis of Postural Sway

Test of Stationality for the Time Series

Stationality means that there is no growth or decline in the time series. The stationality of the series (X_t; t = 1,2, . . ., N) can be tested as follows. Divide the series (X_t) into n equal time segments and compute mean value ($\overline{X}_j$; j = 1,2, . . ., n) and variance S_x^2 (j); j = 1,2, . . ., n) for each duration over N/n times as follows,

$$\overline{X}_1, \overline{X}_2, \ldots, \overline{X}_n \tag{1}$$

$$S_x^2 (1), S_x^2 (2), \ldots, S_x^2 (n) \tag{2}$$

where: N = 500, n = 20.

The test of randomness for the series can be accomplished by the turning point method (Kendall, 1973).

Autoregressive Model and Power Spectrum

Consider that the series (X_t) meets the autoregressive process (AR) of order p such that

$$\widetilde{X}_t = a_1\widetilde{X}_{t-1} + a_2\widetilde{X}_{t-2} + \ldots + a_p\widetilde{X}_{t-p} + Z_t \tag{3}$$

$$\widetilde{X}_t = X_t - X, X = E (X_t) \tag{4}$$

where Z_t is residual.

The AR parameters may be estimated by solving the Yule-Walker equations:

$$\sum_{i=0}^{m} \hat{a}_i(m) R_{xx}(k-i) = 0 \qquad k = 1,2,\ldots, m \qquad (5)$$

$$R_{xx}(1) = C_{xx}(\ell)/C_{xx}(0), \qquad C_{xx}(\ell) = 1/N \sum_{t=1}^{N-\ell} \tilde{X}_t \cdot \tilde{X}_{t+\ell} \quad (6)$$

Akaike's (1970) minimization of final prediction error criterion was used to determine the order.

Using $\hat{a}_i$ the power spectrum is estimated by

$$P_X(f) = \Delta t \cdot 2S_Z^2 / \mid 1 - \sum_{k=1}^{p} \hat{a}_k \cdot \exp(-j2\pi f \Delta tk) \mid^2 \quad 0 \leqslant f \leqslant 1/2\Delta t \qquad (7)$$

(Δt; sampling interval)

$$S_Z^2 = C_{XX}(0) - \sum_{i=1}^{p} \hat{a}_i \cdot C_{XX}(i) \qquad (8)$$

Results

Standard Deviation of Fluctuation

Figure 1 shows the standard deviation of fluctuation for the sagittal and frontal plane with eyes open and eyes closed, respectively. The broken lines show equivalent inclinations.

In these figures, most of the points are distributed above the broken lines. This indicates that in general, sagittal fluctuation is larger than frontal, and with the eyes closed these fluctuations are larger than when open. No difference between normals and patients is evident in these figures.

Stability of Time Series

In a total time series, the same time series occurred 64.1% of the time in mean value, and 93.6% in standard deviation.

Figure 2 shows the sequence of $(\bar{X}_j/\bar{X}_1)$ for a mean value $(\bar{X}_j)$ and of (S_j/S_1) for a standard deviation (S_j) of sway about the center of gravity,

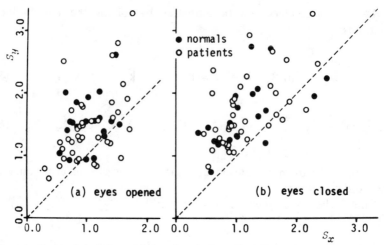

Figure 1 — Standard deviation of fluctuations in the sagittal (S_y) and frontal (S_x) planes to 30 female mental patients and 11 normal male subjects.

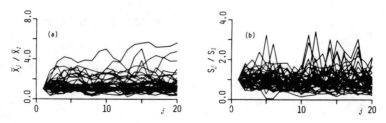

Figure 2 — The transition of mean values and standard deviations about the center of gravity over 39 mental patients.

respectively, for 39 mental patients with eyes open.

If a time series is a stable sequence, the sequence ($\bar{X}_j$), (S_j) must be horizontal along the time axis, j. In other words, the sequence fluctuates around a constant value.

Order of Series with Autoregressive Process

Figure 3 illustrates the distribution of the order of the series with the autoregressive process. This figure shows that the system order range is from 1 to 8, and has a tendency to be of a higher order with the eyes closed as compared to with the eyes open, especially in the frontal plane.

Influence on Postural Sway of Respiration

Under Condition 1, the results of the spectral analysis of postural sway show a peak appeared at near 0.25 Hz for some subjects. This frequency

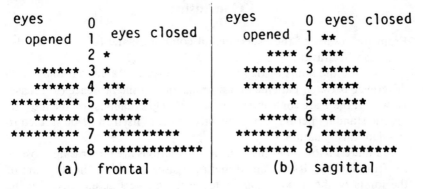

Figure 3 — Distribution of the order of the time series with autoregressive process over 39 mental patients.

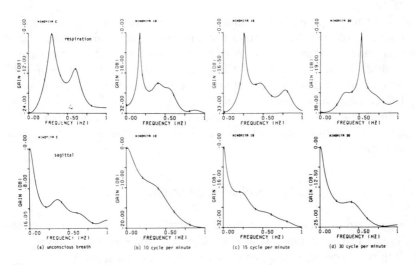

Figure 4 — Spectra of respiration (upper row) and postural sway (lower row) in case of unconscious breath control and conscious breathing with a designated frequency.

is equivalent to 15 cycles/min and corresponds to the frequency of unconscious breathing. To study in detail the influence on postural sway of respiration, Condition 2 was performed.

Figure 4 shows the spectra of postural sway (upper row) and respiration (lower row), (a) breathing without conscious control, and (b), (c), (d) are cases of controlled breathing with frequencies of 10, 15, and 30 cycles/min, respectively.

As seen in these figures, the peak of postural sway shows a spectral shift with a change in respiration. From this result, it can be considered that postural sway is influenced by respiration.

Conclusion

From the results of the experimental conditions, the following conclusions can be drawn:

1. Within the limits of these analyses, it can be stated that there is no difference between patients and normals due to the presence of disease. However, there are some examples such as a gradual posture incline during the standing period. Consequently, posture itself must be studied to evaluate this situation.

2. Figure 1 shows that the sagittal fluctuations are larger than those in the frontal plane. It is assumed that the reason for this is the structure of the joints of the human body. Furthermore, the fluctuations with the eyes closed are larger than with the eyes open. Therefore, this shows that some information needed to control the body is obtained through the use of the eyes.

3. From Experimental Condition 2 it is clear that respiration influences postural sway. Furthermore, postural sway is also influenced by heart rate variability as well as respiration. However, it is impossible to separate the influence of heart rate variability from that of respiration because of the influence of respiration on heart rate variability (Sayers, 1973; Takata and Watanabe, 1978).

In each spectrum of postural sway, the gain was very high at low frequencies, which indicates that postural sway possesses a very long period or trend.

Acknowledgments

Appreciation is extended to the Data Station of Toyota Technical College and to the Nagoya University Computation Center for use of their facilities to perform the numerical analysis for this study.

References

AKAIKE, H. 1970. Statistical predictor identification. Ann. Inst. Math. 22:203-217.

BRÄUER, D., and Seidel, H. 1981. The autoregressive structure of postural sway. In: A. Morecki, K. Fidelius, K. Kedzior, and A. Wit (eds.), Biomechanics VIIA, pp. 155-160. University Park Press, Baltimore.

KENDALL, M.G. 1973. Time Series. Griffin, London.

SAYERS, B. McA. 1973. Analysis of heart variability. Ergonomics 16:17-32.

TAKATA, K., and Watanabe, Y. 1978. Time series analysis on R-R intervals of ECG using an autoregressive model. Proceedings of the International Conference on Cybernetics and Society, Tokyo-Kyoto, pp. 99-104.

Analysis of Body Sway on Seesaw Board

Yoshio Mizuno
Daido Institute of Technology, Aichi, Japan

Ryoichi Hayashi, Akihide Miyake, and Satoru Watanabe
Gifu University School of Medicine, Gifu, Japan

It is well known that a human standing upright is influenced by proprioceptive, vestibular, and visual information. The roles of these three afferent systems concerning postural control have been studied by many investigators (Nashner, 1972; Lestienne et al., 1977; Mauritz and Dietz, 1980). We have investigated the dynamic balancing mechanisms of body sway by means of the seesaw board, and it has been found that the body sway during standing on the seesaw board increased more than that on a flat plate, thereby facilitating the analysis of the characteristics of body sway (Mizuno and Hayashi, 1979).

In order to investigate the influence of vision on postural sway, body sway and EMGs of leg muscles were recorded and compared under two different visual environmental conditions.

Methods

Six healthy subjects were examined when standing on the seesaw board (seesaw) or on the flat plate with their eyes open under the light (LR) or in dark room (DR) conditions. The seesaw consisted of a platform with a curved base (weight 4.25 to 5.96 kg; height 8 cm). Different seesaws varying in the radius of base curvature (30 cm, 50 cm, and 70 cm) were used. The seesaw was set on a gravicorder, and was movable only in the anterior-posterior direction. The angle of inclination of the seesaw positioned upon the gravicorder was measured by a potentiometer joined to a side of the seesaw through a pantograph (see Figure 1A). The subject's normal position while standing upright on the flat plate was traced on a

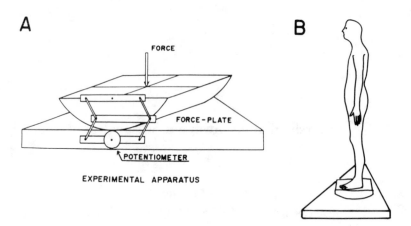

Figure 1 — Experimental apparatus. The seesaw was set on a gravicorder, and was movable only in the anterior-posterior direction. The angle of inclination of the seesaw positioned upon the gravicorder was measured by a potentiometer attached to a side of the seesaw through a pantograph (Figure 1A). Subjects were standing on the seesaw with bare feet (Figure 1B).

paper. The position of the subject's center of gravity was also plotted when the eyes were open. The subjects were asked to stand upright and adjust the position of their centers of gravity to the center of the seesaw (see Figure 1B). They were asked to maintain a normal standing (NS) posture; thereafter they were asked to assume forward leaning (FL) or backward leaning (BL) postures for 90 sec intervals on the seesaw as well as on the flat plate. In the standing upright position, the anterior-posterior direction of body sway (A-P sway), the right-left direction of body sway, and the vertical force were measured by a gravicorder (Sanyo Electric Co.). The power spectra of the A-P sway and the vertical force were calculated by a FFT method with a sampling time of 40 msec in the ranges of 0.1 to 10.0 Hz. EMGs of leg muscles were recorded from tibialis anterior (TA), gastrocnemius (GC) and soleus (GS) with surface electrodes. All data were monitored by a polygraph (13 channel Nihon Kohden K.K.) and were recorded simultaneously on the data recorder (14 channel TEAC R-270A).

Results

When the condition was changed from LR to DR, a marked enhancement of body sway was observed, as is shown in Figure 2A. This enhancement of body sway was developed more clearly on the seesaw, and an example of the results on the seesaw with radius of 50 cm is shown in Figure 2B. When the subject's posture altered in darkness from NS to FL

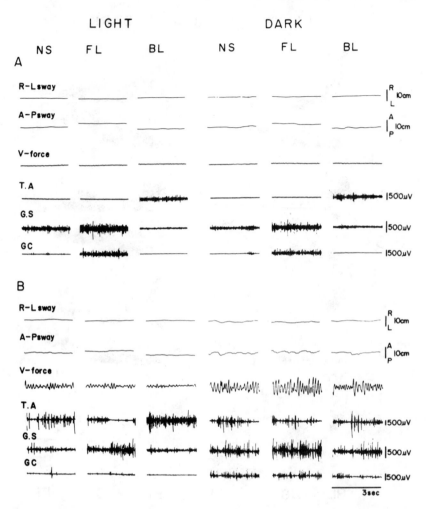

Figure 2 — Recordings during standing on the flat plate (A) and that on the seesaw (B). Subjects maintained normal standing (NS), forward leaning (FL), and backward leaning (BL) in the light condition and in the dark condition. EMGs were recorded from the left leg muscles. T.A. = tibialis anterior; G.S. = soleus; GC = gastrocnemius; LIGHT = in the light condition; DARK = in the dark condition.

or to BL, this enhancement was also found, but the shift of the position of the center of gravity was less in DR than in LR (see Figures 2A, 2B).

When examining the EMGs of the leg muscles under both conditions of LR and DR, it was observed that the tonic activities of GS and GC discharged during FL, while that of TA discharged during BL (see Figure 2A). These leg muscle activities increased more on the seesaw than on the flat plate (see Figure 2B). During FL or BL on the seesaw, phasic activities were intermingled with tonic ones, and the antagonistic muscle

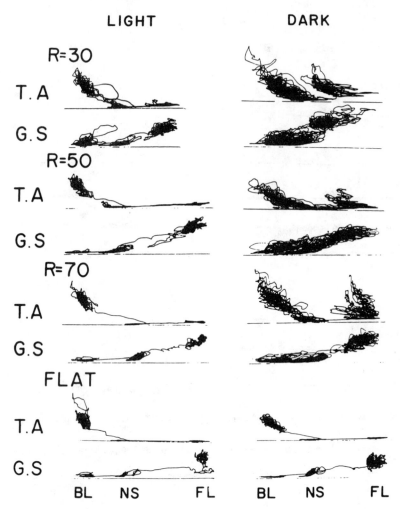

Figure 3—Relationships between the position of the center of gravity and the activities of leg muscles. This figure shows the correlation as Lissajous figures, in which the ordinate is the muscle activity and the abscissa the position of the center of gravity. R expresses curvature radius in cm. Abbreviations are the same as in Figure 2.

discharges represented a characteristic co-contraction phenomenon with those of the agonist muscle. The co-contraction of the antagonist under the DR condition was increased more than under the LR condition. Correlations of muscle activities and the position of the center of gravity expressed by the Lissajous figure are shown in Figure 3. In the DR condition, the intensity of co-contraction of the leg muscles was observed more clearly during FL or BL on the seesaw, when a smaller radius was used.

In contrast to those on the flat plate, the power spectra of body sway

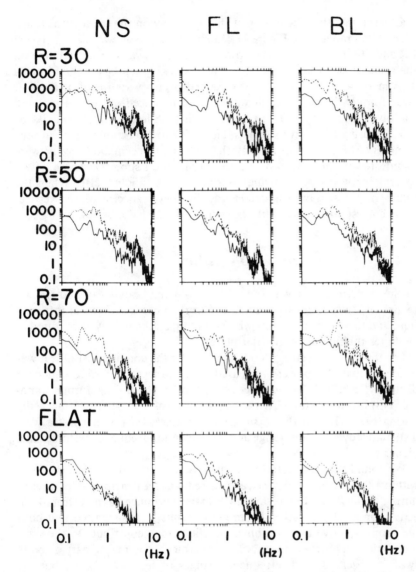

Figure 4—Power spectra of the A-P sway. These power spectra of the A-P sway were calculated by a FFT method with sampling time of 40 msec, and ranged from 0.1 Hz to 10.0 Hz. Solid lines show the powers in the light condition and dotted lines those in the dark condition. Abbreviations are the same as in Figure 2.

(A-P sway) on the seesaw showed a marked increase in amplitude for all ranges (see Figure 4). In LR, it was found that the relationship of the power to the radius of the seesaw was nearly in inverse proportion, namely the power of the A-P sway was increased with a decreasing radius of the seesaw. In DR, this inverse proportion was not observed distinctly, due to the fact that the power of the A-P sway on the seesaw was much larger in all ranges (see Figure 4). When a comparison was made between LR and DR conditions of the power of the A-P sway during standing on the same seesaw radius, the power in DR was much larger than in LR. This increase of the power was observed over a wide range of the A-P sway frequencies. This result was most pronounced on the seesaw, and was observed clearly in 0.2 to 1.0 Hz during NS in five subjects, in 0.2 to 3.0 Hz during FL in five subjects, and in 0.2 to 3.0 Hz during BL in four subjects. The power spectra of the vertical force in body sway were manifested in the ranges of 4.0 to 7.0 Hz in all subjects. These power spectra increased in particular during FL (see Figure 2B). When the radius of the seesaw was decreased, the power spectra of the vertical force corresponded well with the activity of the leg muscles.

Discussion

In this experiment, it was found that both the body sway and the activities of leg muscles were enhanced in the absence of visual information, and these enhancements while standing on the seesaw were found to be larger than those on the flat plate.

For the power spectra of body sway under the condition of DR, it was observed that the powers increased in all ranges, and were enhanced distinctly, especially in the ranges of 0.2 to 1.0 Hz, during normal standing (NS) on the seesaw. This enhancement was also observed markedly during FL or BL on the seesaw, but the increased power spectra were distributed over a wider range and the maximum frequency occurred at a higher point than in NS.

We found the manifestation of the power spectra in the vertical force within 4.0 to 7.0 Hz, and it was considered that the distribution of powers during FL or BL on the seesaw in DR may be produced as an effect of the activity in the leg muscles. From these results, it is suggested that both the proprioceptive and vestibular functions may contribute to the sway much more than they do in NS. Furthermore, the major part of body sway is observed in the frequency ranges of less than 1 Hz, but the mechanism of its appearance is not clear. In this experiment, we found that the power spectra in the ranges 0.2 to 1.0 Hz during standing on the seesaw clearly differed in LR and in DR. Therefore, it was concluded that the visual information suppressed body sway in the low frequency range during standing on the seesaw.

In an unstable condition when the subjects stood on the seesaw, an enhanced body sway, an increase of tonic and phasic leg muscle activities, and a phasic co-contraction of the antagonistic muscles were observed. From these results, it was suggested that for the unstable condition, the dynamic factor rather than the static one took part in the control of posture. However, visual information plays an important role in the control of body sway, although it may not be of concern in high frequency ranges of the dynamic control.

Summary

Body sway and the activities of leg muscles were analyzed when the subjects stood on the seesaw in the absence of the visual information.

The results were as follows:

1. In the absence of visual information and while standing on the seesaw and on the flat plate, body sway and the activities of leg muscles increased markedly.

2. With respect to the power spectra of anterior-posterior sway during standing on the seesaw, the powers in the dark condition (absence of visual information) were larger over a relatively wide range as compared to the light condition. Especially remarkable were the enhanced powers in the ranges of 0.2 to 1.0 Hz.

3. A good correlation between the position of the center of gravity and the tonic activities of leg muscles was found not only in the light condition but also in the dark condition. When standing on the seesaw of the smaller radius, the phasic activities intermingled with the tonic activities and also with the co-contraction of the antagonistic muscles.

4. These results, obtained by using the seesaw, showed that the absence of visual information caused an enhancement of body sway, and this was particularly apparent in the component less than 1 Hz. It may be concluded, therefore, that the visual suppression of body sway took place mainly in the low frequency range.

References

LESTIENNE, F., Soechting, J., and Berthoz, A. 1977. Postural readjustments induced by linear motion of visual scenes. Exp. Brain Res. 28:363-384.

MAURITZ, K.-H., and Dietz, V. 1980. Characteristics of postural instability induced by ischemic blocking of leg afferents. Exp. Brain Res. 38:117-119.

MIZUNO, Y., and Hayashi, R. 1979. Analysis of dynamic balancing in human standing posture. J. Physiol. Soc. Japan. 41:327.

NASHNER, L.M. 1972. Vestibular postural control model. Kybernetic. 10: 106-110.

Postural Effects on the
Aerobic Capacity of Forearm Muscles

T. Yamamoto and K. Fujita
Chukyo University, Aichi, Japan

It is well known that passively tilting the recumbent subject to the head-down position with a tilt table causes a significant increase in forearm blood flow (Roddie and Shepherd, 1956; Mengesha and Bell, 1979). In the head-up position a decrease is also found in forearm blood flow (Brigden et al., 1950; Dornhorst, 1963; Corbett et al., 1971). It remains in doubt, however, whether these cardiovascular responses affect the forearm muscular work capacity. The object of this study was to determine the effects of body tilting on circulatory responses and forearm muscular endurance time.

Methods

Healthy male subjects (n = 34), mainly physical education students, volunteered for this investigation. They ranged in age from 18 to 30 years. Body positions used in this study were three kinds: 1) horizontal-supine (0°), 2) head-down (−40°), and 3) head-up (+40°) positions. After taking a rest in each body position for 10 minutes, the subject performed a hand grip exercise in each body position studied. The exercise was performed by squeezing a handle to which a weight was attached which was equivalent to 1/3 or 1/6 of the maximal grip strength for each subject. The distance of lifting the weight was fixed at 22 mm. The forearm muscular contractions were made rhythmically at a rate of 60/min with a metronome and continued to exhaustion. Forearm and leg blood flow was measured by the venous occlusion method employing a mercury-in-rubber strain-gauge before and after exercise. Blood samples were also obtained from the right cubital vein and brachial artery in 10 selected subjects. Measurements were made of blood lactate and blood gas (Pco_2, Po_2) levels.

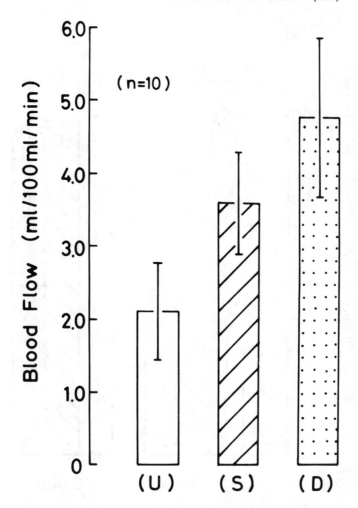

Figure 1–Changes in resting forearm blood flow accompanying body tilting. The symbols 'U', 'S' and 'D' signify head-up, horizontal-supine and head-down positions, respectively. The same symbols apply in all figures. Values are means and standard deviations (n = 10).

Results

Figure 1 shows the changes in resting forearm blood flow related to three different body positions. On the whole, a progressive decrease of the body tilt from +40° (head-up position) to −40° (head-down position) showed a progressive increase in the forearm blood flow of all subjects studied. The mean forearm blood flows were found to be 2.12, 3.61, and 4.79 ml/100 ml tissue/min at +40, 0, and −40°, respectively. They showed significant correlation (P < 0.01 ~ 0.05) with three different angles of passive tilt.

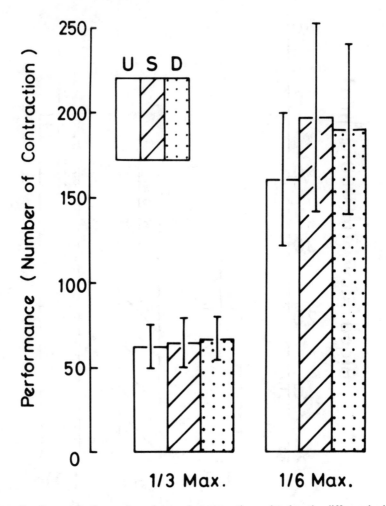

Figure 2 — Changes in the number of muscular contractions related to the different body positions. The numbers of subjects are 14 for the work load of 1/3 maximal grip strength and 17 for 1/6 max. Values are means and standard deviations.

Figure 2 shows the changes in the number of muscular contractions related to both intensity of load and the different body positions. In the work load of heavier weight (1/3 of maximal grip strength), the overall mean number of hand grip contractions was found to be 62.4, 64.6, and 66.5 times at the $+40$, 0, and $-40°$ positions, respectively. They were not significantly different from one another, however. On the other hand, in the work load of lighter weight (1/6 of maximal grip strength), the overall mean number of hand grip contractions was found to be 160.4, 197.0, and 190.4 times at $+40$, 0, and $-40°$ positions, respectively. The overall mean change at head-up tilt was found to be significantly

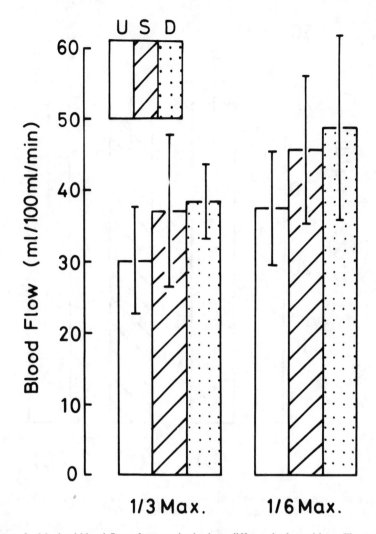

Figure 3—Maximal blood flow after exercise in three different body positions. The numbers of subjects were 5 for the work load of 1/3 maximal grip strength and 7 for 1/6 max. Values are means and standard deviations.

different from the responses at the other two body positions (P < 0.01). The number of contractions in the horizontal-supine and head-down positions, however, were not much different.

Figure 3 shows the changes in maximal forearm blood flow after exercise in three different body positions. In the work load of 1/3 of maximal grip strength, the mean forearm blood flows were found to be 30.10, 37.45, and 38.61 ml/100 ml/min at the + 40, 0, and − 40° positions, respectively. In the work load of 1/6 of maximal grip strength, the mean

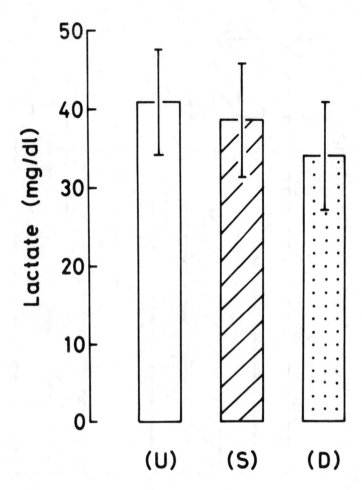

Figure 4 – Comparison of blood lactate immediately after exercise in three different body positions. Values are means and standard deviations (n = 5).

blood flows were 37.85, 45.76, and 48.85 ml/100 ml/min at the +40, 0, and −40° positions, respectively. For the two work load conditions, the mean value in the head-up position was significantly less than those in the other two body positions (P < 0.05). The difference between the mean values in horizontal-supine and head-down positions was not statistically significant, however.

Figure 4 shows the blood lactate immediately after exercise in three different body positions. The mean values were 40.8, 38.6, and 34.1 mg/dl at the +40, 0, and −40° positions, respectively. The significant difference was observed only between the values in head-up and head-down positions (P < 0.05).

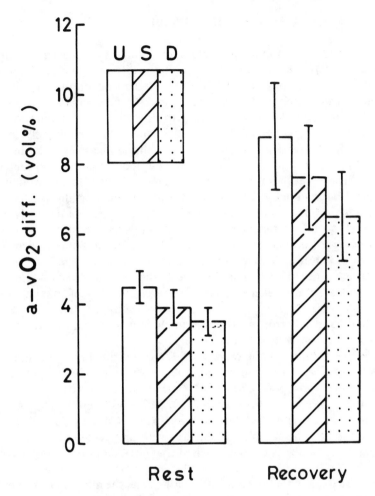

Figure 5 — A-V O₂ difference related to different body positions at rest and after exercise. Values are means and standard deviations (n = 5).

Figure 5 shows the changes in arterial-venous oxygen difference related to three different body positions at rest and after exercise of 1/3 of maximal grip strength. The mean A-V O_2 differences at rest were found to be 37.85, 45.76, and 48.85 vol % at the +40, 0, and −40° positions, respectively. Those after exercise were found to be 8.78, 7.60, and 6.48 vol % at +40, 0, and −40°, respectively. Both at rest and after exercise, the mean A-V O_2 difference in the head-up position was significantly larger than the value in the head-down position ($P < 0.01 \sim 0.05$). The difference between head-up and horizontal-supine positions, or between horizontal-supine and head-down positions, was not statistically significant, however.

Discussion

The performance of muscular endurance (numbers of forearm muscular contractions) in the different body positions differed from one another. That is to say, the postural difference in numbers of contractions was revealed clearly when the lifting weight was 1/6 of maximal grip strength. In the work load of 1/3 of maximal grip strength, the mean durations of work were 62.4, 64.6, and 66.5 at +40, 0, and −40°, respectively. Judging from these results, the effects of body tilt on muscular endurance of one min or so may be slight. The values of blood flow and blood contents in the head-up position, however, were significantly different from those in the other two body positions even in the muscular endurance of one min or so. Roddie and Shepherd (1956) also reported that passively raising the legs of recumbent healthy male subjects, aged 19 to 35 years, caused significant increases in forearm blood flow and the oxygen content of blood from forearm veins draining the muscles. These tendencies observed at rest were similar to those of the present results obtained at rest and even after exercise. As the exercise of 1/3 of maximal grip strength was, however, anaerobic, it was considered that such hemodynamic and hematologic changes might not affect the muscular endurance in each body position.

On the contrary, in the exercise of 1/6 of maximal grip strength, the numbers of forearm muscular contractions at +40° were significantly smaller than those at 0° and −40°. The mean durations of exercise were 160.4, 197.0, and 190.4 sec at +40, 0, and −40°, respectively. As the duration of exercise was about three min or so, the exercise was considered to be more aerobic as compared with the exercise of 1/3 of maximal grip strength. These data suggest that circulatory responses accompanying body tilting have an influence on the aerobic capacity of forearm muscles, and that the longer the work time, the more remarkable the influence.

Acknowledgment

The authors wish to thank the students who acted as subjects and Dr. K. Asahina for his helpful comments.

References

BRIGDEN, W., Howarth, S., and Sharpey-Schafer, E.P. 1950. Postural changes in the peripheral blood-flow of normal subjects with observations on vasovagal fainting reactions as a result of tilting, the lordotic posture, pregnancy and spinal anaesthesia. Clin. Sci. 9:79-90.

CORBETT, J.L., Frankel, H.L., and Harris, P.J. 1971. Cardiovascular responses to tilting in theraplegic man. J. Physiol. London 215:411-431.

DORNHORST, A.C. 1963. Hyperaemia induced by exercise and ischaemia. Br. Med. Bull. 19:137-140.

MENGESHA, Y.A., and Bell, G.H. 1979. Forearm and finger blood flow responses to passive body tilts. J. Appl. Physiol.: Respirat. Environ. Exercise Physiol. 46:288-292.

RODDIE, I.C., and Shepherd, J.T. 1956. The reflex nervous control of human skeletal muscle blood vessels. Clin. Sci. 15:433-440.